C. S. S. R. Kumar, J. Hormes, C. Leuschner (Eds.)

Nanofabrication Towards Biomedical Applications

Further Titles of Interest

Nanofabrication Towards Biomedical Applications

Techniques, Tools, Applications, and Impact

Edited by C. S. S. R. Kumar, J. Hormes, C. Leuschner

WILEY-VCH

WILEY-VCH Verlag GmbH & Co. KGaA

Editors

Dr. Challa S. S. R. Kumar
Center for Advanced Microstructures and Devices
Louisiana State University
6980 Jefferson Highway
Baton Rouge, LA 70806
USA
ckumar1@lsu.edu

Prof. Dr. Josef Hormes
Center for Advanced Microstructures and Devices
Louisiana State University
6980 Jefferson Highway
Baton Rouge, LA 70806
USA
hormes@lsu.edu

Prof. Dr. Carola Leuschner
Reproductive Biotechnology Laboratory
Pennington Biomedical Research Centre
Louisiana State University
6400 Perkins Road
Baton Rouge, LA 70808
USA
leuschc@pbrc.edu

Library of Congress Card No.:
applied for

British Library Cataloguing-in-Publication Data
A catalogue record for this book is available from the British Library.

**Bibliographic information published by
Die Deutsche Bibliothek**
Die Deutsche Bibliothek lists this publication in the Deutsche Nationalbibliografie; detailed bibliographic data is available in the Internet at <http://dnb.ddb.de>.

© 2005 WILEY-VCH Verlag GmbH & Co. KGaA, Weinheim

Printed in the Federal Republic of Germany.

Printed on acid-free paper.

Typesetting   Kühn & Weyh, Satz und Medien, Freiburg
Printing   Strauss GmbH, Mörlenbach
Bookbinding   Litges & Dopf Buchbinderei GmbH, Heppenheim

ISBN-13   978-3-527-31115-6
ISBN-10   3-527-31115-7

Foreword

Nanobiotechnology: Hype, Hope and the Next Small Thing is the title of one of the chapters in this book. This title suggests that the applications of nanotechnology in biology and medicine are still in a somewhat uncertain future, but the contrary is also true: there are already several products, such as zinc oxide nanoparticles in sun cream or nano-silver as a coating material for home appliances to destroy bacteria and prevent them from spreading, that are available on the market. Other, even more exciting applications are in the testing phase, for example, using magnetic nanoparticles for a targeted hyperthermia treatment of brain cancer. There are of course also applications that might become reality in the far future – though there are always surprises possible in nanotechnology, e.g., implantable pumps the size of a molecule that deliver medicines with a precise dose when and where needed, or the possibility to remove a damaged part of a cell and replacing it with a biological machine. These applications are some of the goals stated in the National Institute of Health roadmap for nanomedicine, which was established in spring 2003. This initiative is again part of a larger US National Nanotechnology Initiative (NNI), for which the President's budget will provide about $1 bn for 2005 for projects coordinated by at least ten different federal agencies.

The book aptly named *Nanofabrication Towards Biomedical Applications* is timely as the contributions are all written by experts in their field, summarizing the present status of influence of nanotechnology in biology, biotechnology, medicine, education, economy, society and industry. I am particularly impressed with the judicious combination of chapters covering technical aspects of the various fields of nanobiology and nanomedicine from synthesis and characterization of nanosystems to practical applications, and the societal and educational impact of the emerging new technologies. Thus, this book gives an excellent overview for non-specialists by providing an up-to-date review of the existing literature in addition to providing new insights for interested scientists, giving a jump-start into this emerging research area. I hope this book will stimulate many scientists to start research in these exciting and important directions. I am particularly pleased to recognize the efforts of

Nanofabrication Towards Biomedical Applications. C. S. S. R. Kumar, J. Hormes, C. Leuschner (Eds.)
Copyright © 2005 WILEY-VCH Verlag GmbH & Co. KGaA, Weinheim
ISBN 3-527-31115-7

the Center for Advanced Microstructures and Devices (CAMD) and of the Pennington Biomedical Research Center (PBRC) in taking a lead to spread the influence of biomedical nanotechnology, and I am convinced that the book will be a valuable tool in the hands of all those interested in discovering new paths and opportunities in this fascinating new field.

William L. Jenkins
President, Louisiana State University

Contents

Nanofabrication Towards Biomedical Applications. C. S. S. R. Kumar, J. Hormes, C. Leuschner (Eds.)
Copyright © 2005 WILEY-VCH Verlag GmbH & Co. KGaA, Weinheim
ISBN 3-527-31115-7

Preface

Within a short span of a decade nanotechnology has evolved into a truly interdisciplinary technology touching every traditional scientific discipline. The effect of nanotechnology on biomedical fields has been somewhat slower and is just beginning to gain importance as seen from a recent search on research publications. Of the total number of nanotechnology related publications which are approximately 2500 in the year 2002-2004, only about 10% of them were related to biomedical sciences. Even though, the effect of nanotechnology on biomedical field is slow, it is bound to gain momentum in the years to come as all biological systems embody nanotechnological principles. Slowly but surely, nanomaterials and nanodevices are being developed that have design features on a molecular scale and have the potential to interact directly with cells and macromolecules. The nanoscientific tools that are currently well understood and those that will be developed in future are likely to have an enormous impact on biology, biotechnology and medicine. Similarly, understanding of biology with the help of nanotechnology will enable the production of biomimetic materials with nanoscale architecture. The comparable size scale of nanomaterials and biological materials, such as antibodies and proteins, facilitates the use of these materials for biological and medical applications. Also, in recent years the biomedical community has discovered that the distinctive physical characteristics and novel properties of nanoparticles such as their extraordinarily high surface area to volume ratio, tunable optical emission, magnetic behavior, and others can be exploited for uses ranging from drug delivery to biosensors.

Viewing from the point of biomedical researchers, it is very difficult to fathom out relevant literature and suitable information on nanotechnological tools that would have profound impact on biomedical research as most of the literature is published in physico-chemical journals. It is our endeavor to support the biomedical community by providing the required information on nanotechnology under one umbrella. We are pleased to introduce to our readers a book that covers various facets of nanofabrication which we hope will help biologists and medical researchers. The book covers not only the scientific aspects of nanofabrication tools for biomedical research but also the implications of this new area of research on education, industry and society at large. Our aim is to provide as comprehensive perspective as possible to our readers who are interested in learning, practicing and teaching nanotechnological tools for biomedical fields. We, therefore, designed the contents of the

Nanofabrication Towards Biomedical Applications. C. S. S. R. Kumar, J. Hormes, C. Leuschner (Eds.)
Copyright © 2005 WILEY-VCH Verlag GmbH & Co. KGaA, Weinheim
ISBN 3-527-31115-7

book to have four major sections: (1) Synthetic aspects of nanomaterials, (2) Characterization techniques for nanomaterials (3) Application of nanotechnological tools in biomedical field and (4) Educational, economical and societal implications.

The first section of the book provides information about the fabrication tools for nanomaterials. Fabrication of nanomaterials is by now a very well developed area of research and it is impossible to cover all aspects. Traditionally, synthetic approaches to nanomaterials have been divided into two categories: "top-down" and "bottom-up". "Top-down" practitioners attempt to stretch existing technology to engineer devices with ever-smaller design features. "Bottom-up" researchers attempt to build nanomaterials and devices one molecule/atom at a time, much in the way that living organisms synthesize macromolecules. Therefore, in this volume we made an attempt to explore wet chemical methods for fabrication of metallic nanoparticles, synthetic approaches to carbon nanotubes, and approaches to building of nanostructured materials from low-dimensional building blocks. A fascinating account of bio-mimetic approaches to building materials from nanostructures is dealt in two chapters – "Nanostructured collagen mimics in tissue engineering" and "Molecular biomimetics: Building materials the nature's way, one molecule at a time". We hope to cover other synthetic aspects in subsequent volumes.

The second section of the book covers tools that are currently available for characterization of nanomaterials and is anticipated to give biomedical researchers an opportunity to learn not only basics of some of the very important techniques such as X-ray absorption spectroscopy and X-ray diffraction, transmission electron microscopy, or electron diffraction, but also help in developing an understanding of how these techniques can be utilized to enhance their own research. Also included in this section is a chapter entitled, "Single-molecule detection and manipulation in nanotechnology and biology" which we hope provides our readers up-to-date information about the opportunities that currently exist and future perspectives on tools for visualizing the world at the molecular and nanoscopic level. "Nanotechnologies for Cellular and Molecular Imaging by MRI" is one of the chapters that is anticipated to give our readers an insight into diagnosis and characterization of atherosclerotic plaques. In this section again, there are many more characterization tools and novel detection methods that have been deliberately left behind to be covered in subsequent volumes.

The third section offers examples of how nanotechnological tools are being utilized in biomedical research. While the chapter entitled, "Nanoparticles for Cancer drug delivery" provides a state-of-the-art information on various types of nanoparticles that are currently under development for cancer therapy, a more specific approach using metal nanoshells is described in the chapter-diagnostic and therapeutic application of metal nanoshells. This particular section introduces our readers to other important areas of biomedical research such as gene delivery, and biological agent decontamination that were positively affected by nanotechnology. We do realize that there are many more applications and subject areas in biomedical research that continue to be impacted by nanotechnology. It is impossible to cover all of them in one book, but we hope to be able to cover as many examples as possible by following up with further volumes dedicated to nanofabrication for biomedical applications, which are currently being planned.

The final section and the most important one in our opinion brings out the impact of biomedical nanotechnology on education, society and industry. There is no doubt that nanotechnology is going to significantly affect these important facets of our lives and it is our mission to ensure that researchers working in the area of biomedical nanotechnology become aware of these implications as early as possible. While the chapter, "too small to see" enlightens the readers on how educators are trying to grapple with a situation to educate the new generation about nanotechnology, the chapter aptly titled as "nanobiomedical technology: financial, legal, clinical, political, ethical and societal challenges to implementation" introduces to the reader various global challenges to the implementation of this new technology.

A book series of this magnitude is impossible without the unwavering support from the authors who have taken time of their busy schedule to submit their manuscripts on time and we are indebted to them. We gratefully acknowledge the support from Wiley VCH, in particular to Martin Ottmar, who has been working closely with us to make this first volume of the book series a reality. The Center for Advanced Microstructures and Devices and the Pennington Biomedical Research Center are two unique institutions in Louisiana, USA, who have been providing innumerable opportunities to their employees to excel and we cherish this support and encouragement. Finally, we are indebted to our families for their trust and support in addition to bearing our long absences from our family chores.

Baton Rouge, November 2004
Challa Kumar, Josef Hormes, and Carola Leuschner

List of Contributors

Pulickel M. Ajayan
Rensselaer Polytechnic Institute
Department of Materials Science and
Engineering
Troy, NY 12180
USA

Jennifer Barton
Electrical and Computer Engineering
University of Arizona
1230 Speedway Blvd.
Tucson, AZ 85721
USA

Carl A. Batt
Cornell University
Food Science Department
312 Stocking Hall
Ithaca, NY 14853
USA

Helmut Bönnemann
Max-Planck-Institut für Kohlen-
forschung
Heterogene Katalyse
Kaiser-Wilhelm-Platz 1
D-45470 Mülheim an der Ruhr
Germany

Shelton D. Caruthers
Washington University
School of Medicine
660 S. Euclid Avenue
St. Louis, MO 63110
USA
and
Philips Medical Systems
Cleveland, Ohio
USA

Alex Chen
Rutgers, The State University of New
Jersey
Department of Chemistry
73 Warren Street
Newark, NJ 07102
USA

Daniel T. Chiu
University of Washington
Department of Chemistry
P.O. Box 351700
Seattle, WA 98195-1700
USA

Rebekah Drezek
Rice University
Department of Bioengineering
Houston, TX 77005
USA

Nanofabrication Towards Biomedical Applications. C. S. S. R. Kumar, J. Hormes, C. Leuschner (Eds.)
Copyright © 2005 WILEY-VCH Verlag GmbH & Co. KGaA, Weinheim
ISBN 3-527-31115-7

Steven A. Edwards
S.A. Edwards and Associates
Christiana, TN 37037
USA

Maria P. Gil
Department of Chemical & Bio-
molecular Engineering
300 Lindy Boggs Center
Tulane University
New Orleans, LA 70118
USA

Naomi Halas
Rice University
Departments of Electrical and
Computer Engineering
Houston, TX 77005
USA

Jeffrey D. Hartgerink
Departments of Chemistry and
Bioengineering
Rice University
6100 Main St.
Houston, TX 77005
USA

Huixin He
Rutgers, The State University of
New Jersey
Department of Chemistry
Newark, NJ 07102
USA

Leon Hirsch
Rice University
Department of Bioengineering
Houston, TX 77005
USA

Gavin D.M. Jeffries
University of Washington
Department of Chemistry
P.O. Box 351700
Seattle, WA 98195-1700
USA

Michael D. Kaminski
Nanoscale Engineering Group
Chemical Engineering Division
Argonne National Laboratory
9700 South Cass Avenue
Argonne, IL 60439
USA

Kenneth J. Klabunde
Department of Chemistry
Kansas State University
111 Willard Hall
Manhattan, KS 66505
USA

Challa Kumar
Center for Advanced Microstructures
and Devices
Louisiana State University
6980 Jefferson Hwy.
Baton Rouge, LA 70806
USA

Christopher L. Kuyper
University of Washington
Department of Chemistry
P.O. Box 351700
Seattle, WA 98195-1700
USA

Gregory M. Lanza
School of Medicine
Washington University
660 S. Euclid Avenue
St. Louis, MO 63110
USA

Min-Ho Lee
Rice University
Department of Bioengineering
Houston, TX 77005
USA

Carola Leuschner
Pennington Biomedical Research
Center
6400 Perkins Road
Baton Rouge, LA 70808
USA

Alex Lin
Rice University
Department of Bioengineering
Houston, TX 77005
USA

Christopher Loo
Baylor College of Medicine
Rice University
Department of Bioengineering
Houston, TX 77005
USA

Robert M. Lorenz
University of Washington
Department of Chemistry
P.O. Box 351700
Seattle, WA 98195-1700
USA

Guang Lu
Department of Chemical & Bio-
molecular Engineering
300 Lindy Boggs Center
Tulane University
New Orleans, LA 70118
USA

Yunfeng Lu
Department of Chemical & Bio-
molecular Engineering
300 Lindy Boggs Center
Tulane University
New Orleans, LA 70118
USA

Hartwig Modrow
Physikalisches Institut der
Rheinischen Friedrich-Wilhelms-
Universität Bonn
Nußallee 12
53115 Bonn
Germany

Sergey E. Paramonov
Departments of Chemistry and Bio-
engineering
Rice University
6100 Main St.
Houston, TX 77005
USA

C. K. S. Pillai
Regional Research Laboratory
Polymer Division
Thiruvananthapuram 695019
India

Ryan M. Richards
International University Bremen
Campus-Ring 8, Res III, 116
28759 Bremen
Germany

Axel J. Rosengart
Departments of Neurology and Neuro-
surgery
The University of Chicago and Pritzker
School of Medicine
and
Neuroscience Critical Care Bio-
engineering
Argonne National Laboratory
5841 South Maryland Ave, MC 2030
Chicago, IL, 60637
USA

Latha M. Santhakumaran
University of Medicine and Dentistry of
New Jersey
Robert Wood Johnson Medical School
Department of Medicine
125 Paterson Street, CAB 7090
New Brunswick, NJ 08903
USA

Mehmet Sarikaya
Materials Science & Engineering
University of Washington
Seattle, WA 98195
USA
and
Molecular Biology and Genetics
Istanbul Technical University
Maslak, Istanbul
Turkey

Keith Sheppard
Columbia University
Teachers College
525 West 120th Street, Box 210
New York City, NY 10027
USA

Douglas Spencer
Edu, Inc.
6900-29 Daniels Parkway
Fort Meyers, FL 33912
Florida, 33901
USA

Peter K. Stoimenov
Department of Chemistry,
Kansas State University
111 Willard Hall
Manhattan, KS 66505
USA
Current address:
University of California at
Santa Barbara
Department of Chemistry and Bio-
chemistry
Santa Barbara, CA 93106
USA

Candan Tamerler
Materials Science & Engineering
University of Washington
Seattle, WA 98195
USA
and
Molecular Biology and Genetics
Istanbul Technical University
Maslak, Istanbul
Turkey

Thresia Thomas
University of Medicine and Dentistry of
New Jersey
Robert Wood Johnson Medical School
Department of Environmental and
Occupational Medicine
125 Paterson Street, CAB 7090
New Brunswick, NJ 08903
USA

T. J. Thomas
University of Medicine and Dentistry of
New Jersey
Robert Wood Johnson Medical School
Department of Medicine
125 Paterson Street, CAB 7090
New Brunswick, NJ 08903
USA

Robert Vajtai
Rensselaer Polytechnic Institute
Rensselaer Nanotechnology Center
Troy, NY 12180
USA

Anna M. Waldron
Cornell University
Nanobiotechnology Center
350 Duffield Hall
Ithaca, NY 14853
USA

Donghai Wang
Department of Chemical & Bio-
molecular Engineering
300 Lindy Boggs Center
Tulane University
New Orleans, LA 70118
USA

Bingqing Wei
Louisiana State University
Department of Electrical and
Computer Engineering and Center for
Computation and Technology
EE Building, South Campus Drive
Baton Rouge, LA 70803
USA

Jennifer West
Rice University
Department of Bioengineering
Houston, TX 77005
USA

Samuel A. Wickline
School of Medicine
Washington University
St. Louis, MO 63110
USA

Patrick M. Winter
Washington University
School of Medicine
St. Louis, MO 63110
USA

Jian Min (Jim) Zuo
Department of Material Science and
Engineering and F. Seitz Materials
Research Laboratory
University of Illinois at Urbana-
Champaign
1304 West Green Street
Urbana, IL 61801
USA

I
Fabrication of Nanomaterials

Nanofabrication Towards Biomedical Applications. C. S. S. R. Kumar, J. Hormes, C. Leuschner (Eds.)
Copyright © 2005 WILEY-VCH Verlag GmbH & Co. KGaA, Weinheim
ISBN 3-527-31115-7

1
Synthetic Approaches to Metallic Nanomaterials

Ryan Richards and Helmut Bönnemann

1.1
Introduction

In recent years research involving nanoparticles and nanoscale materials has generated a great deal of interest from scientists and engineers of nearly all disciplines. This interest has been generated in large part by reports that a number of physical properties including optical and magnetic properties, specific heats, melting points, and surface reactivities are size-dependent. These size-dependent properties are widely believed to be a result of the high ratio of surface to bulk atoms as well as the bridging state they represent between atomic and bulk materials. In the nanoscale regime, materials (especially metals and metal oxides) can be thought of as neither atomic species which can be represented by well defined molecular orbitals, nor as standard bulk materials which are represented by electronic band structures, but rather by size-dependent broadened energy states. Because metallic particles are of great importance industrially, an understanding of their properties from small clusters to bulk materials is essential. Although these nanoscale colloidal metals are of interest to scientists of many disciplines, methods for their preparation and chemical applications are primarily the focus of chemists.

Originally called gold sols, colloidal metals first generated interest because of their intensive colors, which enabled them to be used as pigments for glass or ceramics. Nanoparticulate metal colloids are generally defined as isolable particles between 1 and 50 nm in size that are prevented from agglomerating by protecting shells. Depending on the protection shell used they can be redispersed in water ("hydrosols") or organic solvents ("organosols"). The number of potential applications for these colloidal particles is growing rapidly because of the unique electronic structure of the nano-sized metal particles and their extremely large surface areas. A considerable body of knowledge has been gained about these materials throughout the last few decades, and the reader is directed to the numerous books and review articles in the literature which cover these subjects in detail [1–12, 19–26]. This contribution will be focused towards presenting an overview of the synthetic methods used to prepare metallic nanomaterials, factors influencing size and shape, and a survey of potential applications in materials science and biology. Although not covered here, the area of biodirected syntheses is an emerging area of extreme interest [13–18].

Nanofabrication Towards Biomedical Applications C. S. S. R. Kumar, J. Hormes, C. Leuschner (Eds.)
Copyright © 2005 WILEY-VCH Verlag GmbH & Co. KGaA, Weinheim
ISBN 3-527-31115-7

1.2
Wet Chemical Preparations

Nanostructured metal colloids have been obtained by both the so-called "top down" and "bottom up" methods. A typical "top down" method for example involves the mechanical grinding of bulk metals and subsequent stabilization of the resulting nanosized metal particles by the addition of colloidal protecting agents [27, 28]. Metal vapor techniques have also provided chemists with a very versatile route for the production of a wide range of nanostructured metal colloids on a preparative laboratory scale [29–34]. Use of metal vapor techniques is limited because the operation of the apparatus is demanding and it is difficult to obtain a narrow particle size distribution. The "bottom up" methods of wet chemical nanoparticle preparation rely on the chemical reduction of metal salts, electrochemical pathways, or the

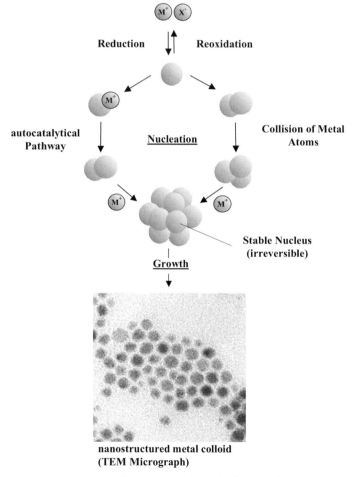

nanostructured metal colloid
(TEM Micrograph)

Figure 1.1. Formation of nanostructured metal colloids via the "salt reduction" method. (Adapted from Ref. [4].)

controlled decomposition of metastable organometallic compounds. A large variety of stabilizers, e.g., donor ligands, polymers, and surfactants, are used to control the growth of the primarily formed nanoclusters and to prevent them from agglomerating. The chemical reduction of transition metal salts in the presence of stabilizing agents to generate zerovalent metal colloids in aqueous or organic media was first published in 1857 by Faraday [35], and this approach has become one of the most common and powerful synthetic methods in this field [10, 11, 36]. The first reproducible standard recipes for the preparation of metal colloids (e.g., for 20 nm gold by reduction of [AuCl$_4^-$] with sodium citrate) were established by Turkevich [1–3]. Based on nucleation, growth, and agglomeration he also proposed a mechanism for the stepwise formation of nanoclusters which in essence is still valid. Data from modern analytical techniques and more recent thermodynamic and kinetic results have been used to refine this model as illustrated in Fig. 1.1 [31–38].

The metal salt is reduced to give zerovalent metal atoms in the embryonic stage of nucleation [37]. These can collide in solution with further metal ions, metal atoms, or clusters to form an irreversible "seed" of stable metal nuclei. Depending on the difference of the redox potentials between the metal salt and the reducing agent applied, and the strength of the metal–metal bonds, the diameter of the "seed" nuclei can be well below 1 nm.

Nanostructured colloidal metals require protective agents for stabilization and to prevent agglomeration. The two basic modes of stabilization which have been distinguished are electrostatic and steric (Fig. 1.2) [36]. Electrostatic stabilization [see Fig. 1.2(a)] involves the coulombic repulsion between the particles caused by the electrical double layer formed by ions adsorbed at the particle surface (e.g., sodium citrate) and the corresponding counterions. As an example, gold sols are prepared by the reduction of [AuCl$_4^-$] with sodium citrate [1–3]. By coordinating sterically demanding organic molecules that act as protective shields on the metallic surface, steric stabilization [Fig. 1.2(b)] is achieved. In this way nanometallic cores are separated

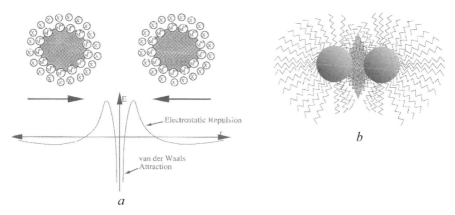

Figure 1.2. (a) Electrostatic stabilization of nanostructured metal colloids. (Scheme adapted from Ref. [36].) (b) Steric stabilization of nanostructured metal colloids. (Scheme adapted from Ref. [36].)

from each other, and agglomeration is prevented. The main classes of protective groups selected from the literature are: polymers and block copolymers [45–48]; P, N, S donors (e.g., phosphines, amines, thioethers) [6, 65–90]; solvents such as THF [6, 91], THF/MeOH [92], or propylene carbonate [93]; long chain alcohols [49–64, 94]; surfactants [6, 7, 9, 21, 22, 93, 95–106]; and organometallics [107–110]. In general, lipophilic protective agents give metal colloids that are soluble in organic media ("organosols") while hydrophilic agents yield water-soluble colloids ("hydrosols"). In Pd organosols stabilized by tetraalkylammonium halides the metal core is protected by a monolayer of the surfactant coat (Fig. 1.3) [111].

Metal hydrosols, in contrast, are stabilized by zwitterionic surfactants which are able to self-aggregate, and are enclosed in organic double layers. After the application of uranylacetate as a contrasting agent, the transmission electron micrographs show that the colloidal Pt particles (average size = 2.8 nm) are surrounded by a double layer zone of the zwitterionic carboxybetaine (3–5 nm). The hydrophilic head group of the betaine interacts with the charged metal surface and the lipophilic tail is associated with the tail of a second surfactant molecule, resulting in the formation a hydrophilic outer sphere (see Fig. 1.4) [112]. Pt or Pt/Au particles can be hosted in the hydrophobic holes of nonionic surfactants, e.g., polyethylene monolaurate [113, 114].

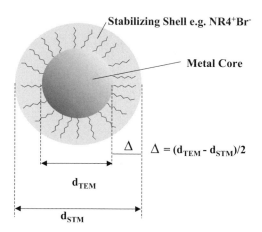

$$\Delta = (d_{TEM} - d_{STM})/2$$

Figure 1.3. Differential transmission electron microscopy/ scanning transmission electron microscopy (TEM/STEM) study of a Pd organosol showing that the metal core (size = d_{TEM}) is surrounded by a monolayer of the surfactant (thickness $\Delta = (d_{TEM} - d_{STM})2$). (Adapted from Ref. [9].)

1.3
Reducing Agents

The type of reducing agent employed has been found to greatly affect the resulting particles. It has been experimentally verified in the case of silver that stronger reducing agents produce smaller nuclei in the "seed" [37]. During the so-called "ripening"

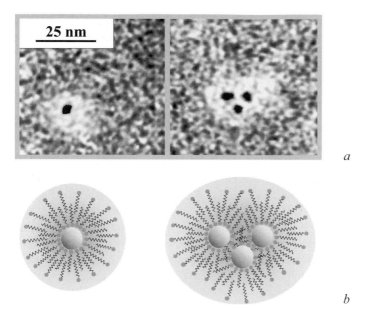

Figure 1.4. (a) TEM micrographs of colloidal Pt particles (single and aggregated, average core size = 2.8 nm) stabilized by carboxybetaine 12 (3–5 nm, contrasted with uranylacetate against the carbon substrate). (b) Schematic model of the hydrosol stabilization by a double layer of the zwitterionic carboxybetaine 12 (= lipophilic alkyl chain; ●ⅶⅶ = hydrophilic, zwitterionic head group). (Adapted from Ref. [4].)

process these nuclei grow to yield colloidal metal particles in the size range of 1–50 nm which have a narrow size distribution. It was assumed that the mechanism for the particle formation is an agglomeration of zerovalent nuclei in the "seed" or – alternatively – collisions of already formed nuclei with reduced metal atoms. The stepwise reductive formation of Ag_3^+ and Ag_4^+ clusters by spectroscopic methods has been followed by Henglein's group [38]. Their results strongly suggest that an autocatalytic pathway is involved in which metal ions are adsorbed and successively reduced at the zerovalent cluster surface. The formation of colloidal Cu protected by cationic surfactants (NR_4^+) has been investigated by in situ X-ray absorption spectroscopy which demonstrated the formation of an intermediate Cu^+ state prior to the nucleation of the particles [41]. It is now generally accepted that the size of the resulting metal colloid is determined by the relative rates of nucleation and particle growth, although the processes taking place during nucleation and particle growth cannot be analyzed separately.

The salt reduction method has the main advantage that in the liquid phase it is reproducible and it allows colloidal nanoparticles with a narrow size distribution to be prepared on the multigram scale. The classical Faraday route via the reduction of $[AuCl_4]^-$ with sodium citrate for example, is still used to prepare standard 20-nm gold sols for histological staining applications [1, 115]. Wet chemical reduction procedures have been applied in the last 20 years or so to combine practically all transition metals with different types of stabilizers, and the whole range of chemical reducing agents has successfully been applied. In 1981, Schmid et al. established the "diborane-as-reductant route" for the synthesis of $Au_{55}(PPh_3)_{12}Cl_6$ (1.4 nm), a full shell ("magic number") nanocluster stabilized by phosphine ligands [57–72]. Clusters of Au_{55} were uniformly formed when a stream of B_2H_6 was carefully introduced into a Au^{III} ion solution. The "diborane route" for $M_{55}L_{12}Cl_n$ nanoclusters was recently reviewed by Finke et al. [11]. Bimetallic nanoclusters that were made accessible by this method have been thoroughly characterized [65–80]. The phosphane ligands may be exchanged in the Au_{55} nanoclusters quantitatively using silsesquioxanes, which causes important changes in the physical and chemical behavior of the gold clusters [80]. The synthesis and general chemistry of nanosized silica-coated metal particles has been elaborated by Mulvaney et al. [80]. The "alcohol reduction process" described by Hirai and Toshima et al. [10, 45–48] is widely applicable to the preparation of colloidal precious metals stabilized by organic polymers such as poly(vinylpyrrolidone) (PVP), poly(vinyl alcohol) (PVA), and poly(methylvinyl ether). Alcohols containing α-hydrogen atoms are oxidized to the corresponding carbonyl compound (e.g., methanol to formaldehyde) during the salt reduction. The method for preparing bimetallic nanoparticles via the coreduction of mixed ions has been evaluated in a recent review [10]. Recently, it has been demonstrated that through the appropriate choice of reduction temperature and acetate ion concentration, ruthenium nanoparticles prepared by the reduction of $RuCl_3$ in a liquid polyol could be monodispersely prepared with sizes in the 1–6 nm range [116]. Hydrogen has been used as an efficient reducing agent for the preparation of electrostatically stabilized metal sols and of polymer-stabilized hydrosols of Pd, Pt, Rh, and Ir [117–121]. Moiseev's giant Pd cluster [Fig. 1.5(a)] [81–86], Finke's polyoxoanion, and tetrabutyl-ammonium-stabilized transition-metal nanoclusters [Fig. 1.5(b)] [11, 40, 122–126] were also prepared by the hydrogen reduction pathway.

Finke et al. have recently reviewed the characterization of Moiseev's "giant" cationic Pd clusters [81–86] [Fig. 1.5(a)] [idealized formula $Pd_{\approx561}L_{\approx60}(OAc)_{\approx180}$ (L= phenanthroline, bipyridine)] and their catalytic properties [11]. The results of a combination of modern instrumental analysis methods applied to Finke's nanoclusters have also recently been carefully discussed [11].

Using CO, formic acid or sodium formate, formaldehyde, and benzaldehyde as reductants, colloidal Pt in water [2, 127] was obtained [128]. Silanes have been found to be effective for the reductive preparation of Pt sols [129, 130]. Duff, Johnson, and Baiker et al. have successfully introduced tetrakis(hydroxymethyl)phosphoniumchloride (THPC) as a reducing agent, which allows the size- and morphology-selective synthesis of Ag, Cu, Pt, and Au nanoparticles from their corresponding metal salts [131–136]. Further, hydrazine [137], hydroxylamine [138], and electrons trapped in, for example, $K^+[(crown)_2K]^-$ [139], have also been successfully applied as reductants. In addition, BH_4^- has been found to be a powerful and valuable reagent for the salt reduction method. A disadvantage, however, is that transition metal borides are often found along with the nanometallic particles [140, 141]. Tetraalkylammonium hydrotriorganoborates [6, 7, 9, 21, 95–97] offer a wide range of applications in the wet chemical reduction of transition metal salts. The reductant $[BEt_3H^-]$ is

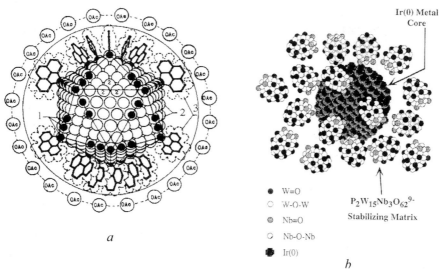

● W=O
○ W-O-W
◉ Nb=O
○ Nb-O-Nb
◐ Ir(0)

$P_2W_{15}Nb_3O_{62}{}^{9-}$
Stabilizing Matrix

a

b

Figure 1.5. (a) Idealized model of Moiseev's "giant palladium cluster" $Pd_{\approx561}Phen_{\approx60}(OAc)_{\approx180}$ (Phen = phenanthroline) (adapted from Ref. [4]). (b) Idealized model of a Finke type Ir(0) nanocluster $P_2W_{15}Nb_3O_{62}{}^9$ – and Bu_4N^+ stabilized $Ir(0)_{\approx300}$. (Adapted from Ref. [4].)

combined with the stabilizing agent (e.g. NR_4^+) in this case. The surface-active NR_4^+ salts are formed immediately at the reduction center at high local concentration and prevent particle aggregation. Trialkylboron is recovered unchanged from the reaction and there are no borides contaminating the products. Most recently it has been demonstrated that the chain length of the alkyl group in the tetraalkylammonium plays a critical role in the stabilization of various metal colloids [142].

$$MX_v + NR_4(BEt_3H) \longrightarrow M_{colloid} + v\,NR_4X + v\,BEt_3 + v/2\,H_2 \uparrow \qquad (1)$$

where M = metals of groups 6–11; X = Cl, Br; v = 1,2,3; and R = alkyl, C_6–C_{20}. The NR_4^+-stabilized metal "raw" colloids as synthesized typically contain 6–12 wt% of metal. "Purified" transition metal colloids containing ca. 70–85 wt% of metal are obtained by work-up with ethanol or ether and subsequent reprecipitation by a solvent of different polarity (see Tab. 9 in Ref. [6]). When NR_4X is coupled to the metal salt prior to the reduction step the pre-preparation of $[NR_4^+\ BEt_3H^-]$ can be avoided. Transition metal nanoparticles stabilized by $NR_4^+X^-$ can also be obtained from NR_4X-transition metal double salts. A number of conventional reducing agents may be applied since the local concentration of the protecting group is sufficiently high to give Eq. (2) [7, 21].

$$(NR_4)_w\ MX_vY_w + v\,Red \longrightarrow M_{colloid} + v\,RedX + w\,NR_4Y \qquad (2)$$

where M = metals; Red = H_2, HCOOH, K, Zn, LiH, $LiBEt_3H$, $NaBEt_3H$, $KBEt_3H$; X,Y = Cl, Br; v, w = 1–3 and R = alkyl, C_6–C_{12}.

The scope and limitations of this method have been evaluated in a recent review [11]. Isolable metal colloids of the zerovalent early transition metals which are stabilized only with THF have been prepared via the $[BEt_3H^-]$ reduction of the preformed THF adducts of $TiBr_4$, [Eq. (3)] $ZrBr_4$, VBr_3, $NbCl_4$, and $MnBr_2$ [Eq. (3)].

$$x \cdot [TiBr_4 \cdot 2\,THF + x \cdot 4\,K[BEt_3H] \xrightarrow{\quad THF,\ 2h,\ 20\,°C \quad} \tag{3}$$

$$[Ti \cdot 0.5\,THF]_x + x \cdot 4\,BEt_3 + x \cdot 4\,KBr\downarrow + x \cdot 4\,H_2\uparrow$$

The results are summarized in Tab. 1.1.

Table 1.1. THF-stabilized organosols of early transition metals.

Product	Starting material	Reducing agent	T (°C)	T (h)	Metal content (%)	Size (nm)
[Ti · 0.5THF]	$TiBr_4 \cdot 2THF$	$K[BEt_3H]$	rt	6	43.5	(<0.8)
[Zr · 0.4THF]	$ZrBr_4 \cdot 2THF$	$K[BEt_3H]$	rt	6	42	–
[V · 0.3THF]	$VBr_3 \cdot 3THF$	$K[BEt_3H]$	rt	2	51	–
[Nb . 0.3THF]	$NbCl_4 \cdot 2THF$	$K[BEt_3H]$	rt	4	48	–
[Mn · 0.3THF]	$MnBr_2 \cdot 2THF$	$K[BEt_3H]$	50	3	70	1–2.5

Detailed studies of [Ti · 0.5 THF] [91] show that it consists of Ti_{13} clusters in the zerovalent state, stabilized by six intact THF molecules (Fig. 1.6).

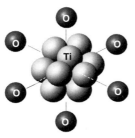

Figure 1.6. Ti_{13} cluster stabilized by six THF-O atoms in an octahedral configuration [7].

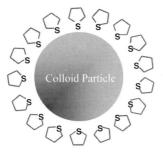

Figure 1.7. Organosols stabilized by tetrahydrothiophene. For M = Ti, V: decomposition. For M= Mn, Pd, Pt: stable colloids.

By analogy, [Mn · 0.3 THF] particles (1–2.5 nm) were prepared [143] and the physical properties studied [144]. In the case of Mn, Pd, and Pt organosols the THF in Eq. (3) was successfully replaced by tetrahydrothiophene (THT); but attempts to stabilize Ti and V this way led to decomposition (Fig. 1.7) [7].

Figure 1.8 gives an overview of the [BEt$_3$H$^-$] method. The advantages of this method may be summarized as follows:

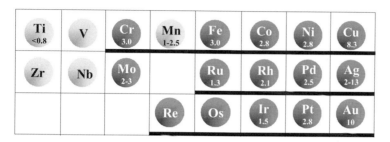

▬▬▬ Nanometal powders

◯ THF-stabilized nanometals

⬤ NR$_4^+$-stabilized nanometals

Figure 1.8. Nanopowders and nanostructured metal colloids accessible via the [BEt$_3$H$^-$] reduction method (including the mean particle sizes obtained). (Adapted from Ref. [7].)

- The method is generally applicable to salts of metals in groups 4–11 in the periodic table.
- It yields extraordinarily stable metal colloids that are easy to isolate as dry powders.
- The particle size distribution is nearly monodisperse.
- Bimetallic colloids are easily accessible by coreduction of different metal salts.
- The synthesis is suitable for multigram preparations and easy to scale up.

One of the drawbacks of this method, however, is that the particle size of the resulting sols cannot be varied by altering the reaction conditions. Using betaines instead of NR$_4^+$ salts as the protecting group in Eq. (1), highly water-soluble hydrosols, particularly those of zerovalent precious metals, were made accessible. A wide variety of hydrophilic surfactants may be used in Eq. (2) [7, 21, 96]. Reetz and Maase et al. have reported a new method for the size- and morphology-selective preparation of metal colloids using tetraalkylammonium carboxylates of the type NR$_4^+$R'CO$_2^-$ (R = octyl, R' = alkyl, aryl, H) both as the reducing agent and the stabilizer [Eq. (4)] [145–147].

$$M^+ + R_4N^+ R'CO_2^- \xrightarrow{50-90\,°C} M^0 (R_4 NR'CO_2)_x + CO_2 + R'\text{-}R \qquad (4)$$

where R = octyl, R′= alkyl, aryl, H. The resulting particle sizes were found to correlate with the electronic nature of the R′ group in the carboxylate. Electron donors produce small nanoclusters while electron-withdrawing substituents R′, in contrast, yield larger particles. For example, Pd particles of 2.2 nm size were found when $Pd(NO_3)_2$ was treated with an excess of tetra(n-octyl)ammonium-carboxylate bearing $R′ = (CH_3)_3CCO_2^-$ as the substituent. The particle size was found to be 5.4 nm with $R′ = Cl_2CHCO_2^-$ (an electron-withdrawing substituent). Bimetallic colloids of the following were obtained with tetra(n-octyl) ammonium formiate as the reductant: Pd/Pt (2.2 nm), Pd/Sn (4.4 nm), Pd/Au (3.3 nm), Pd/Rh (1.8 nm), Pt/Ru (1.7 nm), and Pd/Cu (2.2 nm). The shape of the particles was also found to depend on the reductant: with tetra(n-octyl) ammonium glycolate reduction of $Pd(NO_3)_2$ a significant amount of trigonal particles were detected in the resulting Pd colloid. Recent work in our group has shown that organoaluminum compounds can be used for the "reductive stabilization" of mono- and bimetallic nanoparticles [see Eq. (5) and Tab. 1.2] [107–108].

Table 1.2. Mono- and bimetallic nanocolloids prepared via the organo-aluminum route.

Metal salt	Reducing agent		Solvent Toluene	Conditions		Product	Metal content wt.%	Particle size
	g/mmol	g/mmol	ml	t [°C]	t [h]	m [g]		F [nm]
Ni(acac)$_2$	0.275/1	Al(i-but)$_3$ 0.594/3	100	20	10	0.85	Ni: 13.8	2–4
Fe(acac)$_2$	2.54/10	Al(me)$_3$ 2.1/30	100	20	3	2.4	n.d.	
RhCl$_3$	0.77/3.1	Al(oct)$_3$ 4.1/11.1	150	40	18	4.5	Rh: 8.5 Al: 6.7	2–3
Ag-decanoate	9.3/21.5	Al(oct)$_3$ 8.0/21.8	1000	20	36	17.1	Ag: 11.8 Al: 2.7	8–12
Pt(acac)$_2$	1.15/3	Al(me)$_3$ 0.86/7.6	150	20	24	1.45	Pt: 35.8 Al: 15.4	2.5
PtCl$_2$	0.27/1	Al(me)$_3$ 0.34/3	125	40	16	0.47	Pt: 41.1 Al: 15.2	2.0
Pd(acac)$_2$ Pt(acac)$_2$	0.54/1.8 0.09/0.24	Al(et)$_3$ 0.46/4	500	20	2	0.85	Pd: 22 Pt: 5.5 Al: 12.7	3.2
Pt(acac)$_2$ Ru(acac)$_3$	7.86/20 7.96/20	Al(me)$_3$ 8.64/120	400	60	21	17.1	Pt: 20.6 Ru: 10.5 Al: 19.6	1.3
Pt(acac)$_2$ SnCl$_2$	1.15/2.9 0.19/1	Al(me)$_3$ 0.86/12	100	60	2	1.1	Pt: 27.1 Sn: 5.2 Al: 14.4	n.d.

$$MX_n + AlR_3 \xrightarrow{\text{Toluene}} \quad + \quad [R_2Alacac] \tag{5}$$

M = Metals of Groups 6-11 PSE

X = Halogen, Acetylacetonate n = 2-4
R = C_1-C_8-Alkyl
Particle sizes 1-12nm

where M = metals of groups 6–11; X = halogen, acetylacetonate, n = 2–4; R = C_1–C_8-alkyl; particle sizes 1–12 nm.

Colloids of zerovalent elements of groups 6–11 of the periodic table (and also of tin) may be prepared according to Eq. (5), in the form of stable, isolable organosols. The analytical data available suggest that a layer of condensed organoaluminum species protects the transition metal core against aggregation as visualized in Eq. (5). The exact nature of the "backbone" of the colloidal organoaluminum protecting agent has not yet been completely established.

Quantitative protonolysis experiments have detected the presence of unreacted organoaluminum groups (e.g., Al–CH_3, Al–C_2H_5) from the starting material which are still present in the stabilizer. These active Al–C bonds have been used for controlled protonolysis by long-chain alcohols or organic acids ("modifiers") to give al-alkoxide groups in the stabilizer [Eq. (6)].

$$\xrightarrow[\text{- CH}_4]{\text{+ R-OH (Modifier)}} \tag{6}$$

Modifiers: alcohols, carbonic acids, silanols, sugars, polyalcohols, polyvinylpyrrolidone, surfactants, silica, alumina, etc.

The dispersion characteristics of the original sol can be tailored by this "modification" [Eq. (6)] of the organoaluminum protecting shell. A wide variety of dissolubilities of the colloidal metals in hydrophobic and hydrophilic media (including water) has been achieved this way. The active Al–C bonds in the colloidal protecting shell can react with inorganic surfaces bearing –OH, which opens new ways for the heterogeneous catalyst preparation. The particle size of the metal core is not altered during this modification process (Fig. 1.9) [109].

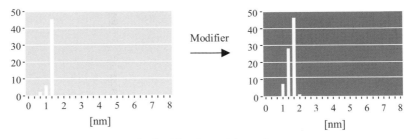

Figure 1.9. Size conservation of colloidal Pt/Ru particles under the hydrophilic modification of the $(CH_3)_n$–Alacac protecting shell using polyethyleneglycol-dodecylether.

1.4
Electrochemical Synthesis

Since 1994 this very versatile preparation route for nanostructured mono- and bimetallic colloids has been further developed by Reetz and his research group [8, 98, 99]. The overall process of electrochemical synthesis [Eq. (7)] can be divided into six elementary steps (see Fig. 1.10).

Anode:		M_{bulk}	$\rightarrow$	$M^{n+} + ne^-$	
Cathode:	$M^{n+} + ne^-$ + stabilizer		$\rightarrow$	$M_{coll}/stabilizer$	(7)
Sum:	M_{bulk} + stabilizer		$\rightarrow$	$M_{coll}/stabilizer$	

1. Oxidative dissolution of the sacrificial Met_{bulk} anode
2. Migration of Met^{n+} ions to the cathode
3. Reductive formation of zerovalent metal atoms at the cathode
4. Formation of metal particles by nucleation and growth
5. Arrest of the growth process and stabilization of the particles by colloidal protecting agents, e.g., tetraalkylammonium ions
6. Precipitation of the nanostructured metal colloids.

Advantages of the electrochemical pathway are that contamination with byproducts resulting from chemical reduction agents is avoided, and that the products are easily isolated from the precipitate. The electrochemical preparation also provides size-selective particle formation. Experiments using Pd as the sacrificial anode in the electrochemical cell to give $(C_8H_{17})_4N^+Br^-$-stabilized $Pd^{(0)}$ particles indicate that the particle size depends on the current density applied: high current densities led to small Pd particles (1.4 nm); low current densities, in contrast, gave larger particles (4.8 nm) [98]. As was seen in a careful analysis of tetraalkylammonium-stabilized Pd

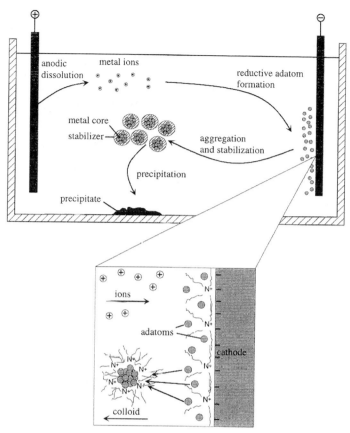

Figure 1.10. Electrochemical formation of NR$_4^+$Cl$^-$-stabilized nanometal. (Adapted from Ref. [9].)

and Ni with a combination of transmission electron microscopy (TEM) and small angle X-ray scattering (SAXS), particle size is not controlled by a single cause but rather can be adjusted by varying the following parameters:

- The distance between the electrodes
- The reaction time and temperature
- The polarity of the solvent.

Through the use of electrochemical synthesis nearly monodisperse Pd$^{(0)}$ particles with sizes between 1 and 6 nm can be obtained. It was also shown that the size of NR$_4^+$-stabilized Ni$^{(0)}$ particles [100] can be adjusted at will. The electrochemical method [98–105] [Eq. (7)] has been successfully applied to prepare a number of monometallic organosols and hydrosols, e.g., of Pd, Ni, Co, Fe, Ti, Ag, and Au on a scale of several hundred milligrams (yields >95%). Using the electrochemical pathway, solvent-stabilized (propylene carbonate) Pd particles (8–10 nm) have also been obtained [93]. If two sacrificial Met$_{bulk}$ anodes are used in a single electrolysis cell, bimetallic nanocolloids (Pd/Ni, Fe/Co, Fe/Ni) are accessible [103]. In the cases of Pt, Rh, Ru, and Mo, which are anodically less readily soluble, the corresponding metal salts were electrochemically reduced at the cathode (see lower part of Fig. 1.10 and Tab. 1.3).

Table 1.3. Electrochemically prepared metallic colloids

Metal salt	d(nm)	Element analysis
PtCl$_2$	2.5[b]	51.21% Pt
PtCl$_2$	5.0[c]	59.71% Pt
RhCl$_3 \cdot$ x H$_2$O	2.5	26.35% Rh
RuCl$_3 \cdot$ x H$_2$O	3.5	38.55% Ru
OsCl$_3$	2.0	37.88%Os
Pd(OAc)$_2$	2.5	54.40% Pd
Mo$_2$(OAc)$_4$	5.0	36.97% Mo
PtCl$_2$ + RuCl$_3 \cdot$ xH$_2$0	2.5	41.79% Pt + 23.63% Rh[d]

[a] Based on stabilizer-containing material.
[b] Current density: 5.00 mA cm^{-2}.
[c] Current density: 0.05 mA cm^{-2}.
[d] Pt-Ru dimetallic cluster.

Tetraalkylammonium-acetate was used both as the supporting electrolyte and the stabilizer in a Kolbe electrolysis at the anode [see Eq. (8)] [104].

$$\text{Cathode:} \quad Pt^{2+} + 2e \longrightarrow Pt^0 \qquad\qquad (8)$$
$$\text{Anode:} \quad 2\,CH_3CO_2 \longrightarrow 2\,CH_3CO_2 + 2e^-$$

Bimetallic nanocolloids can be prepared by combining the electrochemical methods described in Eqs. (7) and (8) (see Tab. 1.4) [104].

Table 1.4. Bimetallic colloids prepared electrochemically

Anode	Metal salt	d(nm)	Stoich. Energy disperse X-ray analysis
Sn	PtCl$_2$	3.0	Pt$_{50}$Sn$_{50}$
Cu	Pd(OAc)$_2$[a]	2.5	Cu$_{44}$Pd$_{56}$
Pd	PtCl$_2$	3.5	Pd$_{50}$Pt$_{50}$

[a] Electrolyte: 0.1M [(n-octyl)$_4$N]OAc/THF.

By modifying the electrochemical method, the synthesis of layered bimetallic nanocolloids (e.g., Pt/Pd) was achieved [100, 105]. A preformed (Oct)$_4$NBr-stabilized Pt colloid core (size: 3.8 nm) was electrolyzed in 0.1 M (Oct)$_4$NBr/THF solution with Pd as the sacrificial anode (Fig. 1.11).

The preformed Pt core may be regarded as a "living metal polymer" on which the Pd atoms are deposited to give "onion-type" bimetallic nanoparticles (5 nm).

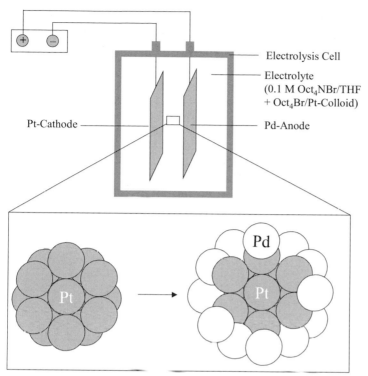

Figure 1.11. Modified electrolysis cell for the preparation of layered bimetallic Pt/Pd nanocolloids. (Adapted from Ref. [100].)

1.5
Decomposition of Low-Valency Transition Metal Complexes

Short-lived nucleation particles of zerovalent metals in solution which may be stabilized by colloidal protecting agents are formed by decomposition of low-valency organometallic complexes and several organic derivatives of the transition metals under the action of heat, light, or ultrasound. Thermolysis [148–153], for example, leads to the rapid decomposition of Co carbonyls to give colloidal Co in organic solutions [148, 149]. Thermolysis of labile precious metal salts in the absence of stabilizers yields colloidal Pd, Pt, and bimetallic Pd/Cu nanoparticles [150] with a broad size distribution. In the presence of stabilizing polymers, such as PVP, these results were greatly improved [151]. Recently, heating in a simple household microwave oven was proposed to prepare nanosized metal particles and colloids [152, 153]. The electromagnetic waves heat the substrate uniformly, leading to more homogeneous nucleation and a shorter aggregation time.

Sonochemical decomposition methods have been successfully developed by Suslick et al. [154] and Gedanken et al. [155–157] and have yielded Fe, Mo_2C, Ni, Pd, and Ag nanoparticles in various stabilizing environments.

By the controlled chemical decomposition of zerovalent transition metal complexes on the addition of CO or H_2 in the presence of appropriate stabilizers, isolable yields of colloidal product in multigram amounts can be prepared [87–90, 158–168]. Bradley and Chaudret et al. [87–90, 159–165] have demonstrated the use of low-valency transition metal olefin complexes as a very clean source for the preparation of nanostructured mono- and bimetallic colloids. Micelles, inverse micelles, and encapsulation methods have also been successfully employed for the preparation of nanoparticulate colloids [38, 39, 94]. It is also worth mentioning that, although beyond the focus of this article, a number of nanoparticulate metal oxide systems have been successfully developed [7, 167–172].

The radiolytic synthesis of mixed Au(III)/Pd(II) solutions has been studied at different dose rates [173]. It was found that at low dose rates, a bilayered cluster with an Au core/Pd shell predominates due to intermetal electron transfer from Pd atoms to Au ions, resulting first in the reduction of the latter to form the core of the particle and then in Pd ion reduction to form the shell. However, at high dose rates when the ion reduction is faster than a possible intermetal electron transfer, genuine alloyed clusters are formed.

1.6
Particle Size Separations

When the particle size deviates less than 15% from the average value, metal colloid sols are generally addressed as "monodisperse." Histograms with a standard deviation σ from the mean particle size of approximately 20% are described as showing a "narrow size distribution." The kinetics of the particle nucleation from atomic units and of the subsequent growth process cannot be observed directly by physical methods. The two primary tools available to the preparative chemist to control the particle size in practice are size-selective separation [51, 174, 175] and size-selective synthesis [41–56, 90, 135–137, 165–181].

So-called *size-selective precipitation* (SPP) was predominantly developed by Pileni [50]. Monodisperse silver particles (2.3 nm, $\sigma = 15\%$) were precipitated from a polydisperse silver colloid solution in hexane by the addition of pyridine in three iterative steps. Recently the two-dimensional "crystallization" of truly monodisperse Au_{55} clusters has been reported by Schmid et al. [174]. Chromatographic separation methods have thus far proven unsuccessful because the colloid was decomposed after the colloidal protecting shell had been stripped off [145]. Cölfen and Pauck have developed size-selective ultracentrifuge separation of Pt colloids [175]. However, although this elegant separation method gives true monodisperse metal colloids, it still provides only milligram-scale samples. Turkevich et al. were the first to describe size-selective colloid synthesis [1, 2]. They were able to vary the particle size of colloidal Pd between 0.55 and 4.5 nm using the salt reduction method. The crucial parameters were the amount of the reducing agent applied, and the pH value. According to the literature on the process of nucleation and particle growth, the essential factors which control the particle size are the strength of the metal–metal bond [48], the molar ratio of metal salt, colloidal stabilizer, reduction agent [1, 128, 135, 176–193], the extent of conversion or the reaction time [128], the temperature applied [1, 177, 189], and the pressure [177]. The preparation of nearly monodisperse nanostructured metal colloids using the salt reduction pathway is well documented in the literature. The "control," i.e., the variation of particle sizes (and shapes), in wet chemical colloid synthesis in practice is left to the intuition of the chemist. At present the most rational method for selecting the particle size is offered by the electrochemical synthesis of Reetz and coworkers. The authors have obtained at will almost monodisperse samples of colloidal Pd and Ni between 1 and 6 nm using variable current densities and suitable adjustment of further essential parameters [98–105]. The resulting particle size in the thermal decomposition method depends on the heat source (see Tab. 1.5) [154]. Size control has also been reported for the sonochemical decomposition method and γ-radiolysis [173, 194, 195].

Table 1.5. Platinum colloids prepared by thermal decomposition methods. (From Ref. [153]).

No.[a]	PVP[b]	NaOH[b]	Average diameter (nm)	Standard deviation (nm)	Relative standard deviation
1	10	0	3.8	0.57	0.15
2	20	0	3.4	0.56	0.16
3	50	0	3.0	0.50	0.17
4	100	0	2.9	0.47	0.16
5	50	2	3.0	0.49	0.16
6	50	4	2.6	0.48	0.18
7	50	6	1.9	0.33	0.17
8	50	8	2.0	0.32	0.16
9	50	10	2.1	0.40	0.19
10	50	0	3.1	1.08	0.35
11	50	8	1.8	0.55	0.31
12	50	0	2.7	0.74	0.27
13	50	8	1.1	0.31	0.28

[a] Nos. 1–9 were prepared by microwave dielectric heating without stirring; nos. 10 and 11 were prepared without stirring, and nos. 12 and 13 were prepared with stirring by oil bath heating.
[b] Data refer to the molar ratios of PVP (as a monomeric unit) and NaOH to Pt respectively.

The domain of preparation methods using constrained environments affords control of the metal particle shape via the preformation of size and the morphology of the products in nano-reaction chambers [49–64]. Recently, the controlled temperature-induced size and shape manipulation of 2- to 6-nm Au particles encapsulated in alkanethiolate monolayers has been reported [62]. The use of near-infrared laser light has induced an enormous increase in the size of thiol-passivated Au particles up to ca. 200 nm [62]. A new medium-energy ion scattering (MEIS) simulation program has successfully been applied to the composition and average particle size analysis of Pt-Rh/α–Al$_2$O$_3$[63].

1.7
Potential Applications in Materials Science

It is expected that metal nanoparticles and their assemblies will have numerous applications in materials science. It has been demonstrated that physical properties including magnetic and optical properties, melting points, specific heats, and surface reactivity are size-dependent. Quantum size effects are related to the "dimensionality" of a system in the nanometer range. "Zero-dimensional" metal particles might still comprise hundreds of atoms. One-dimensional nanoparticle arrangements (cluster wires) are of potential practical interest as semiconducting nanopaths for applications in nanoelectronics. One-dimensional particle arrangements may be induced through host templates. Using vacuum or electrophoretic methods Schmid et al. [196–198] were able to fill the parallel channels of nanoporous alumina membranes with chains/rows of 1.4-nm Au particles giving one-dimensional "quantum wires" consisting of insulated 20- to 100-Au$_{55}$ clusters in a helical array. The diameter of the nanowire could be controlled by varying the pore size.

Interestingly, 1.4-nm Au particles were found to arrange themselves into a linear row when attached to single-stranded DNA oligonucleotides [199, 200]. Driven by the technological significance associated with such architectures, the fabrication of ordered two-dimensional nanoparticle arrays has been successfully achieved by several research groups whose work has recently been reviewed [201]. Planar arrays of uniform metal nanoparticles would allow the design of new "supercomputers" with a superior data storage capacity. Langmuir–Blodgett films of nanometal systems have frequently been studied in this respect. Starting with nanoparticles of defined nuclearity, two-dimensional lattices of thiolized Au$_{55}$, Pd$_{561}$, and Pd$_{1415}$ have been made [202]. Recently, the first successful preparation of two-dimensional hexagonal and cubic lattices of Au$_{55}$ nanoparticles by self-assembly on polymer films was reported [174]. Simply dipping polyethylenimine-modified surfaces into aqueous solutions of acid-functionalized Au$_{55}$ cluster generates the Au$_{55}$ monolayers shown in Fig. 1.12.

The interactions between the nanoparticles and the surface are obviously strong enough to prevent mechanical removal. Whereas the hexagonal form shown in Fig. 12 (a) is normal for an ordered monolayer, the cubic orientation seen in Fig. 12 (b) is unprecedented. Most of the work published on organized nanometal structures is focused on gold particles and sulfur-containing groups in the various ligands [203–208]. Schiffrin et al. have achieved the self-organization of nanosized gold particles using NR$_4^+$X surfactants [209]. Ramos et al. have recently reported the surfactant-mediated two-dimensional crystallization of colloidal crystals [210]. A potential new route to self-assembly of ordered colloidal structures is through the use of attractive Coulomb interactions between colloidal structures and surfactant structures. Nano-structured palladium clusters, stabilized by a monomolecular coat of tetraalkylam-

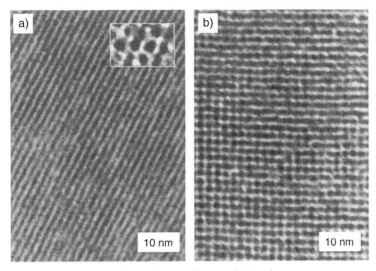

Figure 1.12. Au$_{55}$ monolayers showing a hexagonal (a) and a cubic (b) structure. The insert in (a) shows single clusters in the hexagonal form. (Adapted from Ref. [174].)

monium halide surfactants, self-assemble on carbon surfaces in an ordered manner with the formation of hexagonal close packed (hcp) structures [211–215].

The self-organization of magnetic nanosized cobalt particles was studied by Pileni's group [216, 217]. A comparison of the magnetic properties of deposited cobalt nanoparticles with those dispersed in a solvent indicates a collective flip of magnetization of adjacent particles when they are self-assembled. Mulvaney et al. have described two-dimensional and three-dimensional assemblies of metal core–silica shell nanoparticles in a recent review article [218]. A feature article by Balazs et al. [219] outlines how solid additives can be used to tailor the morphology of binary mixtures containing nanoscopic particles and thereby control the macroscopic properties (e.g., the mechanical integrity) of composites. In addition, computer-aided design has been employed to establish how self-assembled nanostructures can be induced to form arbitrary functional designs on surfaces [220]. Bifunctional spacer molecules such as diamines have been used in attempts to link nanoparticles three-dimensionally [221]. The multilayer deposition of particle arrays on gold has been successfully achieved via the sequential adsorption of dithiol and near-monodisperse nanometal or CdS particles.

Several monometal, bimetal, and metal-semiconductor superlattices have been prepared by dipping a gold substrate into the respective solutions with intermediate steps involving washing and drying [222]. The stepwise three-dimensional assembly of layered gold nanoparticles in porous silica matrices has also been reported [223].

A different field of technological interest stems from the high spin density of nanostructured magnetic metals of the Fe, Co, Ni series [224–225]. THF-stabilized Mn$^{(0)}$ particles which exhibit superparamagnetism below 20K were described as the first example of an antiferromagnetic metal colloid [226].

New strategies utilizing DNA as a construction material for the generation of bio-metallic nanostructures have made it possible to develop larger "nanotechnology devices" (<100 nm) for microelectronic photolithographic applications. DNA is regarded as a promising construction material for the selective positioning of molecular devices because of its recognition capabilities, physicochemical stability, and mechanical rigidity. Seeman was the first to propose DNA for the precise spatial arrangement of three-dimensional networks [227]. Assemblies of DNA-derivatized gold colloids were recently prepared via the DNA hybridization-based self-organization pathway and the resulting defined arrangements of nanometal particles have real applications in laser technology [228–231]. For example, Alivisatos et al. [199] have obtained defined mono-adducts from commercially available 1.4-nm gold clusters where one reactive maleimido group is attached to every particle. These were coupled with thiolated 18-mer oligonucleotides in order to add an individual "codon" sequence. When a single-stranded DNA template containing complementary codons is added, a self-assembly of nanocrystal molecules is observed (Fig. 1.13) [199, 232]. This work has been the subject of recent reviews [232–234].

Niemeyer et al. [233] have recently reported the coupling of metal particles bearing a biotin substituent with the DNA–streptavidin hybrid. The growth of a 12-μm-long, 100- nm wide conductive silver wire has been achieved using a DNA molecule stretched between two gold electrodes as a template [235].

It remains to be seen how the practical applications of these materials will develop over the next few years.

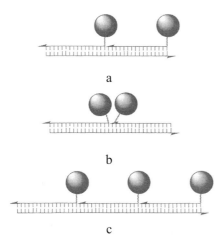

a

b

c

Figure 1.13. Self-organization of conjugates from gold particles (*shaded spheres*) and oligonucleotide codons to supramolecular assemblies by the addition of a template strand. The derivatization of the oligonucleotides in 3′ or 5′ position allows control of the mode: head-to-head (a) or head-to-tail (b) homodimers. The trimer (c) is formed using the complementary sequence in triplicate. (Adapted from Ref. [232].)

References

1 J. Turkevich, P. C. Stevenson, J. Hillier, The nucleation and growth processes in the synthesis of colloidal gold (or is it coagulation of colloidal gold?), *Disc. Faraday Soc.* **1951**, *11*, 55–75.

2 J. Turkevich, G. Kim, Palladium: preparation and catalytic properties of particles of uniform size, *Science* **1970**, *169*, 873.

3 J. Turkevich, Colloidal gold. Part I. Historical and preparative aspects, morphology and structure, *Gold Bulletin* **1985**, *18*, 86–91.

4 H. Bönnemann, R. Richards, Nanoscopic metal particles – synthetic methods and potential applications, *Eur. J. Inorg. Chem.* **2001**, 2455–2480.

5 G. Schmid (ed.), *Clusters and Colloids*, VCH, Weinheim, **1994**.

6 H. Bönnemann, W. Brijoux, R. Brinkmann, et al., Preparation, characterization, and application of fine metal particles and metal colloids using hydrotriorganoborates, *J. Mol. Catal.* **1994**, *86*, 129–177.

7 H. Bönnemann, G. Braun, W. Brijoux, et al., Nanoscale colloidal metals and alloys stabilized by solvents and surfactants. Preparation and use as catalyst precursors, *J. Organometallic Chem.* **1996**, *520*, 143–162.

8 M. T. Reetz, W. Helbig, S. A. Quaiser, Electrochemical methods in the synthesis of nanostructured transition metal clusters, in *Active Metals*, ed. A. Fürstner, VCH, Weinheim, **1996**, 279–297.

9 H. Bönnemann, R. Richards, Nanomaterials as precursors for electrocatalysis, in *Catalysis and Electrocatalysis at Nanoparticle Surfaces*, eds. A. Wieckowski, E. Sarinova, C. Vayenas, Marcel Dekker, New York, **2003**

10 N. Toshima, T. Yonezawa, Bimetallic nanoparticles – novel materials for chemical and physical applications, *New J. Chem.* **1998**, 1179–1201.

11 J. D. Aiken III, R. G. Finke, A review of modern transition-metal nanoclusters: their synthesis, characterization, and applications in catalysis, *J. Mol. Catal. A* **1999**, *145*, 1–44.

12 K. Klabunde (ed.), *Nanoscale Materials in Chemistry*, Wiley Interscience, New York, **2001**.

13 E. Dujardin, S. Mann, Bio-inspired materials chemistry, *Adv. Mater.* **2002**, *14*, 775–788.

14 M. Knez, A. Bittner, F. Boes, et al., Biotemplate synthesis of 3-nm nickel and cobalt nanowires, *Nano Lett.* **2003**, *3(8)*, 1079–1082.

15 E. Dujardin, C. Peet, G. Stubbs, et al., Organization of metallic nanoparticles using tobacco mosaic virus templates, *Nano Lett.* **2003**, *3(3)*, 413–417.

16 M. Reches, E. Gazit, Casting metal nanowires within discrete self-assembled peptide nanotubes, *Science* **2003**, *300*, 625–627.

17 S. Behrens, K. Rahn, W. Habicht, et al., Nanoscale particle arrays induced by highly ordered protein assemblies, *Adv. Mater.* **2002**, *14*, 1621–1625.

18 W. Shenton, T. Douglas, M. Young, et al., Inorganic-organic nanotube composites from template mineralization of tobacco mosaic virus, *Adv. Mater.* **1999**, *11*, 253–256.

19 G. Schmid, in *Applied Homogeneous Catalysis with Organometallic Compounds*, eds. B. Cornils, W. A. Herrmann, Wiley-VCH, Weinheim, **1996**, 636–644, Vol. 2.

20 W. A. Herrmann, B. Cornils, in *Applied Homogeneous Catalysis with Organometallic Compounds*, eds. B. Cornils, W. A. Herrmann, Wiley-VCH, Weinheim, **1996**, 1171–1172, Vol. 2.

21 H. Bönnemann, W. Brijoux, Surfactant-stabilized nanosized colloidal metals and alloys as catalyst precursors, in *Advanced Catalysts and Nanostructured Materials*, ed. W. Moser, Academic Press, San Diego, **1996**, 165–196.

22 H. Bönnemann, W. Brijoux, Potential applications of nanostructured metal colloids, in *Metal Clusters in Chemistry*, eds. P. Braunstein, L. A. Oro, P. R. Raithby, Wiley-VCH, Weinheim, **1999**, 913–931, Vol. 2.

23 T. J. Schmidt, M. Noeske, H. A. Gasteiger, et al., Electrocatalytic activity of PtRu alloy colloids for CO and CO/H_2 electrooxidation: stripping voltammetry and rotating disk measurements, *Langmuir* **1997**, *13*, 2591–2595.

24 E. Auer, W. Behl, T. Lehmann, U. Stenke (to A. G. Degussa), CO tolerant anode catalyst for PEM fuel cells and method for its production, EP 09 24 784 A1 (June 23, 1999).

25 M. Götz, H. Wendt, Binary and ternary anode catalyst formulations including the elements W, Sn and Mo for PEMFCs operated on methanol or reformate gas, *Electrochim. Acta* **1998**, *43*, 3637–3644.

26 T. J. Schmidt, M. Noeske, H. A. Gasteiger, et al., PtRu alloy colloids as precursors for fuel cell catalysts. A combined XPS, AFM, HRTEM, and RDE study, *J. Electrochem. Soc.* **1998**, *145*, 925–931.

27 E. Gaffet, M. Tachikart, O. El Kedim, R. Rahouadj, Nanostructural materials formation by mechanical alloying: morphologic analysis based on transmission and scanning electron microscopic observations, *Mater. Charact.* **1996**, *36*, 185–190.

28 A. Amulyavichus, A. Daugvila, R. Davidonis, C. Sipavichus, Study of chemical composition of nanostructural materials prepared by laser cutting of metals, *Fizika Metallov I Metallovedenie* **1998**, *85*, 111–117.

29 A. Schalnikoff, R. Roginsky, Eine neue Methode der Herstellung kolloider Lösungen, *Z. Kolloid.* **1927**, *43*, 67–70.

30 J. R. Blackborrow, D. Young, *Metal Vapor Synthesis*, Springer-Verlag, New York, **1979**.

31 K. J. Klabunde, *Free Atoms and Particles*, Academic Press, New York, **1980**.

32 K. J. Klabunde, Y.-X. Li, B.-J. Tan, Solvated metal atom dispersed catalysts, *Chem. Mater.* **1991**, *3*, 30–39.

33 J. S. Bradley, in *Clusters and Colloids*, ed. G. Schmid, VCH, Weinheim, **1994**, p. 477f.

34 K. J. Klabunde, G. C. Cardenas-Trivino, in *Active Metals*, ed. A. Fürstner, VCH, Weinheim, **1996**, 237–278.

35 M. Faraday, The Bakerian lecture: experimental relations of gold (and other metals) to light, *Philos. Trans. R. Soc. London* **1857**, *147*, 145–153.

36 J. S. Bradley, in *Clusters and Colloids*, ed. G. Schmid, VCH, Weinheim, **1994**, 469–473.

37 T. Leisner, C. Rosche, S. Wolf, et al., The catalytic role of small-coinage clusters in photography, *Surf. Rev. Lett.* **1996**, *3*, 1105–1108.

38 R. Tausch-Treml, A. Henglein, J. Lilie, Reactivity of silver in aqueous solution-2. pulse radiolysis study, *Ber. Bunsen-Ges. Phys. Chem.* **1978**, *82*, 1335–1343.

39 M. Michaelis, A. Henglein, Reduction of Pd(II) in aqueous solution-stabilization and reactions of an intermediate cluster and Pd colloid formation, *J. Phys. Chem.* **1992**, *96*, 4719–4724.

40 M. A. Watzky, R. G. Finke, Transition metal nanocluster formation kinetic and mechanistic studies. A new mechanism when hydrogen is the reductant: slow, continuous nucleation and fast autocatalytic surface growth, *J. Am. Chem. Soc.* **1997**, *119*, 10382–10400.

41 J. Rothe, J. Hormes, H. Bönnemann, et al., In situ X-ray absorption spectroscopy investigation during the formation of colloidal copper, *J. Am. Chem. Soc.* **1998**, *120*, 6019–6023.

42 A.I. Kirkland, P.P Edwards, D.A. Jefferson, D.G. Duff, in *Annual Reports on the Progress of Chemistry C*, Royal Society of Chemistry, Cambridge, **1990**, 247–305, Vol. 87.

43 S. Ozkar, R. Finke, Transition-metal nanocluster stabilization fundamental studies: hydrogen phosphate as a simple, effective, readily available, robust, and previously unappreciated stabilizer for well-formed, isolable, and redissolvable Ir(0) and other transition-metal nanoclusters, *Langmuir* **2003**, *19*, 6247–6260.

44 B. Hornstein, R. Finke, Transition-metal nanocluster kinetic and mechanistic studies emphasizing nanocluster agglomeration: demonstration of a kinetic method that allows monitoring of all three phases of nanocluster formation and aging, *Chem. Mater.* **2004**, *16*, 139–150.

45 H. Hirai, Y. Nakao, N. Toshima, K. Adachi, Colloidal rhodium in poly-vinylalcohol as hydrogenation catalyst of olefins, *Chem Lett.* **1976**, *9*, 905.

46 H. Hirai, Y. Nakao, N. Toshima, Colloidal rhodium in poly(vinylpyrrolidone) as hydrogenation catalysts for internal olefins, *Chem. Lett.* **1978**, *5*, 545.

47 H. Hirai, Y. Nakao, N. Toshima, Preparation of colloidal rhodium in polyvinyl alcohol by reduction with methanol, *J. Macromol. Sci. Chem.* **1978**, *A12*, 1117.

48 H. Hirai, Y. Nakao, N. Toshima, Preparation of colloidal transition metals in polymers by reduction with alcohols or ethers, *J. Macromol. Sci. Chem.* **1979**, *A13*, 727.

49 J. Tanori, M.P. Pileni, Control of the shape of copper metallic particles by using a colloidal system as template, *Langmuir* **1997**, *13*, 639–646.

50 M. P. Pileni, Nanosized particles made in colloidal assemblies, *Langmuir* **1997**, *13*, 3266–3276.

51 M. Antonietti, C. Göltner, Superstructures of functional colloids: chemistry on the nanometer scale, *Angew. Chem. Int. Ed. Engl.* **1997**, *36*, 911–928.

52 M. P. Pileni, Colloidal self-assemblies used as templates to control size, shape and self-organization of nanoparticles, *Supramol. Sci.* **1998**, *5*, 321–329.

53 M. P. Pileni, Self-organization of magnetic nanosized cobalt particles, *Adv. Mater.* **1998**, *10*, 259–261.

54 M. Antonietti, Functional colloids: structure formation and chemistry on a nanometer scale, *Chem.-Ing. Tech.* **1996**, *68*, 518–523.

55 S. Förster, *Ber. Bunsen-Ges.* **1997**, *101*, 1671–1678.

56 J. J. Storhoff, R. C. Mucic, C. A. Mirkin, Strategies for organizing nanoparticles into aggregate structures and functional materials, *J. Cluster Sci.* **1997**, *8*, 179–216.

57 M. Möller, J. P. Spatz, Mineralization of nanoparticles in block copolymer micelles, *Curr. Opin. Colloid Interface Sci.* **1997**, *2*, 177–187.

58 G.B. Sergeev, M.A. Petrukhina, Encapsulation of small metal particles in solid organic matrixes, *Prog. Solid State Chem.* **1997**, *24*, 183–211.

59 J.P. Wilcoxon, P. Provencio, Use of surfactant micelles to control the structural phase of nanosize iron clusters, *J. Phys. Chem. B* **1999**, *103*, 9809–9812.

60 T. Miyao, N. Toyoizumi, S. Okuda, et al., Preparation of Pt/SiO_2 ultrafine particles in reversed micelles and their catalytic activity, *Chem. Lett.* **1999**, *10*, 1125–26.

61 S. T. Selvan, M. Nogami, A. Nakamura, Y. Hamanaka, A facile sol-gel method for the encapsulation of gold nanoclusters in silica gels and their optical properties, *J. Non-Crystalline Solids* **1999**, *255*, 254–258.

62 M. M. Maye, W. Theng, F. L. Leibowitz, et al., Heating-induced evolution of thiolate-encapsulated gold nanoparticles: A strategy for size and shape manipulations, *Langmuir* **2000**, *16*, 490–497.

63 Y. Niidome, A. Hori, T. Sato, S. Yamada, Enormous size growth of thiol-passivated gold nanoparticles induced by near-IR laser light, *Chem. Lett.* **2000**, *4*, 310–311.

64 I. Konomi, S. Hyodo, T. Motohiro, Simulation of MEIS spectra for quantitative understanding of average size, composition, and size distribution of Pt-Rh alloy nanoparticles, *J. Catal.* **2000**, *192*, 11–17.

65 G. Schmid, R. Pfeil, R. Boese, et al., Au55[P(C6H5)3]12Cl6 a gold cluster of exceptional size, *Chem. Ber.* **1981**, *114*, 3634–3642.

66 G. Schmid, Metal clusters and cluster metals, *Polyhedron* **1988**, *7*, 2321–2329.

67 L. J. de Jongh, J. A. O. de Aguiar, H. B. Brom, et al., Physical properties of high nuclearity metal cluster compounds, *Z. Phys. D. At. Mol. Clusters* **1989**, *12*, 445–450.

68 G. Schmid, B. Morum, J. Malm, Pt-309 a 4 shell platinum cluster, *Angew. Chem. Int. Ed. Engl.* **1989**, *28*, 778–780.

69 G. Schmid, N. Klein, L. Korste, Large transition metal clusters, ligand exchange reactions on $Au_{55}(PPh_3)_{12}Cl_6$ formation of a water soluble Au_{55} cluster, *Polyhedron* **1988**, *7*, 605–608.

70 T. Tominaga, S. Tenma, H. Watanabe, et al., Tracer diffusion of a ligand-stabilized two-shell gold cluster, *Chem. Lett.* **1996**, *12*, 1033.

71 G. Schmid, Large clusters and colloids-metals in the embryonic state, *Chem. Rev.* **1992**, *92*, 1709–1727.

72 H. A. Wicrenga, L. Soethout, I. W. Gerritsen, et al., Direct imaging of $Pd_{561}(phen)38+/-2ON$ and $Au_{55}(PPh_3)_{12}Cl_6$ clusters using scanning tunnelling microscopy, *Adv. Mater.* **1990**, *2*, 482.

73 R. Houbertz, T. Feigenspan, F. Mielke, et al., STM investigations on compact Au_{55} cluster pellets, *Europhys. Lett,* **1994**, *28*, 641.

74 G. Schmid, A. Lehnert, The complexation of gold colloids, *Angew. Chem. Int. Ed. Engl.* **1989**, *28*, 780–781.

75 G. Schmid, V. Maihack, F. Lantermann, S. Peschel, Ligand stabilized metal clusters and colloids: properties and applications, *J. Chem. Soc. Dalton Trans.* **1996**, 589–595.

76 G. Schmid, H. West, J.-O. Malm, et al., Catalytic properties of layered gold-palladium colloids, *Chem. Eur. J.* **1996**, *2*, 1099.

77 U. Simon, R. Flesch, H. Wiggers, et al., Chemical tailoring of the charging energy in metal cluster arrangements by use of bifunctional spacer molecules, *J. Mater. Chem.* **1998**, *8*, 517–518.

78 G. Schmid, S. Peschel, Preparation and scanning probe microscopic characterization of monolayers of ligand-stabilized transition metal clusters and colloids, *New J. Chem.* **1998**, *22*, 669–675.

79 G. Schmid, R. Pugin, J.-O. Malm, J.-O. Bovin, Silsesquioxanes as ligands for gold clusters, *Eur. J. Inorg. Chem.* **1998**, *6*, 813–817.

80 M. Giersig, L. M. Liz-Tarzan, T. Ung, et al., Chemistry of nanosized silica-coated metal particles em study, *Ber. Bunsenges. Phys. Chem.* **1997**, *101*,1617–1620.

81 M. N. Vargaftik. V. P. Zargorodnikov , I. P. Stolarov, et al., Giant palladium clusters as catalysts of oxidative reactions of olefins and alcohols, *J. Mol. Catal.* **1989**, *53*, 315–349.

82 M. N. Vargaftik, V. P. Zargorodnikov, I. P. Stolyarov, et al., A novel giant palladium cluster, *Chem. Commun.* **1985**, *14*, 937–939.

83 V. V. Volkov, G. Van Tendeloo, G. A. Tsirkov, et al., Long- and short-distance ordering of the metal cores of giant Pd clusters, *J. Cryst. Growth* **1996**, *163*, 377.

84 I. I. Moiseev, M. N. Vargaftik, V. V. Volkov, et al., Palladium 561 giant clusters – chemical aspects of self-organization on a nano level, *Mend. Commun.* **1995**, *3*, 87–89.

85 V. Oleshko, V. Volkov, W. Jacob, et al., High resolution electron microscopy and energy electron loss spectroscopy of giant palladium clusters, *Z. Phys. D* **1995**, *34*, 283.

86 I. I. Moiseev, M. N. Vargaftik, T. V. Chernysheva, et al., Catalysis with a palladium giant cluster: phenol oxidative carbonylation to diphenyl carbonate conjugated with reductive nitrobenzene conversion, *J. Mol. Catal. A: Chem.* **1996**, *108*, 77.

87 C. Amiens, D. de Caro, B. Chaudret, J. S. Bradley, Selected synthesis, characterization and spectroscopic studies on a novel class of reduced platinum and palladium particles stabilized by carbonyl and phosphine ligands, *J. Am. Chem. Soc.* **1993**, *115*, 11638–11939.

88 D. deCaro, H. Wally, C. Amiens, B. Chaudret, Synthesis and spectroscopic properties of a novel class of copper particles stabilized by triphenylphosphine, *J. Chem. Soc., Chem. Comm.* **1994**, *16*, 1891–1892.

89 A. Rodriguez, C. Amiens, B. Chaudret, et al., Synthesis and isolation of cuboctahedral and icosahedral platinum nanoparticles. ligand-dependent structures, *Chem. Mater.* **1996**, *8*, 1978–1986.

90 M. Bardaji, O. Vidoni, A. Rodriguez, et al., Synthesis of platinum nanoparticles stabilized by CO and tetrahydrothiophene. Facile conversion to molecular species, *New J. Chem.* **1997**, *21*, 1243–1249.

91 R. Franke, J. Rothe, J. Pollmann, et al., A study of the electronic and geometric structure of colloidal Ti0.0.5THF, *J. Amer. Chem. Soc.* **1996**, *118*, 12090–12097.

92 O. Vidoni, K. Philippot, C. Amiens, et al., Broadening the aldolase catalytic antibody repertoire by combining reactive immunization and transition state theory: new enantio- and diastereoselectivities, *Angew. Chem. Int. Ed. Engl.* **1999**, *38*, 3736–3738.

93 M. T. Reetz, G. Lohmer, Propylene carbonate stabilized nanostructured palladium clusters as catalysts in Heck reactions, *Chem. Commun. (Cambridge)*, **1996**, 1921–1922.

94 D. Mandler, I. Willner, Photohydrogenation of acetylenes in water oil 2-phase systems-application of novel metal colloids and mechanistic aspects of the process, *J. Phys. Chem.* **1987**, *91*, 3600–3605.

95 U.S. Pat. 5,580,492 (Aug. 26, 1993), H. Bönnemann, W. Brijoux, T. Joussen (to Studiengesellschaft Kohle mbH).

96 H. Bönnemann, W. Brijoux, R. Brinkmann, et al., Production of colloidal transition metals in organic phase and their use as catalysts, *Angew. Chem. Int. Ed.* **1991**, *30*, 1344–1346.

97 U.S. Pat. 849,482 (Aug. 29, 1997), H. Bönnemann, W. Brijoux, R. Brinkmann, J. Richter (to Studiengesellschaft Kohle mbH).

98 M. T. Reetz, W. Helbig, Size selective synthesis of nanostructured transition metal clusters, *J. Am. Chem. Soc.* **1994**, *116*, 7401–7402.

99 U.S. Pat. 5,620,564 (Apr. 15, 1997) and U.S. Pat. 5,925,463 (Jul. 20, 1999), (M. T. Reetz, W. Helbig, S. Quaiser (to Studiengesellschaft Kohle).

100 M. A. Winter, PhD Thesis, **1998**, Verlag Mainz, Aachen, ISBN 3–89653–355.

101 J. A. Becker, R. Schäfer, W. Festag, et al., Electrochemical growth of superparamagnetic cobalt clusters, *J. Chem. Phys.* **1995**, *103*, 2520–2527.

102 M. T. Reetz, S. A. Quaiser, C. Merk, Electrochemical preparation of nanostructured titanium clusters: Characterization and use in McMurry type coupling reactions, *Chem. Ber.* **1996**, *129*, 741–743.

103 M. T. Reetz, W. Helbig, S. A. Quaiser, Electrochemical preparation of nanostructural bimetallic clusters, *Chem. Mater.* **1995**, *7*, 2227–2228.

104 M. T. Reetz, S. A. Quaiser, *Angew. Chem.* **1995**, *107*, 2461–2463; A new method for the preparation of nanostructured metal clusters, *Angew. Chem. Int. Ed. Engl.* **1995**, *34*, 2240.

105 U. Kolb, S. A. Quaiser, M. Winter, M. T. Reetz, Investigation of tetraalkylammonium bromide stabilized palladium/platinum bimetallic clusters using extended X-ray absorption fine structure spectroscopy, *Chem. Mater.***1996**, *8*, 1889–1894.

106 J. Kiwi, M. Grätzel, Protection, size factors and reaction dynamics of colloidal redox catalysts mediating light induced hydrogen evolution from water, *J. Am. Chem. Soc.* **1979**, *101*, 7214–7217.

107 H. Bönnemann, W. Brijoux, R. Brinkmann, et al., The reductive stabilization of nanometal colloids by organo-aluminum compounds, *Rev. Roum. Chim.* **1999**, *44*, 1003–1010.

108 J. Sinzig, L. J. De Jongh, H. Bönnemann, et al., Antiferromagnetism of colloidal [MN0.0.3THF]x, *Appl. Organomet. Chem.*, **1998**, *12*, 387–391.

109 WO 99/59713 (November 25, 1999), H. Bönnemann, W. Brijoux, R. Brinkmann (to Studiengesellschaft Kohle).

110 J. S. Bradley, E. W. Hill, M. E. Leonowicz, H. Witzke, Clusters, colloids and catalysis, *J. Mol. Catal.* **1987**, *41*, 59–74.

111 M. T. Reetz, W. Helbig, S. A. Quaiser, et al., Visualization of surfactants on nanostructured palladium clusters by a combination of STM and high-resolution TEM, *Science* **1995**, *267*, 367–369.

112 A. Schulze Tilling, Bimetallische Edelmetallkolloid als Precursor für Kohlenhydratoxidationskatalysatoren, PhD Thesis, RWTH Aachen, **1996**.

113 T. Yonezawa, N. Toshima, Polymer protected and micelle protected gold platinum bimetallic systems, *J. Mol. Catal.* **1993**, *83*, 167–181.

114 N. Toshima, T. Takahashi, H. Hirai, Colloidal platinum catalysts prepared by hydrogen reduction and photoreduction in the presence of surfactant, *Chem. Lett.* **1985**, *8*, 1245–1248.

115 J. S. Bradley, in *Clusters and Colloids*, ed. G. Schmid, VCH, Weinheim, **1994**, 471.

116 G. Viau, R. Brayner, L. Poul, et al., Ruthenium nanoparticles: Size, shape, and self-assemblies, *Chem. Mater.* **2003**, *15*, 486–494.

117 L. D. Rampino, F. F. Nord, Preparation of palladium and platinum synthetic high polymer catalysts and the relationship between particle size and rate of hydrogenation, *J. Am. Chem. Soc.* **1941**, *63*, 2745–2749.

118 L. D. Rampino, F. F. Nord, Applicability of palladium synthetic high polymer catalysts, *J. Am. Chem. Soc.* **1941**, *63*, 3268.

119 L. D. Rampino, F. F. Nord, Systematic studies on palladium high synthetic polymer catalysts, *J. Am. Chem. Soc.* **1943**, *65*, 2121–2125.

120 L. Hernandez, F. F. Nord, Interpretation of the mechanism of catalytic reductions with colloidal rhodium in the liquid phase, *J. Colloid. Sci.* **1948**, *3*, 363–375.

121 W. P. Dunsworth, F. F. Nord, Investigations on the mechanism of catalytic hydrogenations XV. Studies with colloidal iridium *J. Am. Chem. Soc.* **1950**, *72*, 4197–4198.

122 Y. Lin, R. G. Finke, Novel polyoxoanion- and Bu4N+-stabilized, isolable, and redissolvable, 20–30-.ANG. Ir300–900 nanoclusters: The kinetically controlled synthesis, characterization, and mechanism of formation of organic solvent-soluble, reproducible size, and reproducible catalytic activity metal nanoclusters, *J. Am. Chem. Soc.* **1994**, *116*, 8335–8353.

123 Y. Lin, R. G. Finke, A more general approach to distinguishing homogeneous from heterogeneous catalysis: discovery of polyoxoanion and Bu4N+ stabilized, isolable and redissolvable, high reactivity Ir approx 190–450 nanocluster catalysts *Inorg. Chem.* **1994**, *33*, 4891–4910.

124 T. Nagata, M. Pohl, H. Weiner, R. G. Finke, Polyoxoanion-supported organometallic complexes: Carbonyls of rhenium(I), iridium(I), and rhodium(I) that are soluble analogs of solid-oxide-supported M(CO)n+ and that exhibit novel M(CO)n+ mobility, *Inorg. Chem.* **1997**, *36*, 1366.

125 J. D. Aiken III, R. G. Finke, Nanocluster formation synthetic, kinetic, and mechanistic studies. The detection of, and then methods to avoid, hydrogen mass-transfer limitations in the synthesis of polyoxoanion- and tetrabutylammonium-stabilized, near-monodisperse 40+6 .ANG. Rh(0) nanoclusters, *J. Am. Chem. Soc.* **1998**, *120*, 9545–9554.

126 J. D. Aiken III, R. G. Finke, Polyoxoanion- and tetrabutylammonium-stabilized, near-monodisperse, 40 +/− 6 Å Rh(0) similar to (1500) to Rh(0) similar to (3700) nanoclusters: Synthesis, characterization, and hydrogenation catalysis, *Chem. Mater.* **1999**, *11*, 1035–1047.

127 M. R. Mucalo, R. P. Cooney, FTIR spectra of carbon monoxide adsorbed on platinum sols, *Chem. Commun.* **1989**, *2*, 94–95.

128 K. Meguro, M. Torizuka, K. Esumi, The preparation of organo colloidal precious metal particles, *Bull. Chem. Soc. Jpn.* **1988**, *61*, 341–345.

129 L. N. Lewis, N. Lewis, Platinum-catalyzed hydrosilylation – colloid formation as the essential step, *J. Am. Chem. Soc.*, **1986**, *108*, 7228–7231.

130 L. N. Lewis, R. Uriarte, N. Lewis, Metal colloid morphology and catalytic activity – further proof of the intermediacy of colloids in the platinum catalyzed hydrosilation reaction, *J. Catal.* **1991**, *1*, 67–74.

131 A. C. Curtis, D. G. Duff, P. P. Edwards, et al., The morphology and microstructure of colloidal silver and gold, *Angew. Chem. Int. Ed. Engl.* **1987**, *26*, 676.

132 D. G. Duff, A. C. Curtis, P. P. Edwards, et al., The microstructure of colloidal silver- evidence of a tetrahedral growth sequence, *Chem. Commun.*, **1987**, *16*, 1264.

133 A. C. Curtis, D. G. Duff, P. P. Edwards, et al., A morphology-selective copper organosol, *Angew. Chem. Int. Ed. Engl.* **1988**, *27*, 1530.

134 D. G. Duff, P. P. Edwards, J. Evans, et al., A joint structural characterization of colloidal platinum by EXAFS and high-resolution electron microscopy, *Angew. Chem. Int. Ed. Engl.* **1989**, *28*, 590.

135 D. G. Duff, A. Baiker, P. P. Edwards, A new hydrosol of gold clusters. 1. Formation and particle size variation, *Langmuir* **1993**, *9*, 2301–2309.

136 W. Vogel, D. G. Duff, A. Baiker, X-ray structure of a new hydrosol of gold clusters, *Langmuir* **1995**, *11*, 401–404.

137 P. R. van Rheenen, M. J. McKelvey, W. S. Glaunsinger, Synthesis and characterization of small platinum particles formed by the chemical reduction of chloroplatinic acid, *J. Solid State Chem.* **1987**, *67*, 151–169.

138 D. G. Duff, A. Baiker, in *Preparation of Catalysts VI*, ed. G. Poncelet, J. Martens, B. Delmon, P. A. Jacobs, P. Grange, Elsevier Science, Amsterdam, **1995**, 505–512.

139 K.-L. Tsai, J. L. Dye, Synthesis, properties, and characterization of nanometer-size metal particles of homogeneous reduction with alkalides and electrides in aprotic solvents, *Chem. Mater.* **1993**, *5*, 540–546.

140 J. van Wonterghem, S. Mørup, C. J. W. Koch, et al., Formation of ultra-fine amorphous alloy particles by reduction in aqueous-solution, *Nature*, *322*, **1986**, 622.

141 G. N. Glavee, K. J. Klabunde, C. M. Sorensen, G. C. Hadjipanayis, Sodium-borohydride reduction of cobalt ions in non-aqueous media – formation of ultra-fine particles (nanoscale) of cobalt metal, *Inorg. Chem.* **1993**, *32*, 474–477.

142 H. Modrow, S. Bucher, J. Hormes, et al., Model for chainlength-dependent core-surfactant interaction in N(Alkyl)4Cl-stabilized colloidal metal particles obtained from X-ray absorption spectroscopy, *J. Phys. Chem. B* **2003**, *107*, 3684–3689.

143 R. Franke, J. Rothe, R. Becker, et al., A x-ray photoelectron and x-ray absorption spectroscopic study of colloidal [Mn0.0.3THF]x, *Adv. Mater.* **1998**, *10*, 126–132.

144 J. Sinzig, L. J. de Jongh, H. Bönnemann, et al., Antiferromagnetism of colloidal [MN0.0.3THF]x, *Appl. Organomet. Chem.* **1998**, *12*, 387–391.

145 M. Maase, Neue Methoden zur Größen- und formselektiven Darstellung von Metallkolloiden, PhD Thesis, Verlag Mainz, Aachen, **1999**, ISBN 3-89653-463-7.

146 M. T. Reetz, M. Maase, Redox-controlled size-selective fabrication of nanostructured transition metal colloids *Adv. Mater.* **1999**, *11*, 773–777.

147 J. S. Bradley, B. Tesche, W. Busser, et al., Surface spectroscopic study of the stabilization mechanism for shape-selectively synthesized nanostructured transition metal colloids, *J. Am. Chem. Soc.* **2000**, *122*, 4631–4636.

148 P. H. Hess, P. H. Parker, Polymers for stabilization of colloidal cobalt particles, *J. Appl. Polymer. Sci.* **1966**, *10*, 1915–1927.

149 J.R. Thomas, preparation and magnetic properties of colloidal cobalt particles, *J. Appl. Phys.* **1966**, *37*, 2914–2915.

150 K. Esumi, T. Tano, K. Torigue, K. Meguro, Preparation and characterization of bimetallic Pd-Cu colloids by thermal decomposition of their acetate compounds in organic-solvents, *Chem. Mater.* **1990**, *2*, 564.

151 J. S. Bradley, E. W. Hill, C. Klein, et al., Synthesis of monodispersed bimetallic palladium copper nanoscale colloids, *Chem. Mater.* *5*, **1993**, 254–256.

152 Y. Wada, H. Kuramoto, T. Sakata, et al., Preparation of nano-sized nickel metal particles by microwave irradiation, *Chem. Lett.* **1999**, *7*, 607–608.

153 W. Yu, W. Tu, H. Liu, Synthesis of nanoscale platinum colloids by microwave dielectric heating, *Langmuir* **1999**, *15*, 6–9.

154 K. S. Suslick, T. Hyeon, M. Fang, A. Cichowlas, in *Advanced Catalysts and Nanostructured Materials*, Chapter 8, ed. W. Moser, Academic Press, San Diego, **1996**, 197–212.

155 A. Dhas, A. Gedanken, Sonochemical preparation and properties of nanostructured palladium metallic clusters, *J. Mater. Chem.* **1998**, *8*, 445–450.

156 Y. Koltypin, A. Fernandez, C. Rojas, et al., Encapsulation of nickel nanoparticles in carbon obtained by the sonochemical decomposition of Ni(C8H12)2, *Chem. Mater.* **1999**, *11*, 1331–1335.

157 R. A. Salkar, P. Jeevanandam, S. T. Aruna, et al., The sonochemical preparation of amorphous silver nanoparticles, *J. Mater. Chem.* **1999**, *9*, 1333–1335.

158 F. Ciardelli, P. Pertici, Structure and reactivity of aromatic polymers ruthenium catalysts, *Z. Naturforsch. B* **1985**, *40*, 133–140.

159 J. S. Bradley, E. W. Hill, S. Behal, et al., Preparation and characterization of organosols of monodispersed nanoscale palladium – particle-size effects in the binding geometry of adsorbed carbon-monoxide, *Chem. Mater.* **1992**, *4*, 1234–1239.

160 A. Duteil, R. Quéau, B. Chaudret, et al., Preparation of organic solutions or solid films of small particles of ruthenium, palladium, and platinum from organometallic precursors in the presence of cellulose derivatives, *Chem. Mater.* **1993**, *5*, 341–347.

161 D. deCaro, V. Agelou, A. Dutcil, et al., Preparation from organometallic precursors, characterization and some reactivity of copper and gold colloids sterically protected by nitrocellulose, poly(vinylpyrrolidone) or poly(dimethylphenylene oxide), *New J. Chem.* **1995**, *19*, 1265–1274.

162 F. Dassenoy, K. Philippot, T. Ould Ely, et al., Platinum nanoparticles stabilized by CO and octanethiol ligands or polymers: FT-IR, NMR, HREM and WAXS studies, *New J. Chem.* **1998**, *19*, 703–711.

163 J. Osuna, D. deCaro, C. Amiens, et al., Synthesis, characterization, and magnetic properties of cobalt nanoparticles from an organometallic precursor, *J. Phys. Chem.* **1996**, *100*, 14571–14574.

164 T. Ould-Ely, C. Amiens, B. Chaudret, et al., Synthesis of nickel nanoparticles. Influence of aggregation induced by modification of poly(vinylpyrrolidone) chain length on their magnetic properties, *Chem. Mater.* **1999**, *11*, 526–529.

165 J. S. Bradley, E. W. Hill, B. Chaudret, A. Duteil, Surface chemistry on colloidal metals. Reversible adsorbate-induced surface composition changes in colloidal palladium-copper alloys, *Langmuir* **1995**, *11*, 693–695.

166 F. Dassenoy, M.-J. Casanove, P. Lecante, et al., Experimental evidence of structural evolution in ultrafine cobalt particles stabilized in different polymers – from a polytetrahedral arrangement to the hexagonal structure, *J. Chem. Phys.*, **2000**, *112*, 8137–8145.

167 M. T. Reetz, S. Quaiser, M. Winter, et al., Nanostructured metal oxide clusters by oxidation of stabilized metal clusters with air, *Angew. Chem. Ind. Ed.* **1996**, *35*, 2092–2094.

168 M. Verelst, T. Ould Ely, C. Amiens, et al., Synthesis and characterization of CoO, Co3O4, and mixed Co/CoO nanoparticles, *Chem. Mater.* **1999**, *11*, 2702–2708.

169 C.-B. Wang, C.-T. Yeh, Effects of particle size on the progressive oxidation of nanometer platinum, *J. Catal.* **1998**, *178*, 450–456.

170 M. T. Reetz, M. Koch (to Studiengesellschaft Kohle mbH), Water soluble nanostructured metal oxide colloids and method for preparing the same, PCT/EP 99/08594 (November 9, 1999)

171 M. T. Reetz, M. Koch, Water-soluble colloidal Adams catalyst: preparation and use in catalysis, *J. Am. Chem. Soc.* **1999**, *121*, 7933–7934.

172 M. Koch, Wasserlösliche metall- und metalloxide-kolloids: Synthese, Charakterisierung und katalytische Anwendungen, PhD Thesis, Verlag Mainz, Aachen, **1999**, ISBN 3-89653-514-5.

173 H. Remita, A. Etcheberry, J. Belloni, Dose rate effect on bimetallic gold-palladium cluster structure, *J. Phys. Chem. B* **2003**, *107*, 31–36.

174 G. Schmid, M. Bäumle, N. Beyer, Ordered two-dimensional monolayers of Au55 clusters, *Angew. Chem. Int. Ed.* **2000**, *39*, 181–183.

175 H. Cölfen, T. Pauck, Determination of particle size distributions with angstrom resolution, *Colloid Polym. Sci.* **1997**, *275*, 175–180.

176 T. Teranishi, M. Miyake, Size control of palladium nanoparticles and their crystal structures, *Chem. Mater.* **1998**, *10*, 594–600.

177 E. Papirer, P. Horny, H. Balard, et al., The preparation of a ferrofluid by decomposition of dicobalt octacarbonyl 2. Nucleation and growth of particles, *J. Colloid Interface Sci.* **1983**, *94*, 220–228.

178 T. Teranishi, M. Hosoe, M. Miyake, Formation of monodispersed ultrafine platinum particles and their electrophoretic deposition on electrodes, *Adv. Mater.* **1997**, *9*, 65–67.

179 T. Teranishi, I. Kiyokawa, M. Miyake, Synthesis of monodisperse gold nanoparticles using linear polymers as protective agents, *Adv. Mater.* **1998**, *10*, 596–599.

180 K. Esumi, H. Ishizuka, S. Masayoshi, et al., Preparation of bimetallic Pd-Pt colloids in organic solvent by solvent-extraction reduction, *Langmuir* **1991**, *7*, 457–459.

181 R. G. DiScipio, Preparation of colloidal gold particles of various sizes using sodium borohydride and sodium cyanoborohydride, *Anal. Biochem.* **1996**, *236*, 168–170.

182 M. A. Watzky, R.G. Finke, Nanocluster size-control and "magic number" investigations. Experimental tests of the "living-metal polymer" concept and of mechanism-based size-control predictions leading to the syntheses of iridium(0) nanoclusters centering about four sequential magic numbers, *Chem. Mater.* **1997**, *9*, 3083–3095.

183 T. Yonezawa, M. Sutoh, T. Kunitake, Practical preparation of size-controlled gold nanoparticles in water, *Chem. Lett.* **1997**, *7*, 619–620.

184 G. Frens, Interacting particles in contact, *Faraday Discussion* **1990**, *90*, 143–151.

185 X. Zhai, E. Efrima, Silver colloids and interfacial colloids-adsorption of alizarin yellow 2G and its effect on colloidal nucleation, *Langmuir* **1997**, *13*, 420–425.

186 D. V. Leff, P. C. Ohara, J. R. Heath, W. Gelbart, Thermodynamic control of gold nanocrystal size: experiment and theory, *J. Phys. Chem.* **1995**, *99*, 7036–7041.

187 C. H. Chew, J. F. Deng, H. H. Huang, et al., Photochemical formation of silver nanoparticles in poly(N-vinylpyrrolidone), *Langmuir* **1996**, *12*, 909–912.

188 G. Braun, H. Bönnemann, Enantioselective hydrogenations on platinum colloids, *Angew. Chem. Int. Ed. Engl.* **1996**, *35*, 1992–1994.

189 M. B. Mohamed, Zhong L. Wang, M. A. El-Sayed, Temperature-dependent size-controlled nucleation and growth of gold nanoclusters, *J. Phys. Chem. A* **1999**, *103*, 10255–10259.

190 T. Teranishi, M. Miyake, Novel synthesis of monodispersed Pd/Ni nanoparticles, *Chem. Mater.* **1999**, *11*, 3414–3416.

191 G. W. Busser, J. G. van Ommen, J. A. Lercher, in: *Advanced Catalysts and Nanostructured Materials*, ed. W. Moser, Academic Press, San Diego, **1996**, 231–230.

192 M. Antonietti, F. Gröhn, J. Hartmann, L. Bronstein, Nonclassical shapes of noblemetal colloids by synthesis in microgel nanoreactors, *Angew. Chem. Int Ed. Engl.* **1997**, *36*, 2080.

193 K. Torigoe, K. Esumi, Preparation of bimetallic Ag-Pd colloids form silver(I) bis(oxalate)palladate(II), *Langmuir* **1993**, *9*, 1664–1667.

194 J.-S. Jeon, C.-S. Yeh, Studies of silver nano-particles by laser ablation method, *J. Chin. Chem. Soc.* **1998**, *45*, 721–726.

195 J. Bosbach, D. Martin, F. Stietz, et al., Laser induced manipulation of the size and shape of small metal particles: Towards monodisperse clusters on surfaces, *Eur. Phys. J. D* **1999**, *9*, 613–617.

196 G. Hornyak, M. Kröll, R. Pugin, et al., Gold clusters and colloids in alumina nanotubes, *Chem. Eur. J.* **1997**, *3*, 1951.

197 T. Hanaoka, H.-P. Kormann, M. Kröll, et al., Three- dimensional assemblies of gold colloids in nanoporous alumina membranes, *Eur. J. Inorg. Chem.* **1998**, *6*, 807–812.

198 G. Schmid, The role of big metal clusters in nanoscience, *J. Chem. Soc., Dalton Trans.*, **1998**, *7*, 1077–1082.

199 A. P. Alivisatos, K. P. Johnson, X. Peng, et al., Organization of 'nanocrystal molecules' using DNA, *Nature*, **1996**, *382*, 609.

200 A. Freund, T. Lehmann, K.-A. Starz, G. Heinz, R. Schwarz (to Degussa AG), Platinum-aluminium alloy containing catalysts for fuel cell and process for their prepartion, EP 0 743 092 A1 (November 20, 1996)

201 S. Chen, Two-dimensional crosslinked nanoparticle networks, *Adv. Mater.* **2000**, *12*, 186–189.

202 C. N. R. Rao, G. U. Kulkarni, P. J. Thomas, Metal nanoparticles and their assemblies, *Chem. Soc. Rev.*, **2000**, *29*, 27–35.

203 M. Sastry, A. Gole, S. R. Sainkar, Formation of patterned, heterocolloidal nanoparticle thin films, *Langmuir* **2000**, *16*, 3553–3556.

204 T. Teranishi, M. Haga, Y. Shiozawa, M. Miyake, Self-organization of Au nanoparticles protected by 2,6-bis(1'-(8-thiooctyl)benzimidazol-2-yl)pyridine, *J. Am. Chem. Soc.* **2000**, *122*, 4237–4238.

205 C.-C. Hsueh, M.-T. Lee, M. S. Freund, G. S. Ferguson, Electrochemically directed self-assembly on gold, *Angew. Chem. Int. Ed.*, **2000**, *39*, 1228–1230.

206 H. X. He, H. Zhang, Q. G. Li, Z. F. Liu, Fabrication of designed architectures of Au nanoparticles on solid substrate with printed self-assembled monolayers as templates, *Langmuir* **2000**, *16*, 3846–3851.

207 A. K. Boal, F. Illhan, J. E. DeRouchey, et al., Self-assembly of nanoparticles into structured spherical and network aggregates, *Nature*, **2000**, *404*, 746.

208 G. Schmid, W. Meyer-Zaika, R. Pugin, et al., Naked Au55 clusters: dramatic effect of a thiol-terminated dendrimer, *Chem. Eur. J.* **2000**, *6*, 1693.

209 J. Fink, C. J. Kiely, D. Bethell, D. J. Schiffrin, Self-organization of nanosized gold particles, *Chem. Mater.* **1998**, *10*, 922–926.

210 L. Ramos, T. Lubensky, N. Dan, et al., Surfactant-mediated two- dimensional crystallization of colloidal crystals, *Science* **1999**, *286*, 2325–2328.

211 J. Eversole, H. P. Broida, Size and shape effects in light scattering from small silver, copper and gold particles, *Phys. Rev.* **1977**, *B15*, 1644–1655.

212 J. Eversole, H. Broida, Electron microscopy of size distribution and growth of small zinc crystals formed by homogeneous nucleation in a flowing inert gas system, *J. Appl. Phys.* **1974**, *45*, 596.

213 N. Wada, Preparation of fine metal particles by means of evaporation in helium gas, *Jpn. J. Appl. Phys.* **1967**, *6*, 553.

214 S. Kasukate, S. Yatsuga, R. Uyeda, Ultrafine metal particles formed by gas evaporation technique, *Jpn. J. Appl. Phys.* **1974**, *13*, 1714.

215 S. Mochizuki, R. Ruppin, Optical spectra of free silver clusters and microcrystals produced by the gas evaporation technique – transition from atom to microcrystal, *J. Phys. Cond. Mater.* **1993**, *5*, 135.

216 C. Petit, A. Taleb, M.-P. Pileni, Self organization of magnetic nanosized cobalt particles, *Adv. Mater.* **1998**, *10*, 259.

217 C. Petit, T. Cren, D. Roditchev, et al., Single electron tunneling through nano-sized cobalt particles, *Adv. Mater.* **1999**, *11*, 1198.

218 P. Mulvaney, L. Liz-Marzan, M. Giersig, T. Ung, Silica encapsulation of quantum dots and metal clusters, *J. Mater. Chem.* **2000**, *10*, 1259–1270.

219 A. C. Balazs, V. V. Ginzburg, F. Qiu, et al., Multi-scale model for binary mixtures containing nanoscopic particles, *J. Phys. Chem. B.* **2000**, *104*, 3411–3422.

220 H. Fan, Y. Lu, A. Stump, et al., Rapid prototyping of patterned functional nanostructures, *Nature* **2000**, *405*, 55–60.

221 G. Schmid, M. Bäumle, M. Geerkens, et al., Current and future applications of nanoclusters, *Chem. Soc. Rev.* **1999**, *28*, 179–185.

222 K. V. Sarathy, P. J. Thomas, G. U. Kulkarni, C. N. R. Rao, Superlattices of metal and metal-semiconductor quantum dots obtained by layer-by-layer deposition of nanoparticle, *J. Phys. Chem. B.* **1999**, *103*, 399.

223 H. Fan, Y. Zhou, G. P. Lopez, Stepwise assembly in three dimensions. Preparation and characterization of layered gold nanoparticles in porous silica matrixes, *Adv. Mater.* **1997**, *9*, 728–731.

224 J. L. Dormann, D. Fiorani, eds., *Magnetic Properties of Fine Particles*, North Holland, Amsterdam, **1992**.

225 L. J. deJongh, D. A. van Leeuwen, J. M. van Ruitenbeek, J. Sinzig, *Nato Aso Ser. C*, **1996**, *484*, 615–643.

226 L. J. de Jongh, J. Sinzig, H. Bönnemann, et al., Antiferromagnetism of colloidal [MN0.0.3THF]x , *Appl. Organomet. Chem.*, **1998**, *12*, 387–391.

227 N. C. Seeman, Nucleic-acid junctions and lattices, *J. Theor. Biol.* **1982**, *99*, 237.

228 C. A. Mirkin, R. L. Letsinger, R. C. Mucic, J. J. Storhoff, A DNA-based method for rationally assembling nanoparticles into macroscopic materials, *Nature* **1996**, *382*, 607–609.

229 J. J. Storhoff, R. C. Mucic, C. A. Mirkin, Programmed materials synthesis with DNA, *Chem. Rev.* **1999**, *99*, 1849–1862.

230 J. J. Storhoff, A. A. Lazarides, R. C. Mucic, et al., What controls the optical properties of DNA-linked gold nanoparticle assemblies?, *J. Am. Chem. Soc.* **2000**, *122*, 4640–4650.

231 C. A. Mirkin, Programming the assembly of two- and three-dimensional architectures with DNA and nanoscale inorganic building blocks, *Inorg. Chem.*, **2000**, *39*, 2258–2272.

232 C. M. Niemeyer, DNA as a material for nanotechnology, *Angew. Chem., Int. Ed. Engl.* **1997**, *36*, 585–587.

233 C. M. Niemeyer, W. Bürger, J. Peplies, Covalent DNA-streptavidin conjugates as building blocks for novel biometallic nanostructures, *Angew. Chem., Int. Ed. Engl.* **1998**, *37*, 2265–2268.

234 C. M. Niemeyer, Progress in "engineering up" nanotechnology devices utilizing DNA as a construction material, *Appl. Phys. A* **1999**, *68*, 119–124.

235 E. Braun, Y. Eichen, U. Sivan, G. Ben-Yoseph, DNA-templated assembly and electrode attachment of a conducting silver wire, *Nature*, **1998**, *391*, 775–778.

2
Synthetic Approaches for Carbon Nanotubes

Bingqing Wei, Robert Vajtai, and Pulickel M. Ajayan

This chapter will briefly introduce different types of nanoscale carbon structures, such as fullerenes, carbon nanotubes, carbon onions, carbon nanofibers, nanodiamonds, and nanoporous activated carbon, as well as their synthesis methods. In particular, synthetic approaches for carbon nanotube production will be described. In any such review, it is important to discuss the various applications of nanotubes. Considering the scope of this book we will just give a brief perspective on one of the key future applications of nanotubes, namely biomedical, after discussing the synthetic routes to creating novel carbon-based nanostructures.

2.1
Introduction

Carbon is one of the most important elements in nature and in the human body. It is found in many different structures with entirely different properties, such as graphite, diamond, and amorphous carbon. The two different isomorphic crystalline structures of carbon are graphite and diamond; the discovery of fullerenes and nanotubes has certainly changed this view and has brought in a whole range of topologically different carbon architectures with nanoscale dimensions. It is worth distinguishing the structural differences among these different carbon structures first, then the state-of-the-art achievements in making them will be introduced in the following sections.

2.1.1
Structure of Carbon Nanomaterials

The carbon atoms in graphite, diamond, and fullerenes (C_{60}) are held together by strong covalent bonds. As demonstrated in Fig. 2.1, it is the lattice arrangement of atoms and the type of bonding that differentiate these forms of elemental carbon. Understanding how carbon atoms are arranged in each type of material allows a better understanding of why these compounds have different properties. Notice that graphite, for example, has large sheets of hexagonal honeycomb lattice and strong bonds in these planes. The sheets interact, but they are so far apart (~ 3.35 Å) and

Nanofabrication Towards Biomedical Applications. C. S. S. R. Kumar, J. Hormes, C. Leuschner (Eds.)
Copyright © 2005 WILEY-VCH Verlag GmbH & Co. KGaA Weinheim
ISBN 3-527-31115-7

the interactions are via weak Van der Waals force. Because the layers of carbon rings can slide over each other, graphite is a good lubricant. Diamond, however, has each carbon atom bonded to four other carbon atoms in a tetrahedral arrangement. Diamond can be cleaved along its planes, but it cannot flake apart into layers because of this tetrahedral arrangement of carbon atoms. C_{60}, the most important member in the fullerene family, is shaped like a soccer ball, often called a buckyball, reminiscent of the stable geodesic domes that were built by the architect Buckminster Fuller. Perhaps the biggest difference of the fullerenes when compared to their crystalline graphite and diamond counterparts is that the fullerenes are molecular, with exact numbers (60, 70, 82, etc.) of carbon atoms. Fullerenes are discrete entities but can be used as the building blocks of lattices. Carbon nanotubes are extensions of the fullerene structure, as we will discuss later in this chapter.

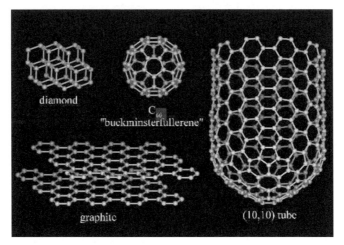

Figure 2.1. Different crystal carbon structures [1].

2.1.2
Wide Range of Properties

Fullerenes and their derivatives, carbon nanotubes, have triggered extensive and attractive research and promise to be one of the key materials in nanotechnology. Unlike other existing materials, carbon nanomaterials have found many scientific and technological applications in diverse areas, as illustrated in Fig. 2.2. Because of the wide range of superior properties (mechanical, electrical, thermal) that are inherent in carbon nanostructures, specifically nanotubes, and due to the simplicity of their structures (making them good model systems to study physics in nanoscale materials), they play an important role in the current rapid expansion of fundamental studies on nanostructures and potential use in nanotechnology. Based on their dimensions, their novel electronic structures, and their controllable chemical functionality, carbon nanotubes and other carbon nanomaterials are expected to be used in several applications, including medical and biomedical areas, such as drug delivery and diagnostic devices. In this chapter we will look at the various ways in which carbon nanostructures can be synthesized, with emphasis on nanotubes, and briefly discuss the limitations in these procedures. From here we start with an introduction of members in the carbon nanomaterials family, including fullerenes, nano-onions, nanofibers, nanotubes, nanoscale diamonds and diamond-like carbon and nanoporous activated-carbon (Section 2.2). After a brief summary in Section 2.3 of the main approaches for the synthesis of carbon nanotubes, which are arc discharge, laser ablation, and chemical vapor deposition (CVD), our recent efforts focusing on controllable growth of nanotube architectures based on substrate-site-selective growth by the CVD technique are summarized in Section 2.4. A perspective of nanotubes in medical and bio-medical applications is also presented in Section 2.5, before the conclusion.

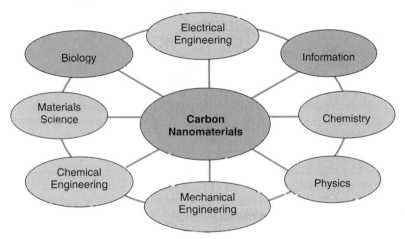

Figure 2.2. Scientific and technological areas to which carbon nanomaterials research has relevance.

2.2
Family of Carbon Nanomaterials

As briefly mentioned at the beginning of this chapter, the most recently discovered form of crystalline carbon is the fullerene family. Fullerenes were discovered in 1985 [2]. The discovery of C_{60} led to the discovery of carbon nanotubes in 1991 [3]. Since then, researchers around the world have been exploring both the basic science and potential applications of these novel materials. In this section, the structures and synthesis techniques of several important carbon nanomaterials will be introduced.

2.2.1
Fullerenes

C_{60} – buckminsterfullerene or the buckyball – involves 60 carbon atoms in a "soccer ball"-shaped structure. The carbon atoms in C_{60} are arranged in a shell that is made of 20 hexagons and 12 pentagons. This is required by Euler's theorem, where 12 pentagons

faces. Rearrangement via the Stone–Wales transformation can create pentagons and heptagons [7]; an appropriate combination of both ensures the uniform spherical curvature of the onion. The shrinkage of the shells creates a surface tension that tends to make the object spherical and generates pressure within the onion.

A simple method for producing high-quality spherical carbon nano-onions in large quantities without the use of expensive vacuum equipment has been reported recently [8]. The nested onion nanoparticles were generated by arc discharge between two graphite electrodes, similar to the one used for C_{60} production but submerged in deionized water, that is, in a nonvacuum environment. After the electric discharge the nano-onions remain afloat on the water surface, while the rest of the carbon structures produced fall to the bottom of the beaker, giving material of high purity. The average diameter of the nano-onions is 25–30 nm (range 5–40 nm), a useful size range for many lubrication applications. Nano-onions have been successfully used as an effective catalyst for an important industrial reaction to convert ethylbenzene into styrene [9]. The synthesis of styrene is one of the ten most important industrial chemical processes, and is commonly conducted using a catalyst of potassium-promoted iron oxide (K–Fe) to effect oxidative dehydrogenation of ethylbenzene. This process is thermodynamically limited to a maximum yield of 50%. Carbon onions, in contrast, produce styrene in 62% yield in these preliminary experiments, and the researchers expect further improvement. They think that the key to a still more active catalyst lies in generating an optimal distribution of active sites on the surface of the carbon nano-onions.

2.2.3
Carbon Nanofibers

The term "carbon nanofiber" summarizes a large family of different filamentous nanocarbons. Carbon nanofibers, like carbon nanotubes, are mainly related to graphite structures and their structures have been well investigated [10–12] during the past 50 years or so.

Carbon fibers (micron size structures) have been around for several decades and are commercially important materials. They can be made in a variety of ways, such as by using organic polymers like polyacrylonitrile or by vapor phase deposition with the assistance of catalyst particles (vapor-grown carbon fibers or VGCFs) [13, 14]. The latter are in structure, morphology and degree of graphitization closer to carbon nanofibers.

Proposed models of carbon fiber growth assume that the process usually consists of two steps: first, a high-aspect core filament grows by catalyst-assisted vapor phase growth, and then a (sometimes only partially) graphitized carbon deposition thickens the template structure up to the final dimensions [15–17]. The initial growth stage has been a difficult process to understand. Very recently, however, state-of-the-art transmission electron microscopy equipment has offered the possibility to track the early stages of the nanofiber growth with sufficient spatial and temporal resolution [18, 19]. As shown in Fig. 2.6, the initial equilibrium shape of the catalyst particle (here a Ni particle) transforms into a highly elongated shape. The elongation of the Ni particle appears to be correlated with the formation of graphene sheets at the graphene–Ni interface with their basal (002) planes oriented parallel to the Ni surface. Hence, the reshaping of the Ni nanocluster assists the alignment of graphene layers into a tubular structure. The elongation of the Ni nanocrystal continues until it reaches a high aspect ratio, before it abruptly contracts to a spherical shape within less than ~0.5 s [Fig. 2.6(h)]. The contraction is attributed to the fact that the increase in the Ni surface energy can no longer be compensated for by the energy gained when binding the graphitic fiber to the Ni surface. The elongation/contraction scenario continues in a periodic manner as the nanofiber grows.

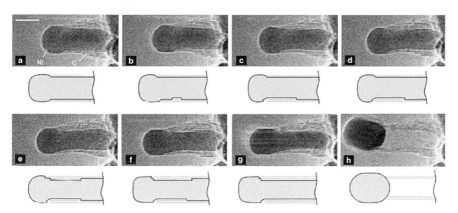

Figure 2.6. Snap images showing a sequence of the growing process of a carbon nanofiber. (From Ref. [19], with permission.)

Carbon nanofibers have various applications and are normally used as fillers to improve the mechanical and thermal properties of composite materials [20]. Other applications include their uses as electrically conducting fillers, as catalyst support, in nanoelectronic devices, as artificial muscles, as field emitters, and in gas and electrochemical energy storage matrices. The production of carbon nanofibers is a mature technology and a 40-ton-per-year commercial plant is already operating in Japan for the production of vapor-grown carbon nanofibers [21].

2.2.4
Carbon Nanotubes

Carbon nanotubes, nanoscale cylinders constructed from sp^2 hybridized carbon bonds, form a hexagonal honeycomb lattice as in graphite. Containing a hollow center, carbon nanotubes are derived from graphene sheets that are rolled up into tubes with a seamless structure [22, 23]. The nanotubes are terminated by fullerene endcaps with six pentagons on both ends. The unique mechanical and electrical properties of carbon nanotubes are directly related to the characteristics of the carbon bonds and their organization into the tube lattice. The nanotubes can be distinguished by the number of layers that make up their cylindrical walls: single-walled nanotubes (SWNTs) and multi-walled nanotubes (MWNTs). Figure 2.7 shows models and high-resolution transmission electron microscopy (HRTEM) images of a SWNT and a MWNT. Along the tube axis, carbon nanotubes show excellent mechanical properties due to the nature of the strong C–C bonding and the seamless structure. The most intriguing property of carbon nanotubes comes from their unique electronic structure. Depending on the diameter and chirality, carbon nanotubes can be either metallic nanowires with quantum transport property or semiconducting nanomaterials with varying band gaps. They also show good thermal and electrical conductivity, chemical stability, and high mechanical strength. Because of these, carbon nanotubes are promising materials for various applications (for examples, see

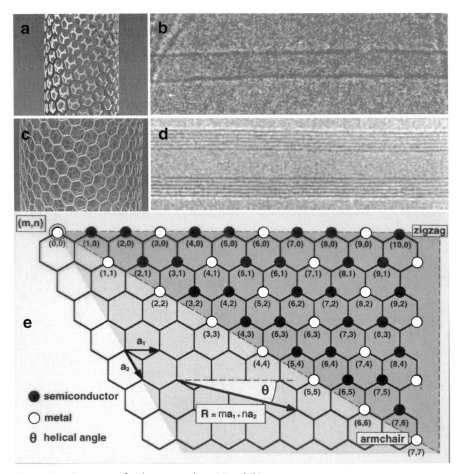

Figure 2.7. Structures of carbon nanotubes. (a) and (b) are SWNT and (c) and (d) are MWNT. (e) Indices with a pair of integers (*n*, *m*) to classify nanotube structures: the white dots denote metallic nanotubes and the black ones are semiconducting nanotubes. (From Ref. [23], with permission.)

Section 2.5) [22–24]. A detailed description of carbon nanotubes synthesis will be presented in Sections 2.3 and 2.4.

2.2.5
Nanoscale Diamonds and Diamond-Like Carbon

Nanocrystalline diamond films are unique new materials with applications in fields as diverse as tribology, cold cathodes, corrosion resistance, electrochemical electrodes, and conformal coatings on microelectromechnical system (MEMS) devices. Man-made diamond crystals were successfully produced in the 1950s by the high-pressure, high-temperature method. An alternative method, CVD of diamond at low pressure (typically with the use of an excited CH_4/H_2 mixture on substrates held at ~700 to 800 °C), has also been applied successfully over the last 15–20 years. With the assistance of a double bias, hot-filament CVD was employed to synthesize nanoscale diamonds (Fig. 2.8) [25].

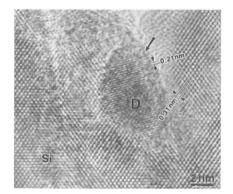

Figure 2.8. HRTEM image of a diamond crystallite (diameter ~6 nm) grown directly on Si with a random alignment. (From Ref. [25], with permission.)

Ultra-nanocrystalline diamond (UNCD) has been synthesized by the modification of the microwave plasma chemical vapor deposition (MPCVD) process [26]. UNCD thin films are synthesized using argon-rich plasmas instead of the hydrogen-rich plasmas normally used to deposit microcrystalline diamond (>1 µm grain size). The use of small amounts of carbon source gases (C_{60}, CH_4, C_2H_2) with argon leads to the formation of C_2 dimers, which are the growth species for all UNCD thin films. UNCD grown from C_2 precursors consists of ultra-small (2–5 nm) grains and atomically abrupt grain boundaries. These films are superior in many ways to traditional microcrystalline diamond films: they are smooth, dense, pinhole-free, and phase-pure, and can be conformally coated on a wide variety of materials and high-aspect-ratio structures. UNCD is finding a wide range of industrial applications in MEMS, as tribo-coatings, as photonic switches in optical cross-connects, as field emission cathodes, as electrochemical electrodes, and as hermetic coatings on bioimplants [26]. With the ability to tailor both the film structure and the electronic properties independently, UNCD can be optimized for several important applications. Most excitingly, the ability to electronically dope the material both n- and p-type opens the door to the next generation of novel high-speed, high-temperature, and even biocompatible electronics. In addition to nanocrystalline diamond films, it worth mentioning another form of carbon nanomaterial, diamond-like carbon, which has been evaluated as a coating to improve biocompatibility of orthopedic and cardiovascular implants [27].

2.2.6
Nanoporous Activated Carbon

The surface area of a solid increases when it becomes nanoporous, improving catalytic, absorbent, and adsorbent properties. Activated carbon is an example of a nanoporous material that, like zeolites, has been in use for a long time. Nanoporous carbon consists of very-high-surface-area carbon with a tunable and very narrow pore size distribution. It is suitable in applications such as energy storage systems, catalytic, specific adsorbents, and gas separation. Recent research has developed a templating technique using silica nanoparticles that can create activated carbon with uniform 8-nm and 12-nm pore sizes. The resulting materials show adsorption greater than 10 times that of commercial activated carbon [28]. High surface area of porous carbon materials is ideal for holding dispersed metal catalyst particles for use in heterogeneous catalysis (Fig. 2.9).

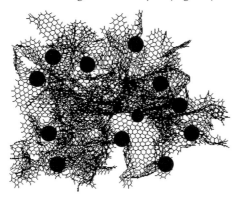

Figure 2.9. Atomic level schematics of nanoporous carbon structure with loaded catalyst. (From Ref. [29], with permission.)

Nanoporous carbon membranes prepared by pyrolysis of poly(furfuryl alcohol) on porous stainless steel disks have been investigated for gas separation [30]. The nanoporous carbon molecular sieve membranes can be prepared with very high size and shape selectivity. This offers the opportunity to extend the range of application of carbon membranes beyond surface selective flow. That nanoporous carbon molecular sieves with pore sizes less than 5 Å can separate nitrogen from oxygen has been known for some time, and is the basis for nitrogen pressure swing adsorption. The separation, done over packed beds of nanoporous carbon, is based on the kinetics of diffusion rather than on the thermodynamics of adsorption.

2.3
Synthesis of Carbon Nanotubes

There are many synthesis methods, such as electric arc, laser ablation chemical vapor deposition, pyrolysis, electrochemical methods, template-based synthesis, flames, and so on, that have been employed to synthesize carbon nanotubes. In this section, three main techniques for the synthesis of carbon nanotubes, both SWNTs and MWNTs, are introduced. They are the arc-discharge method (electric arc), laser ablation (pulsed laser vaporization), and CVD.

2.3.1
Nanotube Growth via the Arc-Discharge Method

Electric arc was the first method reported for producing carbon nanotubes [3] and also the first mass-production technique [31]. For MWNT production two high-purity graphite electrodes are used. During the growth process, nanotubes are formed and deposited on the cathode; the anode is continuously consumed in the process.

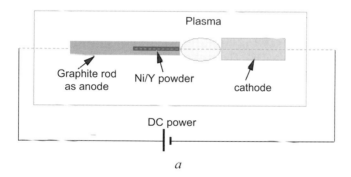

a

b   *c*

Figure 2.10. (a) Apparatus to produce carbon nanotubes by the electric-arc method. The setup produces multi-walled nanotubes (MWNTs) when pure carbon rods are applied as electrodes, and produces single-walled nanotubes (SWNTs) when a metal catalyst is mixed into the core of the anode. (b) Photograph of the cathode deposition; (c) TEM image of the MWNTs.

and nano electromechanical instruments based on these types of three-dimensional networks of carbon nanotubes may be realized in the future by this approach.

2.4.1
Substrate-Site-Selective Growth

In this specific CVD process we placed predefined SiO_2/Si patterns (Fig. 2.12) into a conventional tube furnace. The nanotubes were grown from a mixture of xylene (C_8H_{10}) and ferrocene [$Fe(C_5H_5)_2$]. The reactor is preheated gradually to 800 °C and then a 0.01-g ml^{-1} solution of ferrocene in xylene is preheated to about 150 °C and fed into the reactor. In this reaction ferrocene is the nanotube nucleation initiator and xylene is the carbon source. This precursor combination results in highly selective growth of 20- to 30-nm-diameter MWNTs on the SiO_2 surfaces: no nanotube growth is observed on pristine Si surfaces or on the native oxide layer. The film is comprised of vertically aligned nanotubes with center-to-center average spacing of ~50 nm [41].

To reveal the reason for this strong substrate selectivity we investigated the mechanism of catalyst particle formation on different substrates. Particles are formed and deposited both on silicon and silica surfaces, with a size around 20–40 nm in the silicon oxide region and slightly bigger in the Si region. Measurements made by an

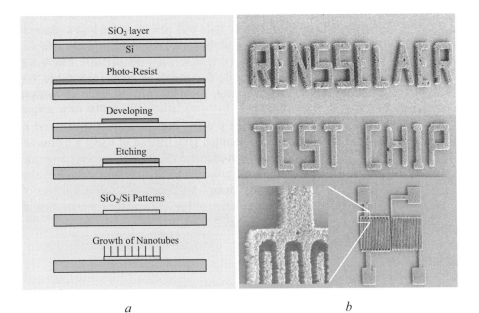

a   *b*

Figure 2.12. The process used to create silica patterns on a silicon substrate. (a) The sequence from top to bottom: generation of a SiO_2 layer on a Si wafer; spin coating of a photoresist; photoexposure through a mask and resist development; etching the exposed SiO_2; photoresist removal and growth of nanotubes on the patterned SiO_2 surface. (b) Patterned MWNT layers.

electron probe microanalyzer and by Auger electron spectroscopy showed that iron and carbon are dispersed over both regions. Cross-sectional TEM observation of the substrates showed that particles on the top of silicon oxide surfaces are pure gamma iron (fcc Fe), which is an active catalyst for carbon nanotube growth. However, on the silicon surface iron silicide is formed by substrate particle interactions, and these silicide phases are not effective catalysts [42].

2.4.2
Three-Dimensional Nanotube Architectures

The site selectivity of the floating catalyst CVD method is a powerful tool to design mesoscale structures similar to ones used as MEMS. To demonstrate this possibility we carried out simultaneous multidirectional growth and multilayered growth of ordered nanotubes [43, 44]. To provide depth for the horizontal growth, thick silica layers were deposited by plasma-enhanced CVD (PECVD) to create high-aspect-ratio silica features. Patterns of Si/SiO$_2$ of various shapes were generated by photolithography followed by a combination of wet and/or dry etching. CVD growth of nanotubes is stimulated in a manner similar to the above-described method using a xylene/ferrocene mixture (Fig. 2.13). With this method nanotube growth in mutually orthogonal directions and growth with oblique inclinations (that is, neither orthogonal nor planar with respect to the substrate plane) has been illustrated [44].

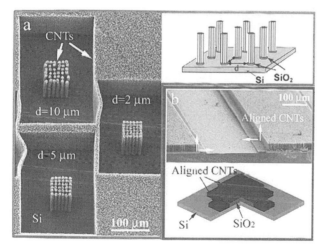

Figure 2.13. Carbon nanotube structures deposited on patterned, three-dimensional silicon/silica templates. The length of the nanotubes in both vertical and horizontal directions is about 60 μm. The thickness of the patterned SiO$_2$ layer was 8.5 μm in the presented experiment. (From Ref. [43], with permission.)

2.4.3
Super-Long SWNT Strands

Mesostructures of SWNTs can also be grown by techniques involving vapor phase catalyst delivery, similar to the above. We have synthesized long strands of ordered SWNTs (Fig. 2.14) by a CVD deposition technique with the floating catalyst method, but in a vertical furnace, using n-hexane as a carbon source [45, 46]. The n-hexane solution with ferrocene content of 0.018 g ml^{-1} and thiophene of 0.4 wt% was introduced into the reactor after heating the reactor to the pyrolysis temperature (~1100 °C), using hydrogen as carrier gas. In the process SWNTs formed in abundance with yields of ~0.5 g h^{-1}. The formation of very long SWNT strands, also in large amounts (approximately 20–30% of the product), is the unique characteristic of this vertical floating process. The SWNT strands consist of smaller-diameter ropes, which also seem very long, almost as long as the strand itself (several inches) and can be manually handled quite easily.

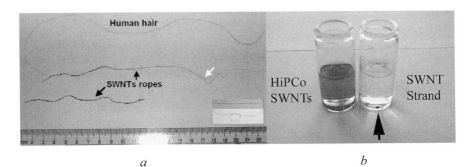

a   *b*

Figure 2.14. (a) Nanotube strand made of the longest SWNT produced to date. (b) Comparison of two samples of similar mass (HiPCO and the long strands) ultrasonicated in ethanol.

Raman spectra taken from different areas along the strands are very similar to each other, indicating homogeneity along the strands. The diameter of SWNTs determined from the frequency of the radial breathing modes (RBM) is in the range of 1.1–1.7 nm, with a dominance of tubes of 1.1 nm, in accordance with our TEM investigations. Measurements were directly conducted on these strands, namely direct tensile test and electrical measurements. The stress in the strands was recorded on individual SWNT strands of centimeter lengths, of diameter 5–20 μm. Modulus values obtained directly from the measurements were 49–77 GPa, smaller than what is predicted for individual nanotubes but still of relatively large magnitude compared to existing fibers. Macroscopic electrical resistivity of nanotube strands was measured from room temperature to 5K using a four-probe method. The crossover temperature from metallic to semiconducting state occurred at about 90K. The metallic resistivity is about six times the value of single bundles reported previously but less than any other macroscopic SWNT structures, suggesting continuous conducting paths along the long nanotube strands.

Double-walled carbon nanotubes (DWNTs) have been brought to the forefront of research in recent years in the nanotube community due to the fact that they are at the frontier between SWNTs and MWNTs and could have the advantages of both. They open a possibility of functionalizing the outer wall, which will ensure connections with the external environment, while retaining the remarkable mechanical and electronic properties of the inner nanotube. This may prove to be very useful for their integration into systems and composites. DWNTs were first reported using pyrolytic organic precursor [47] and recently have been synthesized using the arc-discharge [48] and the CVD techniques [49].

2.5
Perspective on Biomedical Applications

Due to the fact that carbon nanomaterials, particularly carbon nanotubes, have novel structures, small and well-defined dimensions paired with extremely high electrical and thermal conductivity, high mechanical strength and flexibility, as discussed before, various prospective applications of nanotubes have been proposed such as nanoelectronic devices, energy storage, nanocomposites, nanosensors, and so on. A detailed review of this aspect can be found in Refs. [22–24]. Considering the scope of this book we will just give a brief perspective on nanotubes in biomedical applications, one of their key future areas of application.

2.5.1
Imaging and Diagnostics

A striking example is shown in Fig. 2.15, using an individual MWNT attached to the end of a scanning probe microscope tip for imaging biomolecules [50]. The advantage of the nanotube tip is its slenderness and the ability to image features (such as very small, deep surface cracks) which are almost impossible to probe using the

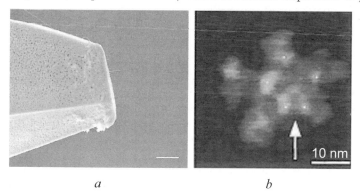

a   *b*

Figure 2.15. Individual MWNT grown directly onto an AFM tip (a) and a picture to demonstrate its lateral resolution (b). (From Ref. [50], with permission.)

larger, blunter-etched Si or metal tips. Biological molecules such as DNA can readily be imaged at higher resolution using nanotube tips than with conventional scanning tunneling microscope tips. MWNT and SWNT tips were used in a tapping mode to image biological molecules such as amyloid-b-protofibrils (related to Alzheimer's disease) with a resolution never achieved before.

Another example of medical diagnostic imaging is x-ray radiation coming from a nanotube x-ray source [51]. The basic design of the x-ray tube has not changed significantly in the last century. The x-ray intensity generated using a carbon nanotube (CNT)-based field-emission cathode is sufficient to image a human extremity (Fig. 2.16). The device can readily produce both continuous and pulsed x-rays with a programmable wave form and repetition rate. Pulsed x-ray with a repetition rate greater than 100 kHz was readily achieved by programming the gate voltage. The CNT-based cold-cathode x-ray technology can potentially lead to portable and miniature x-ray sources for industrial and medical applications.

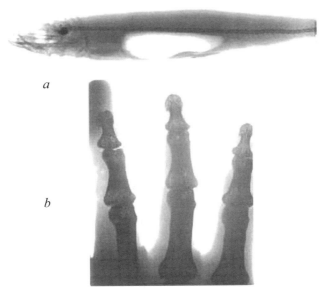

a

b

Figure 2.16. X-ray images of a fish (a) and a humanoid hand (b) taken using the nanotube x-ray source. Detailed bone structures are clearly resolved. (From Ref. [51], with permission.)

2.5.2
Biosensors

Carbon nanotubes can be used for monitoring enzyme activity. Immobilization of proteins on the sidewall of carbon nanotubes through a linking molecule has already been demonstrated [52]. Proteins carrying pH-dependent charged groups that can electrostatically gate a semiconducting SWNT created the possibility to construct a nanosize protein and/or pH sensor. Redox enzymes go through a catalytic reaction cycle where groups in the enzyme temporarily change their charge state and conformational changes occur in the enzyme. This enzymatic activity can potentially be monitored with a nanotube sensor. Figure 2.17 demonstrates the use of individual semiconducting SWNTs as versatile biosensors [53]. Controlled attachment of the redox enzyme glucose oxidase (GOx) to the nanotube sidewall is achieved through a linking molecule and is found to induce a clear change of the conductance. The enzyme-coated tube is found to act as a pH sensor with large and reversible changes in conductance upon changes in pH. Upon addition of glucose, the substrate of GOx, a steplike response can be monitored in real time, indicating that the sensor is capable of measuring enzymatic activity at the level of a single nanotube. This demonstration of nanotube-based biosensors provides a new tool for enzymatic studies and opens the way to biomolecular diagnostics.

Another approach has been explored using enzyme-containing polymer-SWNT composites as unique biocatalytic materials [54]. The biocatalytic composites were prepared by suspending SWNT and α-chymotrypsin (CT) directly into a poly(methyl methacrylate) solution in toluene. The activity of the resulting CT-polymer-nanotube films was observed to be higher than both polymer-CT and polymer-graphite-CT films. SWNT-containing composites show higher enzyme activity than the non-

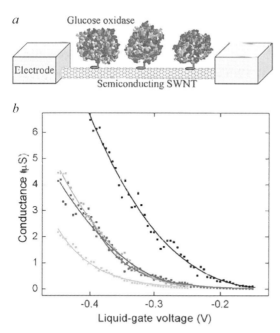

Figure 2.17. (a) Two electrodes connecting a semiconducting SWNT with GOx enzymes immobilized on its surface. (b) Conductance of a semiconducting SWNT as a function of the liquid-gate voltage in water. Data are for the bare SWNT (I), after 2 h in dimethylformamide with/without the linking molecule (II), and after GOx immobilization (III). (From Ref. [53], with permission.)

13 M. Endo, Y. A. Kim, T. Matusita, T. Hayashi, From vapor-grown carbon fibers (VGCFs) to carbon nanotubes, in *Carbon Filaments and Nanotubes: Common Origins, Differing Applications?* eds. L. P. Biro, C. A. Bernardo, G. G. Tibbetts, Ph. Lambin, Kluwer Academic Publishers, Dordrecht, Boston, London, **2000**.

14 M. Bognitzki, W. Czado, T. Frese, et al., Nanostructured fibers via electrospinning, *Adv. Mater.* **2001**, *13*, 70–72.

15 G. G. Tibbetts, Nucleation and growth of carbon filaments and vapor grown carbon fibers, in *Carbon Filaments and Nanotubes: Common Origins, Differing Applications?* eds. L. P. Biro, C. A. Bernardo, G. G. Tibbetts, Ph. Lambin, Kluwer Academic Publishers, Dordrecht, Boston, London, **2000**.

16 R. T. K. Baker, P. S. Harris, The formation of filamentous carbon, in *Chemistry and Physics of Carbon*, eds. P. L. Walker, P. A. Thrower, Dekker, New York, **1978**.

17 M. Endo, K. Takeuchi, K. Kobori, et al., Pyrolytic carbon nanotubes from vapor-grown carbon-fibers, *Carbon* **1995**, *33*, 873–881.

18 P. M. Ajayan, Nanotechnology – how does a nanofibre grow? *Nature* **2004**, *427*, 402–403.

19 S. Helveg, C. López-Cartes, J. Sehested, et al., Atomic-scale imaging of carbon nanofibre growth, *Nature* **2004**, *427*, 426–429.

20 J. P. Issi, B. Nysten, Electrical and thermal transport properties in carbon fibers, in *Carbon Fibers*, eds. J. B. Donnet, S. Rebouillat, T. K. Wang, J. C. M. Peng, Marcel Dekker, New York, **1998**.

21 www.japancorp.net/Article.Asp?Art_ID=5719

22 P. M. Ajayan, Nanotubes from carbon, *Chem. Rev.* **1999**, *99*, 1787–1799.

23 M. S. Dresselhaus, G. Dresselhaus, P. Avouris, *Carbon Nanotubes: Synthesis, Structure, Properties, and Applications*, Springer, Berlin, New York, **2001**.

24 R. H. Baughman, A. A. Zakhidov, W. A. de Heer, Carbon nanotubes – the route toward applications, *Science* **2002**, *297*, 787–792.

25 S. T. Lee, H. Y. Peng, X. T. Zhou, et al., A nucleation site and mechanism leading to epitaxial growth of diamond films, *Science* **2000**, *287*, 104–106.

26 D. M. Gruen, Nanocrystalline diamond films, *Ann. Rev. Mater. Sci.* **1999**, *29*, 211–259.

27 A. Singh, G. Ehteshami, S. Massia, et al., Glial cell and fibroblast cytotoxicity study on plasma-deposited diamond-like carbon coatings, *Biomaterials* **2003**, *24*, 5083–5089.

28 S. J. Han, K. Sohn, T. Hyeon, Fabrication of new nanoporous carbons through silica templates and their application to the adsorption of bulky dyes, *Chem. Mater.* **2000**, *12*, 3337–3341.

29 P. M. Ajayan, I. S. Schadler, P. V. Braun, *Nanocomposite Science and Technology*, Wiley-VCH, Weinheim, Germany, **2003**.

30 M. Acharya, H. C. Foley, Transport in nanoporous carbon membranes: experiments and analysis, *AICHE J.* **2000**, *46*, 911–922.

31 T. W. Ebbesen, P. M. Ajayan, Large-scale synthesis of carbon nanotubes, *Nature* **1992**, *358*, 220–221.

32 S. Iijima, T. Ichihashi, Single-shell carbon nanotubes of 1-nm diameter, *Nature* **1993**, *363*, 603–605.

33 D. S. Bethune, C. H. Kiang, M. S. de Vries, et al., Cobalt-catalyzed growth of carbon nanotubes with single-atomic-layer walls, *Nature* **1993**, *363*, 605–607.

34 C. Journet, W. K. Maser, P. Bernier, et al., Large-scale production of single-walled carbon nanotubes by the electric-arc technique, *Nature* **1997**, *388*, 756–758.

35 L. B. Kiss, R. Vajtai, P. M. Ajayan, Random walk in gas vortices and nanotube self-assembly, *Phys. Status Solidi (b)* **1999**, *214(1)*, R3–R4.

36 T. Guo, P. Nikoleav, A. G. Rinzler, et al., Self-assembly of tubular fullerenes, *J. Phys. Chem.* **1995**, *99*, 10694–10697.

37 A. Thess, R. Lee, P. Nikolaev, et al., Crystalline ropes of metallic carbon nanotubes, *Science* **1996**, *273*, 483–487.

38 P. C. Eklund, B. K. Pradhan, U. J. Kim, et al., Large-scale production of single-walled carbon nanotubes using ultrafast pulses from a free electron laser, *Nano Lett.* **2002**, *2*, 561–566.

39 P. Nikolaev, M. J. Bronikowski, R. K. Bradley, et al., Gas-phase catalytic growth of single-walled carbon nanotubes from carbon monoxide, *Chem. Phys. Lett.* **1999**, *313*, 91–97.

40 Z. J. Zhang, B. Q. Wei, G. Ramanath, P. M. Ajayan, Substrate-site selective growth of aligned carbon nanotubes, *Appl. Phys. Lett.* **2000**, *77*, 3764–3766.

41 J. T. Drotar, B. Q. Wei, Y. P. Zhao, et al., Reflection high-energy electron diffraction from carbon nanotubes, *Phys. Rev. B* **2001**, *64*, 125417.

42 Y. J. Jung, B. Q. Wei, R. Vajtai, et al., Mechanism of selective growth of carbon nanotubes on SiO_2/Si patterns, *Nano Lett.* **2003**, *3*, 561–564.

43 B. Q. Wei, R. Vajtai, Y. Jung, et al., Organized assembly of carbon nanotubes – cunning refinements help to customize the architecture of nanotube structures, *Nature*, **2002**, *416*, 495–496.

44 B. Q. Wei, R. Vajtai, Y. Jung, et al., Assembly of highly organized carbon nanotube architectures by chemical vapor deposition, *Chem. Mater.*, **2003**, *15*, 1598–1606.

45 H. W. Zhu, C. L. Xu, D. H. Wu, et al., Direct synthesis of long single-walled carbon nanotube strands, *Science* **2002**, *296*, 884–886.

46 B. Q. Wei, R. Vajtai, Y. Y. Choi, et al., Structural characterizations of long single-walled carbon nanotube strands, *Nano Lett.* **2002**, *2*, 1105–1107.

47 A. Sarkar, H. W. Kroto, M. Endo, Hemi-toroidal networks in pyrolytic carbon nanotubes, *Carbon* **1995**, *33*, 51–55.

48 J. L. Hutchison, N. A. Kiselev, E. P. Krinichnaya, et al., Double-walled carbon nanotubes fabricated by a hydrogen arc discharge method, *Carbon* **2001**, *39*, 761–770.

49 A. Peigney, P. Coquay, E. Flahaut, et al., A study of the formation of single- and double-walled carbon nanotubes by a CVD method, *J. Phys. Chem. B* **2001**, *105*, 9699–9710.

50 J. H. Hafner, C. L. Cheung, C. M. Lieber, Growth of nanotubes for probe microscopy tips, *Nature* **1999**, *398*, 761.

51 G. Z. Yue, Q. Qiu, B. Gao, et al., Generation of continuous and pulsed diagnostic imaging x-ray radiation using a carbon-nanotube-based field-emission cathode, *Appl. Phys. Lett.* **2002**, *81*, 355–357.

52 R. J. Chen, Y. Zhang, D. Wang, H. Dai, Non-covalent sidewall functionalization of single-walled carbon nanotubes for protein immobilization, *J. Am. Chem. Soc.* **2001**, *123*, 3838.

53 K. Besteman, J.O. Lee, F.G.M. Wiertz, et al., Enzyme-coated carbon nanotubes as single-molecule biosensors, *Nano Lett.* **2003**, *3*, 727–730.

54 K. Rege, N. R. Raravikar, D. Y. Kim, et al., Enzyme-polymer-single walled carbon nanotube composites as biocatalytic films, *Nano Lett.* **2003**, *3*, 829–832.

3
Nanostructured Systems from Low-Dimensional Building Blocks

Donghai Wang, Maria P. Gil, Guang Lu, and Yunfeng Lu

3.1
Introduction

Compared to their bulk and molecular counterparts, materials with nanoscale structure, such as quantum dots, nanocrystals, nanorods, nanowires, nanobelts, nano tubes, and nanomeshes [1–3] show novel electronic, magnetic, optical, thermal, and mechanic properties due to quantum effects. Such unique materials therefore hold great promise for a large spectrum of applications such as energy storage and conversion, information storage and processing, chemical transformations and purifications, chemical and biological sensing, and drug discoveries and screening.

Realization of these promises requires translation of the properties of nanoscale components into the dimensions devices can use. One of the most significant steps is to directly synthesize the nanostructures with designed two-dimensional (2D) networks, three-dimensional (3D) networks, or more complicated hierarchical structures, or to assemble the preformed nanoscale building blocks into such usable structures.

This chapter first provides an overview of the fabrication of nanostructured systems, including: (i) synthesis of 2D/3D nanostructures through self-assembly of the preformed building blocks, (ii) biomimetic assemblies and biomolecular recognition-based assemblies, (iii) template-assisted assemblies, (iv) external field-assisted assemblies, and (v) 2D/3D nanostructures by direct synthesis. Among these methods, the synthesis of nanostructured systems using non-covalent interactions among the building blocks offers a simple and efficient route, since the syntheses of such building blocks are well developed. Beyond this, the biomimetic assembles and bio-recognition-based assemblies utilize the specific non-covalent interactions to form nanostructured systems with better design and control. The template-assisted assembly approach integrates top-down fabrication techniques with bottom-up approaches to fabricate more complicated systems. Due to the weak nature of non-covalent interactions, external-field-induced assembly may be required to assemble the building blocks into oriented systems. Compared with the above assembling approaches, direct synthesis of nanostructured systems is the ultimate goal since it avoids the use of weak non-covalent interactions, providing robust systems with efficient inter- and intra-network communications (e.g., charge, heat, and mass trans-

Nanofabrication Towards Biomedical Applications. C. S. S. R. Kumar, J. Hormes, C. Leuschner (Eds.)
Copyright © 2005 WILEY-VCH Verlag GmbH & Co. KGaA, Weinheim
ISBN 3-527-31115-7

port). After the synthesis review, applications of the nanostructured systems are also presented. Note that current applications are far beyond what we can summarize here. Therefore, only representative applications are presented in this chapter.

3.2
Nanostructured System by Self-Assembly

Among the various synthesis approaches (e.g., lithography and e-beam lithography), the self-assembly approach has emerged as one of the most promising routes to synthesize a large variety of nanostructures. This approach utilizes non-covalent interactions such as van der Waals, hydrogen bonding, electrostatic, capillary force, and π–π interactions organizing various building blocks into 2D, 3D, or more complicated hierarchical structures. These building blocks characterize the nanoscale structures (i.e., they have at least one dimension in the 1–100 nm scale). Usually, the nanoscale building blocks are defined as zero-dimensional (0D) for spherical nanoparticles, one-dimensional (1D) for nanowire, nanorod, nanobelt, and nanotube structures, and 2D or 3D for hierarchical nanostructures. This section reviews the 2D and 3D assemblies using 0D and 1D building blocks.

3.2.1
Nanoparticle Assemblies

Recent research has enabled the versatile syntheses of a large variety of metal, semiconductor, and metal oxide nanocrystals, which can be self-assembled into superstructures through van der Waals [4], hydrophobic [5, 6], electrostatic [7–10], hydrogen bonding [11–13], and covalent bonding interactions [14, 15]. Excellent examples include the self-assembly of Au [4, 16, 17], Ag [18], Co [19], Ag_2S [20], CdS [21, 22], CdSe [23], FePt [6], and $CoPt_3$[24] nanocrystals. For example, binary Au nanoparticles organize themselves into regular AB_2 or AB 2D superstructures, depending on the relative amount of each species and the ratio of particle diameters [16, 25]. Mixing two monodispersed colloidal solutions of $CoPt_3$ nanocrystals with different diameters (4.5 and 2.6 nm diameter), followed by slow evaporation of the solvent, results in the formation of a 3D superlattice of AB_5 type [24]. To date, most of the research is focused on the self-assembly of nanocrystals into monolayers, rings, superlattices, particles, and thin films. By using polyelectrolytes as a mediator, nanocrystals can also be assembled layer by layer into thin films through electrostatic interactions [26–28].

3.2.1.1   **Role of Capping Molecules**
The surface of nanoparticles is often passivated in solutions by functional organic molecules. The stabilizing molecules and the nature of the nanocrystals (e.g., crystalline structure and preferred crystal orientation) may therefore significantly affect the assembling process and resulting superstructure. For instance, 3D assembly of *fcc* Ag nanocrystals often occurs through the preferred interactions of the [100] facets

and [111] facets [4]. The stabilizing molecules used include organic molecules terminated with alkyl, acid, amine [9], pyrrolyl [14], or dithiol [15] groups; iodine [29]; dendrimers [30]; macromolecules such as multivalent polymer receptors [11, 31] and polyelectrolytes [26–28]; and biomolecules [32, 33]. Among these molecules, hydrophobic chain-terminated ligands are the most commonly used capping molecules for the assembly of nanoscale components through hydrophobic or van der Waals interactions. In addition, the use of capping molecules with terminated functional groups enables the manipulation of molecular interactions and thus tunable superstructures. For example, Rotello and colleagues demonstrated the assembly of metal nanocrystals with a functionalized polymer into structured spherical networks through thymine/diaminotriazine multivalent hydrogen bonding (see Fig. 3.1) [34]. Briefly, Au nanocrystals were first modified using thymine-functionalized alkanethiol. Mixing the modified Au nanocrystals with diaminotriazine-functionalized polystyrene creates spherical nanocrystal/polymer network assemblies through hydrogen bonding. The size and structure of these networks can be further tuned at different temperatures. Mirkin and Alivisatos et al. have demonstrated the formation of aggregated metal nanocrystals using DNA as the recognition element [35, 36]. Nanocrystals capsulated with two complementary DNA strands recognize each other and assemble into aggregates based on DNA molecular recognition. Such biologically driven self-assembly of nanoscale components has been extensively investigated using biomolecular recognition/binding (e.g., DNA, antibody–antigen, etc.), replacing hydrophobic, electrostatic, or hydrogen bonding interactions. This will be discussed later in Section 3.4.

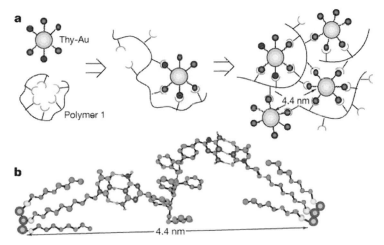

Figure 3.1. Proposed mechanism (a) and structure (b) for self-assembly of Au nanocrystals with functionalized polymer through multivalent hydrogen bonding interactions. (From Ref. [34].)

3.2.1.2 **Multicomponent Assembly**

Assemblies of two or more different nanocrystals provide a new route to control nanocrystal superstructures and properties. Such multicomponent assemblies can be achieved through precisely controlling the sizes and size distribution of the nanocrystals, as well as the assembly conditions. For example, Ag and Au monodispersed nanocrystals with two different particle sizes can self-assemble into regular 2D AB superstructures [16, 25]. Murray and colleagues assembled magnetic (γ-Fe$_2$O$_3$) and semiconducting (PbSe) nanoparticles into large length scale 3D binary superlattices based on hydrophobic and van der Waals interactions [5]. By tuning the particle size and size ratio, the assembled superlattices of magnetic nanocrystals and semiconductor quantum dots can have long-range ordered AB$_2$ and AB$_{13}$ superlattice structure. Figure 3.2 shows transmission electron microscope (TEM) images of self-assembled superlattices of magnetic and semiconductor nanoparticles. These multicomponent nanocrystal assemblies may allow engineering of more complex materials from which new or synergistic properties can emerge. For example, two magnetic nanoparticles with different coercivities and remnant magnetizations can self-assemble and obtain an optimized nanocomposite with larger coercivity and higher remnant magnetization than that composed of either part [37].

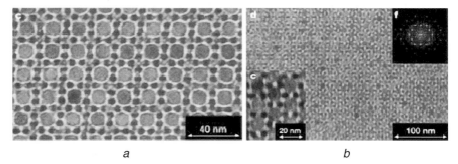

a   *b*

Figure 3.2. Transmission electron microscope (TEM) images of self-assembled superlattices of magnetic (γ-Fe$_2$O$_3$) and semiconductor (PbSe) nanoparticles. (a) TEM image of the [100] plane of superlattice type AB$_{13}$; (b) TEM image of the [001] plane of superlattice type AB$_2$. insets are high-resolution TEM and fast Fourier transform (FFT) images. (From Ref. [5].)

Electrostatic related self-organization has also been applied for two-component self-assembly, such as SiO$_2$-Au [7], SiO$_2$–CdSe [8], Au–CdS [9], and TiO$_2$–CdS [10] systems. Such assemblies often involve the functionalization of one type of colloidal building block with an amine derivative, and a counterpart building block with a carboxylic acid derivative. Mixing the two components results in the spontaneous formation of electrostatically bound mixed-colloidal constructs. The shapes and sizes of these ensembles could then be controlled via variation of the sizes and composition of the building blocks. This electrostatically induced self-organization can be disassembled and reassembled at different pH, regardless of the nature of the nanocrystal cores [8]. This process can also be used to form composite films via layer-by-layer deposition [10]. Bifunctional ligands have also been applied into a Pt–Au assembly process [12].

3.2.2
1D Nanostructure Assemblies

Recently, 1D nanostructures such as nanorods, nanowires, nanotubes, and nano-belts have been widely studied. Current self-assembly processes of 1D nanostructures are mainly focused on functionalized nanorods of Co [38], CuO [39], Au [40], CdSe [41], BaCrO$_4$ [42], and BaWO$_4$ [43]. For example, CdSe nanorods can be macro-scopically aligned in a nematic liquid-crystal phase [44, 45], and superlattice structures of these nanorods formed upon deposition on a substrate are determined by liquid-crystalline phases that are formed prior to complete solvent evaporation [46].

The assembly of 1D nanostructures is also based on non-covalent interactions. For example, El-Sayed et al. reported that gold nanorods with an aspect ratio of 4.6 coated with two different cationic surfactants could assemble into long-range ordered structures on a substrate upon solvent evaporation and subsequent increase in concentration [40]. This ordered packing of the nanorods on the substrate could be a result of the hydrodynamic pressure of solvent stream and lateral capillary forces. As solvent evaporates, the pressure of the thin film on the substrate compared to the pressure in the suspension decreases. This pressure gradient produces a suspension influx from the bulk solution toward the thin film. The solvent flux compensates for the evaporated solvent from the film, and the nanoparticle flux causes particle accumulation and dense assembly therein. In explaining the tendency of nanorods to align parallel to each other, El-Sayed and colleagues claim that it is due to higher lateral capillary forces along the length of the nanorods as compared to the width. Murphy and colleagues [33] investigated self-assembly of higher aspect ratio gold nanorods with one surfactant system and obtained results consistent with El-Sayed's observation.

With the existence of different functional sections or blocks along 1D nanostructures, self-assembly can be guided through block-to-block interactions between different 1D nanoscale components. Based on the discrete surface chemistry of different sections or blocks, various kinds of molecules can be attached to the blocks using well-established methods [e.g., self-assembled monolayers (SAMs)]. For instance, Kovtyukhova and Mallouk developed a selective functionalization of Au/Pt/Au nanowires based on the different reactivity of Pt and Au towards isocyanides and thiols [47]. The mercaptoethyl-amine-bearing gold portion of the nanowires can be tagged with fluorescent indicator molecules to image the spatially localized SAMs along the length of the nanowires. Mirkin and colleagues have also found that multicomponent rods composed of inorganic hydrophilic sections (Au) and organic hydrophobic domains (oxidized polypyrrole) assemble into superstructures as shown in Fig. 3.3 [48]. Due to compositional differences between the inorganic and organic portions, the block ends tend to phase-segregate in a way that aligns the structures. The assembled 3D superstructure can be tuned into bundles, tubes, and sheets by controlling the block numbers and ratio.

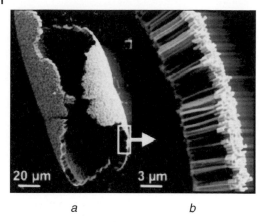

a   b

Figure 3.3. Scanning electron microscope (SEM) images of assemblies of Au-polypyrrole nanorods with a 1:4 block-length ratio (a), and an enlarged view of the boxed area, revealing the highly oriented amphiphilic rods (b). (From Ref. [48].)

As an alternative long-range interaction to capillary forces for nanorod assembly, magnetic interactions can also direct and stabilize the self-assembly of ordered, 3D nanorod structures. Whitesides and colleagues demonstrated that metal nanorods containing alternating sections of ferromagnetic and diamagnetic materials along the nanorods can self-assemble into stable microstructures [49]. Magnetization of the ferromagnetic disk-like sections within individual rods polarizes the sections perpendicular to the physical axis of the rods and promotes lateral interactions that direct the self-assembly of the rods.

3.3
Biomimetic and Biomolecular Recognition Assembly

The above self-assembled systems are driven by nonspecific, non-covalent interactions. Utilization of biomolecules such as DNA, antibodies, viruses, and biotinylated proteins [33, 35, 36, 50, 51] may provide the self-assemblies with biomolecular recognition, better structural control, and responsive functionality.

3.3.1
Assembly by Biomolecular Recognition

3.3.1.1 DNA-Assisted Assembly
Self-recognizing biomolecules such as DNA can direct the assembly of attached nanoscale components. Earlier works by Mirkin and Alivisatos and colleagues have shown that complementary DNA oligonucleotides could be used to direct the assembly of nanocrystals such as Au and CdSe [35, 36]. This method takes advantage of

molecular recognition of the complementary oligonucleotides anchored on the nanocrystals. For example, no recognition occurs when two noncomplementary strands of DNA attached with nanocrystals are mixed together. However, adding a third strand of half-complementary DNA to each of the grafted sequences induces hybridization and drives the self-assembly process. Similarly, by functionalizing two complementary oligonucleotides on two different nanocrystals (size or type), DNA-directed binary nanocrystal assembled networks have been prepared [52]. Such nanocrystal-labeled systems are of great interest for highly selective colorimetric detection for oligonucleotides. This technique has also been extended to assemble nanorods and nanowires. For example, large-scale uniaxial organization of gold nanorods can be achieved by using specific DNA hybridization [32, 53]. Silver nano-wires have also been assembled between two prefabricated leads [54].

Single-stranded DNA molecules self-assemble into DNA-branched motif complexes known as tiles. These tiles carry sticky ends that preferentially match the sticky ends of other tiles, facilitating their assembly into lattices. Li et al. used a line-ar array of DNA triple crossover molecules (TX) to controllably template the self-assembly of streptavidin linear arrays through biotin–streptavidin interactions [55]. This self-assembled DNA–streptavidin lattice was used as a scaffold to precisely position periodically gold nanoparticles within the lattice for potential nanoelectronic applications. DNA tiles can also assemble into 2D lattice sheets with proper design of the sticky ends. These structures can also be used as templates for the synthesis of superstructures of other materials such as nanoparticles, nanorods, and carbon nanotubes [56].

3.3.1.2 **Protein-Assisted Assemblies**

A large number of complementary protein systems offer a new direction for nano-structure synthesis. Mann and colleagues developed protein recognition directed assembly of Au nanocrystals such as antibody/antigen and streptavidin/biotin recognition pairs [33]. Similar to the DNA-directed assembly, nanocrystals functionalized with two complementary proteins may assemble into superstructures based on protein-specific recognition or binding. Shenton et al. attached either IgE or IgG antibodies to the nanocrystals. Subsequent addition of appropriate antigens resulted in the interparticle conjugations shown in Fig. 3.4 [57]. Similarly, Murphy and colleagues found that functionalization of gold nanorods with biotin followed by the addition of streptavidin results in an end-to-end assembly [58], rather than a side-by-side assembly, which is similar to that achieved when using DNA as linkers [32].

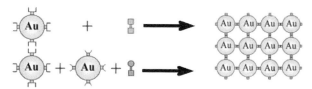

Figure 3.4. Use of surface-attached antibodies and artificial antigens for cross-linking of Au nanoparticles. (From Ref. [57].)

3.3.1.3 **Virus-Assisted Assemblies**

Viruses have also been used as templates for the synthesis of magnetic and semiconductor nanomaterials, as shown in Fig. 3.5. The morphologies achieved (e.g., particles, liquid crystals, films, and fibers) are dependent on the virus structure and the assembly process [50]. For example, Douglas et al. used cowpea chlorotic mottle virus (CCMV), which forms a cage-like protein shell, to grow iron oxide and polyoxometalates around the scaffold [59, 60]. Nanowires of metals, semiconductors, and magnetic materials have also been synthesized using rod-shaped viruses such as M13 bacteriophage and tobacco mosaic viruses as templates. For example, beginning with the ZnS nanoparticles nucleated on the viruses, controlled growth connects these particles and forms nanowires. Interestingly, the specific metal–virus binding directs the nucleation process and creates the nanowires consisting of oriented hexagonal wurtzite ZnS nanocrystals.

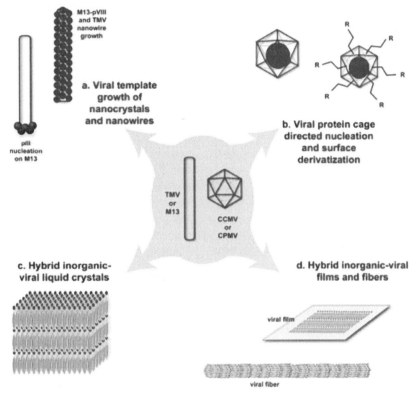

Figure 3.5. Routes for virus-assisted material synthesis. (a) Rod-shaped viruses used to synthesize nanowires. (b) Cages used to synthesize nanoparticles within the interior or the external surface which can be modified to assemble into viral arrays. (c) Rod-shaped viruses for liquid crystal fabrication. (d) Inorganically modified liquid crystals fabricated into viral films and viral fibers. (From Ref. [50].)

3.3.2
Biomimetic Assembly Process

The biomimetic approach is an elegant route to organize nanostructures across various length scales. Research in this area focuses on the production of hierarchical complex nanostructures using self-assembled hybrid organic and inorganic building blocks. As an example, complex $BaSO_4$ morphologies such as bundled fibers, brush-like and cone-shaped structures, and unusual cone-on-cone assemblies were formed in aqueous polyacrylate solutions at room temperature by a simple precipitation reaction [61]. The formation process may be quite complicated. Figure 3.6 shows an example of a nanofilament formation process which involves eight steps: anionic polymer-mediated nanoparticle aggregation and crystallization (steps 1–3); adsorbence of the anionic polymer to positively charged crystal faces (step 4); unidirectional aggregation of the crystalline $BaSO_4$, which forms the nanofilament (steps 5 and 6); secondary side-by-side aggregation, which forms nanofilament bundles (step 7); and, finally, the formation of cone-shaped structure composed of nanofilament bundles is obtained (step 8). With the optimization of variables such as temperature, pH, and concentration, highly ordered funnel-like $BaSO_4$ superstructures were obtained as shown in Fig. 3.7. This complex superstructure experiences remarkable self-similar growth and forms very long $BaSO_4$ fiber bundles with repetitive growth patterns [62].

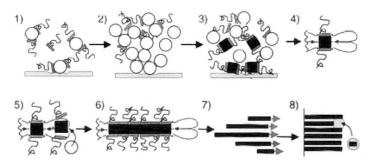

Figure 3.6. Proposed mechanism for the heterogeneous nucleation and growth of fiber bundles. (From Ref. [61].)

Similarly, complex ZnO superstructures have been biomimetically synthesized in solution using citrate ions to control the growth behavior of the crystals [63, 64]. In this approach, ZnO nanocrystals were first prefabricated on the substrate. In the absence of the citrate ions, ZnO grows in rod-like structures along the [001] direction. With the addition of citrate ions, ZnO growth is directed into nanoplates due to specific adsorption of the citrate ions to the [002] surface. By repeating the growth step several times, ZnO superstructures consisting of arrays of oriented ZnO nano-columns intercalated by layers of nanoplates can be obtained (see Fig. 3.8).

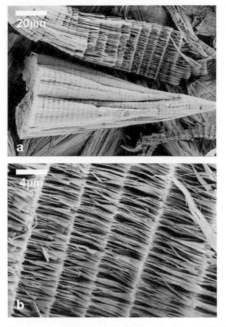

Figure 3.7. SEM images of complex forms of BaSO₄ bundles and superstructures: (a) highly ordered funnel-like superstructures with multiple hierarchical aligned cones, (b) enlarged image of the detailed superstructure with repetitive patterns. (From Ref. [62].)

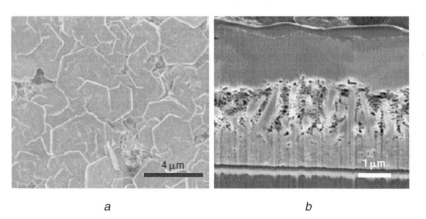

a   b

Figure 3.8. SEM images of plate-like structures on top of ZnO bilayers (a); and column-to-plate transition in the ZnO bilayers (b). (From Ref. [64].)

3.4
Template-Assisted Integration and Assembly

Nanoscale components with uniform size and shape can spontaneously assemble into ordered 2D or 3D aggregates based on different specific, non-covalent interactions. This energy-minimization-driven self-assembly provides a basic and effective bottom-up route to organize nanoscale components into highly ordered and thermodynamically stable structures for macroscopic material applications. However, it is difficult to have effective control over the size, shape, position, and structure of the assemblies for device applications by self-assembly only. For this purpose, the complementary top-down microfabrication strategy has been integrated into the self-assembly of nanoscale components.

3.4.1
Template-Assisted Self-Assembly

Microfabrication techniques such as photolithography, soft lithography, and scanning probe microscopy lithography provide robust routes to pattern solid surfaces with relief structures and different properties at scales ranging from tens of nanometers to hundreds of micrometers. The predetermined structures in these patterned surfaces could serve as templates to direct self-assembly of nanoscale components in the confined spaces and positions, which produce assemblies with controllable sizes, shapes, position, and even assembly structures.

3.4.1.1 Templating with Relief Structures

The microfluidic method is an effective way to control the self-assembly of nanoscale components to obtain aggregates with predetermined size, shape, and position. This is often achieved by filling the fluidic channels by capillary force or other interactions with solutions containing nanoscale components. Ordered channels are usually formed between a patterned poly(dimethylsiloxane) (PDMS) mold and a substrate [65] or between a flat PDMS and a substrate patterned with relief structures [66]. Self-assembly of the building blocks within the confined channels can generate patterned 2D or 3D lattices after evaporation of the solvent. Alternatively, using a liquid dewetting method in the microchannels, Yang et al. demonstrated a novel method to assemble molybdenum selenide molecules into ordered parallel arrays of nanowires along the corners of the channels. Crossbar junctions of nanowires could also be obtained by repeating this process within the aligned channels orthogonal to the preformed nanowires [67]. Ozin et al. also showed that substrates with separated or continuous relief structures can be used as templates to direct the self-assembly of colloidal particles into controllable aggregates using a two-step spin coating method. Briefly, a low spin speed was first used to make colloidal particles fall into the relief structures. A high spin speed was then used to remove excess colloidal particles from the substrate surface [68]. Xia et al. developed another method to assemble colloidal particles into complex aggregates by using a special cell consisting of planar and patterned substrates (see Fig. 3.9). When a suspension of

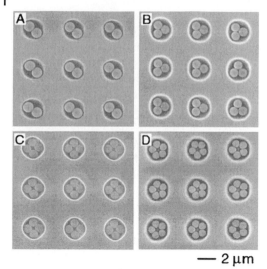

— 2 μm

Figure 3.9. SEM images of ordered arrays of (a) dimer, (b) trimer, (c) square tetramer, and (d) pentagon assemblies of polystyrene colloidal particles with diameters of 1000, 900, 800, and 700 nm, respectively. The arrays were fabricated by template-assisted self-assembly using substrates patterned with an ordered array of 2-μm cylindrical holes. (From Ref. [69].)

colloidal particles flew across the patterned substrate, capillary forces formed at the meniscus between the suspension surface and the substrate pushed the particles into the relief structures and also removed the particles on the top surface. During evaporation of solvent from the relief structures, colloidal particles assembled into complex aggregates relative to the particle size and the dimension of the template [69].

The above approaches are somewhat limited to the formation of concrete close-packed nanoscale component assemblies on substrates. Yang et al. demonstrated the use of substrates with patterned wettability to form organic liquid patterns. Subsequent assembly of the colloidal particles on the liquid-patterned surface generated ordered arrays of voids in the colloidal crystal systems [70]. The formation of specific microstructures within the colloidal crystals is important for the fabrication of opti-

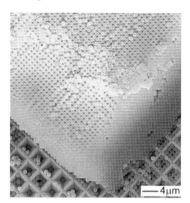

—4μm

Figure 3.10. SEM image of a face-centered cubic structured colloidal crystal assembly with the [100] plane parallel to the substrate. This assembly was fabricated by the crystallization of polystyrene colloidal particles on a substrate patterned with an array of square pyramidal pits. (From Ref. [72].)

cal cavities, waveguides, and photonic chips. Besides control over the size, shape, and position of the aggregates, work demonstrated by Blaaderen, Ozin, and Xia et al. showed that relief structure-assisted self-assembly can also provide effective control over the assembly structures [66, 71, 72]. For example, crystallization of colloidal particles on the substrates patterned with specific structures (e.g., square pyramidal pits) can produce face-centered cubic crystals with preferred orientations (see Fig. 3.10) [66, 71, 72].

3.4.1.2 Templating with Functionalized Patterned Surfaces

The use of substrates with functionalized patterned surfaces provides another route to confine self-assembly of the building blocks. Such patterned surfaces are often prepared through photolithography and microcontact printing techniques. For example, Braun and Hammond et al. studied the self-assembly of charged colloidal particles on substrates with patterned charges [73, 74]. The charged colloidal particles in the suspension were selectively adsorbed to the patterned regions with opposite charges through electrostatic interactions. Similarly, patterned aggregates of nanorods on gold substrates were also achieved [47]. It is also possible to spatially address individual charged colloidal particles on such patterned surfaces using a dip-pen lithography technique and electrostatic interactions [75]. Aksay et al. further explored this technique by combining an external electric field and surface with patterned charge density (see Fig. 3.11) [76].

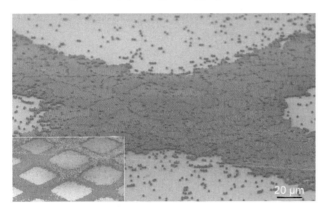

Figure 3.11.   SEM image of ordered colloidal particle aggregates produced by external electric-field-directed assembly of colloidal particles on an indium tin oxide (ITO) electrode. The electrode was patterned with different current density by exposure to ultraviolet light through a mask. (From Ref. [76].)

3.4.2
Patterning of Nanoscale Component Assemblies

The patterning methods discussed above are based on self-assembly of building blocks directed by a patterned surface or structure. Alternatively, patterned structures can be prepared by postpatterning the preformed nanoscale component assemblies using appropriate microfabrication techniques. For example, Brust et al. demonstrated that an electron beam lithography technique can be used to pattern Langmuir-Blodgett (LB) self-assembled monolayers or multilayers of alkanethiol-capped gold nanoparticles [77]. The electron beam-exposed regions of the LB film became insoluble as a result of the cross-linking of the capping alkanethiol molecules. Following removal of the nanoparticles on the unexposed regions, gold nanoparticle assemblies with features as small as 50 nm were generated. A similar method was used to pattern self-assembled films of surfactant-stabilized FePt nanoparticles using pulsed and focused laser beams to carbonize the surfactant into insoluble components [78]. The patterned nanoparticle films were further annealed to convert the FePt nanoparticle assemblies into ferromagnetic, ordered, face-centered tetragonal superlattices [6].

Other techniques developed involve patterning preformed monolayers using soft-lithographic techniques. For example, by transferring the Fe_2O_3 nanoparticle film on the patterned surface of a PDMS stamp, Yang et al. demonstrated that self-assembled nanoparticles could serve as the "ink" and be further transferred onto a silicon wafer using a microcontact printing technique [79]. Yang et al. also developed a lift-up soft lithography technique, in which a single layer of colloidal particles from the top layer of a colloidal crystal film was selectively transferred onto a PDMS stamp surface. Using this method, it is also possible to realize fine control over the microstructures of colloidal films using a layer-by-layer lift-up process [80].

3.5
External-Field-Induced Assembly

Although self-assembly is a spontaneous and thermodynamically favorable procedure, external fields may be required for the assembly of nanoscale building blocks to micro/macro scales because of the weak nature of the non-covalent interactions. External fields, such as electric [81], magnetic [82], electrophoretic [83, 84], and electroosmotic flow [85], shearing, and pressure fields [42, 86] have been applied to align or assemble the nanoscale components into larger length scales.

3.5.1
Flow-Directed Assembly

An excellent example of shearing field-assisted assembly is the alignment and assembly of nanowires into parallel arrays or functional networks achieved by microfluidic flow and surface-patterning techniques [86]. Usually, a higher flow rate leads to a stronger shearing force and therefore better nanowire alignment. The surface coverage of the aligned nanowires can be controlled by flow duration and solution concentrations. The use of patterned surface chemistry enables the nanowires to be patterned and aligned preferentially along the fluidic direction. Parallel or crossed arrays of nanowires can be easily obtained using this technique (see Fig. 3.12) [86].

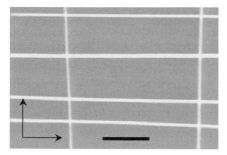

Figure 3.12. A SEM image of a crossed array of InP nanowires fabricated by a two-step flow technique with orthogonal channel direction. The scale bar is 500 nm. (From Ref. [86].)

3.5.2
Electric-Field-Induced Assembly

Electric fields have been extensively used to align nanoscale components such as carbon nanotubes [87], semiconductor nanowires [81], and block copolymer meso-structures [88] perpendicular or parallel to the substrate. Lieber and colleagues found that preformed indium phosphide nanowires prefer to align along the direction of an electric field [81]. These aligned nanowires can bridge two electrodes and form nanowire circuits. The ability of nanowires to align under an electric field is due to the anisotropic polarizability of 1D nanostructures. External electric fields have also been exploited to control carbon nanotube orientations during chemical vapor deposition [87, 89]. It is worth mentioning that carbon nanotubes can also be aligned by mechanical shear [90], anisotropic flow [91], gel extrusion [92, 93], and magnetic field [94].

3.5.3
Electrophoretic Assembly

Electrophoretic deposition is another efficient way of assembling charged building blocks into 2D/3D organized colloidal systems such as micrometer-sized colloid silica [85, 95], polystyrene latexes [76, 96], metal nanocrystals [83, 84, 97–99], and zeolite and diamond nanoparticles [100, 101]. During the deposition process, colloidal particles with surface charges are accelerated in the solution under the applied electric fields and deposited on the target electrodes. The surface charges of the building blocks are dependent on the nature of the materials and the synthesis conditions. For example, zeolite and diamond nanoparticles at acidic conditions may become positively charged and deposit on the cathode [100, 101]. Dense zeolite films and hollow fibers were obtained through manipulating the charge density of the zeolite nanoparticles and electric field strength. Similarly, thiol and cationic surfactant-capped gold nanocrystals were electrophoretically deposited to form 2D and 3D arrays of gold nanocrystal thin films [84, 98]. As-deposited gold nanocrystal films display prominent surface plasmon band absorption and weak IR region band absorption, which indicates that these nanoparticles are well dispersed in the thin films [98]. However, deposition using high concentrations of gold colloidal solutions may result in nanocrystal aggregation, indicated by a broad IR region absorption.

3.5.4
Assembly Using Langmuir–Blodgett Techniques

LB techniques have also been exploited to assemble nanoscale building blocks at liquid/vapor interfaces. The assembly of Au, Ag, and CdS nanoparticles [102, 103] using the LB technique often results in hexagonally closed packed arrays due to iso-tropic interparticle interactions. Compared with nanocrystal assembly, the LB assembly of anisotropic 1D nanostructures often results in more complicated struc-tures. Yang and colleagues investigated the LB assembly of $BaCrO_4$ and $BaWO_4$, and Ag nanowires [42, 43, 104], and discovered that the assembled superstructures can be tuned through controlling the interface pressure. For example, an increase of interface pressure (e.g., by compression of the monolayer) transited the $BaCrO_4$ nanorod superstructures from side-by-side aggregates to nematic and smectic liquid crystalline phases (see Fig. 3.13). However, compression of the LB assembly of Au nanorods with a similar aspect ratio only resulted in side-by-side aggregates, which can be attributed to strong van der Waals interactions and directional capillary forces among the nanorods [105]. Such nanowire or nanoparticle superstructures can be easily transferred to the substrate for further device fabrication. For example, Lieber

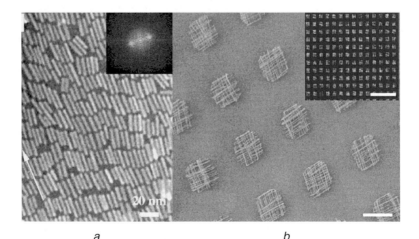

a   b

Figure 3.13. (a) TEM image and FFT image (inset) of LB monolayer of $BaCrO_4$ nanorods with a smectic configuration (from Ref. [42]); (b) a SEM image and optical micrograph (inset) of a patterned multilayered LB film consisting of crossed nanowires. Scale bars are 10 and 100 µm in the SEM image and optical micrograph, respectively (from Ref. [106]).

et al. demonstrated the patterning of single- or multilayered films of nanowires using a combination of LB and photolithography techniques (see Fig. 3.13) [106].

3.6
Direct Synthesis of 2D/3D Nanostructure

Nanoscale building blocks, such as 0D, 1D and 2D nanostructures, are being intensely studied, not only because of their unique properties arising from size effects, but also because of their capability for directing nanosystem integration. As a result, hierarchical assemblies of molecular and nanoscale components into length scales that devices can use are essential for the success of nanotechnology. Most of the existing methods described above utilize non-covalent or external-field interactions, such as self-assembly [5], microfluidic [86], LB [42, 105], and other techniques [81] to assemble the preformed nanoscale building blocks. Due to the weak nature of non-covalent interactions, such assemblies are not favored for device applications, where intensive communication (efficient charge transport) between nanoscale components, chemical and mechanical stability, large surface areas, or good accessibility are required. Direct synthesis of 2D/3D continuous nanoscale arrays or networks at micro/macro scales is therefore paramount for nanoscale components to realize real-world device applications. The following section surveys the most cited synthetic approaches used for direct 2D/3D nanostructure synthesis, templated synthesis, and oriented nanostructure growth.

3.6.1
Templated Synthesis

Most templating growth methods, which involve confined growth of materials within a template (e.g., pores) followed by removal of the template, provide a flexible and affordable synthesis route for a variety of nanostructured materials. Examples of such templates include hard templates (e.g., porous alumina films, track-etched polycarbonate films, and mesoporous silica) [107–111] and soft templates (e.g., liquid crystalline phases and amphiphilic block copolymers) [112–115]. The hard templating approach is conceptually simple to implement; however, the use of porous alumina or polycarbonate membranes as templates usually results in nanowires with polycrystallinity or with large wire diameters (20–1000 nm) that may preclude their quantum confinement effects. Surfactant-templated mesoporous silica contains unique nanoscale pore channels, controllable pore surface chemistry, and well-controlled morphology, providing ideal templates for the synthesis of 2D/3D nanoscale arrays or networks [116, 117]. Since the syntheses of nanowire arrays using porous alumina or track-etched polycarbonate films as templates have been well developed, this section will emphasize templating syntheses using mesoporous silica as templates. Other template synthesis methods based on sacrificial or negative templates will not be involved in this section due to their lack of macroscopic morphology control.

3.6.1.1 **Mesoporous Silica-Templated Synthesis**

Mesoporous silica templates are typically prepared by the co-assembly of silicates and surfactants into highly ordered lyotropic liquid-crystalline (LC) phases. Removal of the surfactant creates mesoporous silica with unimodal 2- to 20-nm-diameter pore channels that are arranged into hexagonal, cubic, or lamellar mesostructures. The template synthesis strategies are therefore based on the confined growth of metals, polymers, alloys, and semiconductors within these pore channels and subsequent silica template removal. The beauty of this approach is the replication of these complicated but well-studied silicate/surfactant LC phases, which allows the synthesis of nanoscale structures with precise structural and compositional control. For example, diameters of nanowires can be tuned from 2 to 20 nm based on the template pore sizes while the nanowire arrangements can be controlled from 2D arrays to 3D networks using templates with 2D and 3D pore channels, respectively. Furthermore, mesoporous silica can be readily made in the forms of powders [116], particles [118], thin films [117, 119], fibers [120], and monoliths [121], which provides the possibility to prepare various nanostructures with different macroscopic morphologies. Besides the precise structural control over multiple-length scales, this method can be extended to synthesize nanowires with various chemical compositions, such as catalytic noble metals and alloys, ferromagnetic metals and alloys, and semiconductors with tunable optical and electrical properties.

To date, various metals, carbon, and semiconductors have been grown confined within mesoporous silica templates using ion exchange and chemical reduction [122], supercritical fluid [123], chemical vapor deposition [124], electroless deposition [125], and electrodeposition techniques [109]. However, most of the nanostructures synthesized usually lack the macroscopic continuity required for device applications [126]. Recently, Lu et al. developed a novel method to continuously grow 2D/3D nanostructures using electrodeposition within mesoporous silica templates [109]. Unlike other techniques, this electrodeposition method can gradually fill the mesoporous channels with a large variety of materials through electrochemical reactions.

Figure 3.14 schematically shows the formation of 3D nanowire networks using mesoporous silica containing 3D interconnective pore channels as templates. As depicted, the synthesis procedure includes the following steps: (i) coating a mesostructured silica film on a conductive substrate through co-assembly of silicate and surfactant [116, 117]; (ii) removing the surfactant LC template to create a silica film with a 3D mesoporous network; (iii) filling the pore channels with metals or semiconductors by electrodeposition; and (iv) removing the silica template to create a replicated mesoporous nanowire network. This method continuously grows various materials within the mesoporous channels from the bottom conductive substrate upward, providing a simple route to synthesize 3D nanowire networks. Such a complete filling process in turn allows a precise mesostructure and macroscopic morphological control. For example, the use of mesoporous thin film and monolith templates respectively allow the synthesis of nanostructures in these forms.

Figure 3.15 shows representative TEM images of Pd nanowires prepared using mesoporous silica thin films with 2D porous channels as templates [109]. As-synthesized nanowires have an average diameter around 8–9 nm, which is similar to the

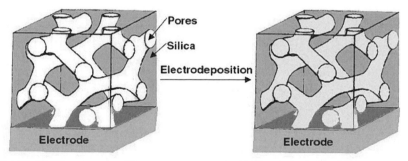

a) 3D Cubic Mesoporous Template   b) 3D Nanowire/silica Nanocomposites

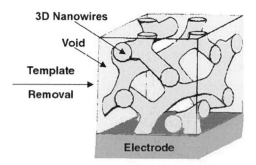

c) 3D Nanowire Network

Figure 3.14. Formation of 3D continuous macroscopic metal or semiconductor nanowire networks using a templated electrodeposition technique. (a) Formation of a cubic mesoporous silica film on an electrode through co-assembly of silicate and surfactant (steps 1 and 2); (b) filling the pore channels with metals or semiconductors by electrodeposition (step 3); (c) removal of the silica template to create a replicated mesoporous nanowire network (step 4).

pore diameter of the template. Due to the defects of the surfactant/silica LC structure, hexagonally arranged mesopore channels may cross and result in interconnected crossed nanowires [see the area marked by an arrow in Fig. 3.15(a)]. This further indicates that the mesostructure of the nanowires can be precisely controlled through replicating the LC mesophases. The high-resolution TEM image and selected-area electron diffraction (SAED) pattern of a single nanowire [Fig. 3.15(b,c)] indicate the formation of local single crystal metal nanowires. This may be due to the unique nucleation and growth of the metal in the confined template. The synthesized nanowires aggregate into bundles and form swirling structures spreading across the film [see Fig. 3.15(d)], which result from the replication of the swirling mesostructure of hexagonally arranged mesoporous silica. X-ray diffraction studies also indicate that these nanowires are organized in hexagonal closed-packed arrays. This example strongly suggests that nanowire thin films with controlled mesostructures can be prepared using electrodeposition within a mesoporous silica template.

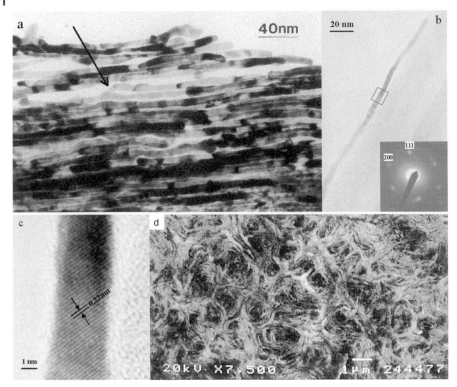

Figure 3.15. TEM and SEM images of Pd nanowires prepared by electrodeposition within a mesoporous silica template. (a) A TEM image of templated electrodeposited Pd nanowires. (b) A single Pd nanowire with the inset showing the selected-area electron diffraction of the single Pd nanowire. (c) Selected-area (see b) high-resolution TEM image of the Pd nanowire showing *fcc* crystalline lattice fringes. (d) Typical SEM top view image of Pd nanowire thin films. (From Ref. [109].)

Furthermore, nanowire networks with hierarchical pore structures can be readily achieved by incorporating colloidal silica porogens with different sizes and shapes into the self-assembled surfactant/silicate templates. For example, hierarchical templates were prepared by introducing colloidal silica spheres (diameters 20–100 nm) as secondary porogens into the surfactant/silicate thin film assemblies. Subsequent electrodeposition and template removal result in hierarchical porous nanowire networks.

3.6.1.2 Direct Nanostructures Synthesis Using Soft Templates

Other than hard templates like mesoporous silica, soft templates such as surfactant liquid crystals [114, 127, 128], block copolymers [115], organic gels [129], and biocellulose substances [130, 131] have also been utilized for direct nanostructure synthesis. For example, electrodeposition using a reverse surfactant LC phase as a template results in the formation of 2D aggregated nanowires with poor structural control [114]. By selectively attaching organic metallic complexes to a specific section of a block copolymer,

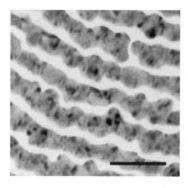

Figure 3.16. TEM image of arrays of silver nanowires in a polystyrene-polymethylacrylate copolymer film formed by the preferential deposition of silver in polystyrene domains. (From Ref. [115].)

ordered assemblies (e.g., lamellar, hexagonal, and cubic mesostructures) spatially define the complexes within the polymers. Further reduction of the metal complexes results in metal nanoparticles embedded within nanoscale domains of the block copolymer matrix [132, 133]. To avoid the complicated synthesis procedure involved in the above methods, metal can be thermally evaporated directly onto self-assembled copolymer films. Due to the different wetting properties [115], metal nanoparticle arrays, chains, or nanowires were preferentially formed within the lamellar phase of one block domain as shown in Fig. 3.16. Similarly, metal complexes can also be evaporated on self-assembled copolymer surfaces. Diffusion and preferred reduction of the metal complexes by the specific block may create three-dimensionally arranged metal nanoparticles within the polymer matrices [115, 134, 135].

Polysaccharides such as dextran have been reported as another type of soft template. Self-supported macroporous frameworks of silver, gold/copper oxide, silver/copper oxide, and silver/titania have been synthesized using such templates [130]. A silver sponge prepared by this technique is shown in Fig. 3.17. The synthesis procedure often involves heating mixtures of metal-salt and dextran to 500–900 °C. Expansion of the dextran molecules during thermal degradation creates porosity within the framework and forms organized metal/metal oxide networks.

Figure 3.17. SEM images of silver sponges. (a) A sponge prepared at 520 °C (scale bar 200 μm). (b) A sponge monolith prepared under the same conditions showing individual Ag rods and continuous pores (scale bar 20 μm). (From Ref. [130].)

3.6.2
Direct Synthesis of Oriented 1D Nanostructure Arrays

Oriented 1D nanostructures, either parallel or perpendicular to the substrate, have broad applications ranging from chemical and biological sensing to nanoelectronic devices, optical emission, display and data storage, and energy conversion and storage such as photovoltaics, batteries, and capacitors. To date, several methods have been developed to synthesize 1D nanoscale components into large-area oriented arrays, such as templated synthesis [110], electrospinning [136,137], seeded growth [64,138], epitaxial growth [139], and nanoimprinting techniques [140].

Synthesis of 1D nanostructured arrays that are parallel to the substrate are often based on the assembly of preformed nanowires through microfluidic, Langmuir-Blodgett (see Section 3.5), nanoimprinting, and electrospinning techniques. Using a nanoimprinting technique, ultra-high-density parallel nanowire arrays can be fabricated by transferring cross-sectional information of layer-by-layer structured films vertically onto a substrate [140]. Electrospinning can also be used in the synthesis of parallel nanowire arrays of polymer and metal oxides [136, 137, 141].

Oriented nanostructures perpendicular to the substrate have recently received much interest due to their open structure, high surface area and porosity, and favorable orientations that may provide enhanced device performance. Besides the traditional templating approach in which materials of interest are synthesized within the pore channels of the anodized alumina and track-etched polycarbonate membranes [110], the templateless approach is emerging as a novel and efficient method of fabricating oriented nanostructured arrays such as carbon nanotubes [142, 143], ZnO [63, 138], TiO_2 [144], MoO_x [145], SiC_xN_y [146], and polyaniline [147]. The following section focuses on templateless synthesis of oriented nanostructures through chemical vapor deposition, hydrothermal, and seeded growth methods.

3.6.2.1 Oriented Arrays by Chemical Vapor Deposition
Chemical vapor deposition is a widely used technique to prepare inorganic oriented nanostructures such as carbon nanotubes [142, 143], ZnO [139], and MoO_x [145]. For example, Yang and colleagues synthesized highly oriented ZnO nanowire arrays via catalytic epitaxial crystal growth on an Au thin-film-coated sapphire substrate [139]. Because of the good epitaxial interface between the [0001] plane of the ZnO nanowire and the [110] plane of the sapphire substrate, the ZnO nanowires achieved are oriented perpendicularly to the substrate along their [0001] direction (see Fig. 3.18). Due to the unique nanoscale structure, ZnO nanowire arrays display photoluminescence with an intensity that can be increased to form lasing action with increasing incident optical pump power. Zhou et al. used a thermal evaporation technique to synthesize well-aligned nanowire arrays of MoO_2 with diameters in the range of 50–120 nm and lengths up to 4 μm [145]. In the synthesis procedure, thermal evaporated Mo reacts with residual oxygen in the vacuum chamber, forming MoO_2 nanowire arrays on the substrate. As-synthesized MoO_2 nanowires were subsequently treated with O_2 to form MoO_3 or with H_2 to form metallic Mo nanowire arrays. This procedure has also been used to synthesize tungsten and tungsten oxide nanowire arrays.

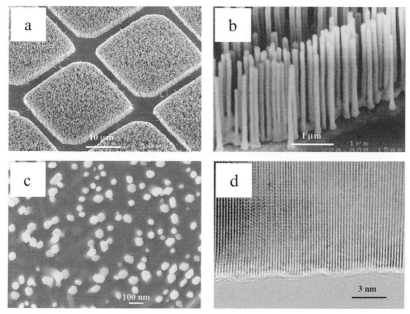

Figure 3.18. (a–c) SEM images of ZnO nanowire arrays on sapphire substrates at different magnifications; (c) gives a top view. (d) A high resolution TEM image of an individual nanowire showing the [0001] growth direction. (From Ref. [139].)

3.6.2.2 Seeded Solution Growth

The seeded solution growth method is used to prepare large-area oriented nanostructured arrays in mild reaction conditions through carefully manipulating the nucleation and growth processes. It has many advantages, such as naturally inspired mild growth, hierarchical control of nanostructures [148], and applicability for a wide range of materials such as metal oxides and polymers [144, 147]. For example, oriented ZnO nanorods or nanowires[64, 138] can be readily synthesized using the seeded growth method. First, ZnO seeding nanocrystals were dip-coated or spin-cast onto a substrate. Subsequent immersing of the coated substrate in an aqueous solution containing zinc nitrate hydrate and organic molecules (e.g., methenamine, diethylenetriamine, and hexamethyltetramine) affords controlled growth of the perpendicularly oriented ZnO crystals along their [0001] direction. During this process, the organic molecules serve as a structural directing agent that directs the crystal growth orientation. For example, Tian et al. synthesized large arrays of oriented helical ZnO nanorods and columns in solution using citrate ions to control the growth behavior of the crystal [63, 64]. Citrate ions specifically adsorb to the [002] surface and force the ZnO crystal to grow into plates. Further spiral growth of the ZnO plates produces oriented helical nanorods and columns in the presence of the ions (see Fig. 3.19).

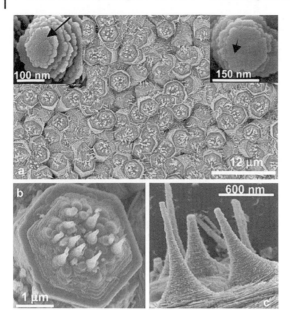

Figure 3.19. SEM micrographs of helical ZnO nanorods on oriented ZnO crystals. (a) Large arrays of well-aligned helical ZnO nanocolumns on top of base ZnO rods. (b) Precisely aligned ZnO nanocolumns on the [002] surface of one ZnO rod. (c) Tilted high-magnification SEM image of arrays of helical nanocolumns on the [002] surface. (From Ref. [63].)

Similarly, oriented TiO_2 nanotubes can be synthesized in an alkaline solution using TiO_2 nanocrystal-seeded substrates and subsequent controlled growth [144]. Such nanotubes are composed of multilayer sheets with inner and outer diameters of 3.7 nm and 12 nm, respectively. Besides the synthesis of metal oxide oriented nanostructures, oriented polymer nanostructures such as polyaniline nanowires ~100 nm in diameter and ~0.8 μm in length can be easily prepared using an electro-deposition approach (see Fig. 3.20) [147]. The synthesis procedure was based on the formation of seeding polyaniline nanoparticles on electrodes by applying a galvano-static current density of 0.08 mA cm^{-2}. Subsequent deposition using lower current

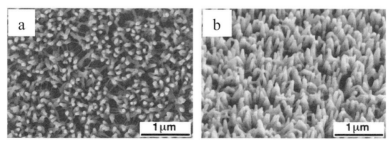

Figure 3.20. SEM micrographs of oriented polyaniline nano-wires on Pt. (a) Top view, (b) tilted view. (From Ref. [147].)

densities grows nanowires that are perpendicular to the electrode. This approach avoids the use of porous templates and grows polymer nanowire arrays on complicated surfaces for biological sensing and other applications [147, 149].

3.7
Applications

3.7.1
Chemical and Biological Sensing Applications

The unique structure and composition of nanoscale systems endows them with unique and superior properties, which may open a new avenue for novel device applications such as energy conversion, information storage, advanced electronic devices, drug screening, and chemical/biological sensing. The following section reviews some representative applications.

Optical, magnetic, and electrical-based detections have been achieved using nanostructured systems. For example, using metal nanocrystals with surface-enhanced Raman optical properties or fluorescent semiconductor nanocrystals, typical optical detection systems have been developed for specific biomedical sensing and imaging applications [150, 151]. Other examples involve the use of nanocrystals attached to high-density oligonucleotides to optically detect nucleic acids or DNA [152, 153]. Magnetic-based detection systems use nanoparticles that serve as magnetic resonance contrast enhancement agents for biological detections [154] and biomolecular separation [155]. Compared with the nanoparticle-based systems, 1D nanostructures such as carbon nanotubes [156, 157], semiconductor nanowires [158], polymer nanowires [159], and metal nanojunctions[160, 161] have been broadly used for electrical conductance-based detections, which will be discussed in detail in the following section.

3.7.1.1 Carbon-Nanotube-Based Sensing
The electrical properties of single-walled carbon nanotubes (SWNTs) are extremely sensitive to chemical gaseous environments. Direct absorption of gases on the nanotubes changes their electronic structure, which can be detected electrically. The large surface areas and charge-sensitive conductance of carbon nanotubes makes them excellent candidate materials for sensing a variety of gases such as NO_2, NH_3, and O_2 [156, 157]. Furthermore, carbon nanotubes have been integrated in field-effect transistors (FET) to detect ammonia [162], aromatic compounds [163], alcohol vapor [164], and specific protein bindings [165]. For example, Star et al. fabricated carbon-nanotube-based FET devices and examined the sensing of monosubstituted benzenes, such as aniline, phenol, anisole, toluene, chlorobenzene, and nitrobenzene (see Fig. 3.21) [165]. They found a linear increase of gate voltage shifting with increasing electron-donating capabilities of the absorbents ($NH_2 > O\text{-}H > OCH_3 > Cl > NO_2$). Similarly, FET devices based on carbon nanotubes attached to biomolecules can be used to detect complementary bio-conjugated molecules [165]

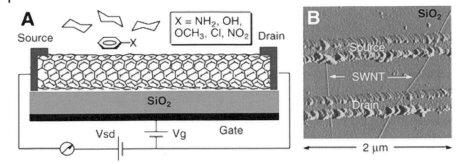

Figure 3.21. (a) Field-effect transistor with a SWNT transducer contacted by two electrodes (source and drain) and a silicon back gate exposed to a cyclohexanc solution of benzene derivatives. (b) An AFM topograph of the nanotube device. (From Ref. [163].)

3.7.1.2 Semiconducting-Nanowire-Based Sensing

Lieber and colleagues were the first to demonstrate the sensitivity of semiconductor nanowires to the local chemical environment. The sensing principle involves monitoring the change in conductivity caused by changes in surface charge of a single nanowire exposed to chemical and biological species [158]. For example, a silicon nanowire device can be used as a pH nanosensor by modifying the silicon oxide surface with 3-aminopropyltriethoxysilane [158]. The amine groups on the nanowire surface can undergo protonation and deprotonation, which causes a change in the surface charge of the nanowire, resulting in a change in conductance. This change is linearly dependent on pH and the response is also reversible. Similarly, the silicon nanowires can be functionalized with biotin. The well-characterized specific binding of biotin–streptavidin monitored by the changes in conductance of the functionalized nanowires enables their use as biomolecular sensors (see Fig. 3.22). The specific binding mechanism can also be used to detect complementary DNA sequences by DNA-functionalized nanowires [166, 167].

Figure 3.22. A biotin-modified silicon nanowire (SiNW) (left) and subsequent binding of streptavidin to the SiNW surface (right). (From Ref. [158].)

Other types of nanoscale devices that have been widely studied are the solid-state SnO_2 based sensors, in which adsorption and desorption of gas molecules on the SnO_2 surface causes a change in resistance. Recently, a SnO_2 nanowire sensor has been reported with improved sensitivity over a thin-film-based sensor [168]. Larger surface areas of the nanowires result in further improvement in both sensitivity and sensor reversibility [169].

Conductive polymer nanowires are another active element used for conductance-based sensing. Specifically, polyaniline has received much attention because of its reversible doping/de-doping chemistry and stable electrical conduction. Single polyaniline/PEO (Poly(ethylene oxide)) nanowire sensors have shown a rapid and reversible resistance change upon exposure to NH_3 gas [170]. In addition, sensors based on polyaniline nanofiber networks display improved sensitivity and stability due to their high surface areas [159].

3.7.1.3 Metallic-Nanowire-Based Sensing

Like semiconductor nanowires, so-called metal quantum wires (extremely small metal nanowires formed by a few metal atoms) can be used to detect very subtle changes in the chemical environment [171]. Adsorption of chemical species scatters the metallic bonded electrons in the quantum wires, decreasing the quantized conductance to a fractional value [172, 173]. Tao and colleagues used this principle to develop nanojunction sensors that measure the change in conductance, which is dependent on the adsorption strength of the molecules to the quantum wire [161]. This method has been used to detect 2,2'-bipyridine, adenine, and mercaptopropionic acid [173, 161]. Trace metal ions can also be detected using the nanojunction electrodes [174]. By sweeping the electrode potentials cathodically or anodically, metal ions can be deposited or dissolved in the nanojunction gap. Deposition of only a few metal ions into the gap can trigger a large change in conductance and provide a very sensitive detection. The voltage at which a point contact is formed and dissolved is specific for each ion, providing a possible route for ion sensing and determination.

Metal mesowires fabricated through step-edge electrodeposition have been reported as excellent gaseous sensors for H_2 and NH_3 [160, 175]. The sensing mechanism is similar to the nanojunction sensors since most of these mesowires are composed of series of connected metal nanoparticles with interparticle boundaries acting as nanojunctions. For example, Pd nanoparticles in mesowires expand and increase the electrical conductivity in the presence of H_2 [160].

3.7.2
Other Applications of Integrated Nanoscale Component Assemblies

Other than sensing applications, integrated nanostructured systems also provide new avenues to other areas such as high-density information storage and other microelectronic applications. For example, nanoscale transistors have been fabricated with individual semiconductor nanowires or carbon nanotubes with performance comparable to or exceeding that of conventional materials [176, 177]. Macroscopic thin-film transistors (TFTs) were fabricated using oriented Si nanowire or CdS nanoribbons as semiconducting channels [178]. These high-performance TFTs were produced on various substrates, including plastics, using a low-temperature assembly process shown in Fig. 3.23. The preformed nanowires were dispersed in solution and assembled onto the surface of the chosen substrate through flow-directed alignment. These thin-film transistors display comparable and in some cases su-

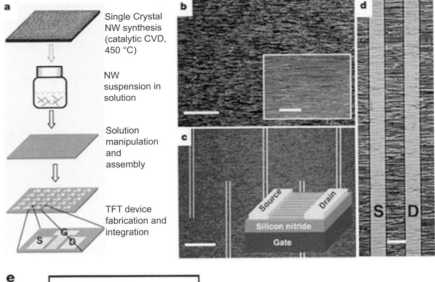

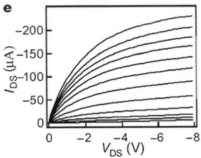

Figure 3.23. Nanowire thin-film transistor (NW-TFT). (a) NW-TFT fabrication process. (b) Optical micrograph of a flow-aligned nanowire thin film (scale bar 80 μm). Inset shows higher magnification (scale bar 20 μm). (c) Optical micrograph of NW-TFTs fabricated on nanowire thin films (scale bar 100 μm). (d) Optical micrograph of a NW-TFT, where parallel arrays of nanowires are clearly seen to bridge the full distance from the source to the drain electrodes (scale bar 5 μm). (e) Drain current (I_{DS}) versus drain–source voltage (V_{DS}) at increasing gate voltages (V_{GS}) in steps of 1 V starting from the top at $V_{GS} = -10$ V. (From Ref. [178].)

perior key transistor parameters for macroelectronic applications when compared to materials currently used for this application.

For high-density information storage applications, the data storage media requires room-temperature ferromagnetic materials with reduced magnetic domain sizes. 2D or 3D single-component superlattices assembled from 2- to 11-nm Co nanoparticles contain small magnetic domain sizes, but are superparamagnetic at room temperature and therefore not suitable for data storage media applications [6, 179, 180]. Skumryev et al. reported that isolated ferromagnetic Co nanoparticles embedded in an antiferromagnetic matrix show ferromagnetic behavior at room temperature, due to magnetic coupling. The extra magnetic anisotropy energy generated exceeds the thermal energy and results in ferromagnetic properties [181]. To develop high-density information storage systems, 2D nanowire arrays have been intensively investigated because of their uniaxial anisotropic magnetization [112, 182]. 3D Co nanowire networks, unlike Co thin films or 2D wire arrays, possess isotropic magnetism with an enhanced coercivity of 255 Oe compared with Co thin films, whose coercivity is on the order of 10 Oe [112].

3.8
Concluding Remarks

Nanotechnology is the latest rage in science and technology. However, the future of nanotechnology depends on the understanding and discovery of materials' properties at the nanoscale, efficient manufacture of nanoscale materials, and, most importantly, device applications of nanoscale materials for real-world applications. The above chapter summarizes the fabrications of nanostructured systems consisting of low-dimensional building blocks, a tiny piece of nanotechnology. However, we truly believe that the combination of the current available top-down fabrication techniques with bottom-up approaches – a harmony between industrial civilization and nature – may guide the research of nanotechnology into a new age.

References

1 Alivisatos, A. P., Semiconductor clusters, nanocrystals, and quantum dots. *Science* (Washington, D. C.), **1996**. *271* (5251), 933–937.

2 Xia, Y., P. Yang, Y. Sun, Y. Wu, B. Mayers, B. Gates, Y. Yin, F. Kim, H. Yan, One-dimensional nanostructures: synthesis, characterization, and applications. *Adv. Mater.* **2003**. *15* (5), 353–389.

3 Corso, M., W. Auwaerter, M. Muntwiler, A. Tamai, T. Greber, J. Osterwalder, Boron nitride nanomesh. *Science* **2004**. *303* (5655), 217–220.

4 Wang, Z. L., Structural analysis of self-assembling nanocrystal superlattices. *Adv. Mater.* **1998**. *10* (1), 13–30.

5 Redl, F. X., K. S. Cho, C.B. Murray, S. O'Brien, Three-dimensional binary superlattices of magnetic nanocrystals and semiconductor quantum dots. *Nature* **2003**. *424* (6943), 968–971.

6 Sun, S., C. B. Murray, D. Weller, L. Folks, A. Moser, Monodisperse FePt nanoparticles and ferromagnetic FePt nanocrystal superlattices. *Science* **2000**. *287* (5460), 1989–1992.

7 Galow, T. H., A. K. Boal, V. M. Rotello, A "building block" approach to mixed-colloid systems through electrostatic self-organization. *Adv. Mater.* **2000**. *12* (8), 576–579.

8 Hiramatsu, H., F. E. Osterloh, pH-Controlled assembly and disassembly of electrostatically linked CdSe-SiO2 and Au-SiO2 nanoparticle clusters. *Langmuir* **2003**. *19* (17), 7003–7011.

9 Kolny, J., A. Kornowski, H. Weller, Self-organization of cadmium sulfide and gold nanoparticles by electrostatic interaction. *Nano Lett.* **2002**. *2* (4), 361–364.

10 Hao, E., B. Yang, J. Zhang, X. Zhang, J. Sun, J. Shen, Assembly of alternating TiO2 and CdS nanoparticle composite films. *J. Mater. Chem.* **1998**. *8* (6), 1327–1328.

11 Boal, A. K., V. M. Rotello, Fabrication and self-optimization of multivalent receptors on nanoparticle scaffolds. *J. Am. Chem. Soc.* **2000**. *122* (4), 734–735.

12 Gomez, S., L. Erades, K. Philippot, B. Chaudret, V. Colliere, O. Balmes, J.-O. Bovin, Platinum colloids stabilized by bifunctional ligands: self-organization and connection to gold. *Chem. Commun.* **2001**(16), 1474–1475.

13 Han, L., J. Luo, N. N. Kariuki, M. M. Maye, V. W. Jones, C. J. Zhong, Novel interparticle spatial properties of hydrogen-bonding mediated nanoparticle assembly. *Chem. Mater.* **2003**. *15* (1), 29–37.

14 Wang, T., D. Zhang, W. Xu, S. Li, D. Zhu, New approach to the assembly of gold nanoparticles: formation of stable gold nanoparticle ensemble with chainlike structures by chemical oxidation in solution. *Langmuir,* **2002**. *18* (22), 8655–8659.

15 Mayer, C. R., S. Neveu, and V. Cabuil, 3D hybrid nanonetworks from gold-functionalized nanoparticles. *Adv. Mater.* **2002**. *14* (8), 595–597.

16 Kiely, C. J., J. Fink, M. Brust, D. Bethell, D. J. Schiffrin, Spontaneous ordering of bimodal ensembles of nanoscopic gold clusters. *Nature* **1998**. *396* (6710), 444–446.

17 Whetten, R. L., J. T. Khoury, M. M. Alvarez, S. Murthy, I. Vezmar, Z. L. Wang, P. W. Stephens, C. L. Cleveland, W. D. Luedtke, U. Landman, Nanocrystal gold molecules. *Adv. Mater.* **1996**. *8* (5), 428–433.

18 Korgel, B.A., S. Fullam, S. Connolly, D. Fitzmaurice, Assembly and self-organization of silver nanocrystal superlattices: ordered "soft spheres". *J. Phys. Chem. B* **1998**. *102* (43), 8379–8388.

19 Puntes, V. F., K. M. Krishnan, P. Alivisatos, Synthesis, self-assembly, and magnetic behavior of a two-dimensional superlattice of single-crystal .vepsiln.-Co nanoparticles. *Appl. Phys. Lett.* **2001**. *78* (15), 2187–2189.

20 Gao, F., Q. Lu, D. Zhao, Controllable assembly of ordered semiconductor Ag2S nanostructures. *Nano Lett.* **2003**. *3* (1), 85–88.

21 Hu, K., M. Brust, A. J. Bard, Characterization and surface charge measurement of self-assembled CdS nanoparticle films. *Chem. Mater.* **1998**. *10* (4), 1160–1165.

22 Nakanishi, T., B. Ohtani, K. Uosaki, Fabrication and characterization of CdS-nanoparticle mono- and multilayers on a self-assembled monolayer of alkanedithiols on gold. *J. Phys. Chem. B* **1998**. *102* (9), 1571–1577.

23 Marx, E., D. S. Ginger, K. Walzer, K. Stokbro, N. C. Greenham, Self-assembled monolayers of CdSe nanocrystals on doped GaAs substrates. *Nano Lett.* **2002**. *2* (8), 911–914.

24 Shevchenko, E. V., D. V. Talapin, A. L. Rogach, A. Kornowski, M. Haase, H. Weller, Colloidal synthesis and self-assembly of CoPt3 nanocrystals. [Erratum for *J. Am. Chem. Soc.* 2002, 124, 11480–11485] *J. Am. Chem. Soc.* **2002**. *124* (46), 13958.

25 Kiely, C. J., J. Fink, J. G. Zheng, M. Brust, D. Bethell, D. J. Schiffrin, Ordered colloidal nanoalloys. *Adv. Mater.* **2000**. *12* (9), 640–643.

26 Ostrander, J. W., A. A. Mamedov, N. A. Kotov, Two modes of linear layer-by-layer growth of nanoparticle-polyelectrolyte multilayers and different interactions in the layer-by-layer deposition. *J. Am. Chem. Soc.* **2001**. *123* (6), 1101–1110.

27 Caruso, F., H. Lichtenfeld, M. Giersig, H. Moehwald, Electrostatic self-assembly of silica nanoparticle-polyelectrolyte multilayers on polystyrene latex particles. *J. Am. Chem. Soc.* **1998**. *120* (33), 8523–8524.

28 Feldheim, D. L., K. C. Grabar, M. J. Natan, T. E. Mallouk, Electron transfer in self-assembled inorganic polyelectrolyte/metal nanoparticle heterostructures. *J. Am. Chem. Soc.* **1996**. *118* (32), 7640–7641.

29 Cheng, W., S. Dong, E. Wang, Iodine-induced gold-nanoparticle fusion/fragmentation/ aggregation and iodine-linked nanostructured assemblies on a glass substrate. *Angew. Chem. Int. Ed.* **2003**. *42* (4), 449–452.

30 Frankamp, B. L., A. K. Boal, V. M. Rotello, Controlled interparticle spacing through self-assembly of au nanoparticles and poly(amidoamine) dendrimers. *J. Am. Chem. Soc.* **2002**. *124* (51), 15146–15147.

31 Fullam, S., H. Rensmo, S. N. Rao, D. Fitzmaurice, Noncovalent self-assembly of silver and gold nanocrystal aggregates in solution. *Chem. Mater.* **2002**. *14* (9), 3643–3650.

32 Dujardin, E., S. Mann, L.-B. Hsin, C. R. C. Wang, DNA-driven self-assembly of gold nanorods. *Chem. Commun.* **2001**(14), 1264–1265.

33 Mann, S., W. Shenton, M. Li, S. Connolly, D. Fitzmaurice, Biologically programmed nanoparticle assembly. *Adv. Mater.* **2000**. *12* (2), 147–150.

34 Boal, A. K., F. Ilhan, J. E. DeRouchey, T. Thurn-Albrecht, T. P. Russell, V. M. Rotello, Self-assembly of nanoparticles into structured spherical and network aggregates. *Nature*, **2000**. *404* (6779), 746–748.

35 Alivisatos, A. P., K. P. Johnsson, X. Peng, T. E. Wilson, C. J. Loweth, M. P. Bruchez, Jr., P. G. Schultz, Organization of 'nanocrystal molecules' using DNA. *Nature* **1996**. *382* (6592), 609–611.

36 Mirkin, C. A., R. L. Letsinger, R. C. Mucic, J. J. Storhoff, A DNA-based method for rationally assembling nanoparticles into macroscopic materials. *Nature* **1996**. *382* (6592), 607–609.

37 Zeng, H., J. Li, J. P. Liu, Z. L. Wang, S. Sun, Exchange-coupled nanocomposite magnets by nanoparticle self-assembly. *Nature* **2002**. *420* (6914), 395–398.

38 Dumestre, F., B. Chaudret, C. Amiens, M. Respaud, P. Fejes, P. Renaud, P. Zurcher, Unprecedented crystalline super-lattices of monodisperse cobalt nanorods. *Angew. Chem. Int. Ed.* **2003**. *42* (42), 5213–5216.

39 Chang, Y., H. C. Zeng, Controlled synthesis and self-assembly of single-crystalline cuo nanorods and nanoribbons. *Crystal Growth Design* **2004**. *4* (2), 397–402.

40 Nikoobakht, B., Z. L. Wang, M. A. El-Sayed, Self-assembly of gold nanorods. *J. Phys. Chem. B* **2000**. *104* (36), 8635–8640.

41 Li, L.-s., J. Walda, L. Manna, A. P. Alivisatos, Semiconductor nanorod liquid crystals. *Nano Lett.* **2002**. *2* (6), 557–560.

42 Kim, F., S. Kwan, J. Akana, P. Yang, Langmuir-Blodgett nanorod assembly. *J. Am. Chem. Soc.* **2001**. *123* (18), 4360–4361.

43 Kwan, S., F. Kim, J. Akana, P. Yang, Synthesis and assembly of BaWO4 nanorods. *Chem. Commun.* **2001**(5), 447–448.

44 Onsager, I., *Ann. N. Y. Acad. Sci.* **1949**. *51* (627).

45 Li, L.-S., A. P. Alivisatos, Semiconductor nanorod liquid crystals and their assembly on a substrate. *Adv. Mater.* **2003**. *15* (5), 408–411.

46 Jana, N. R., L. A. Gearheart, S. O. Obare, C. J. Johnson, K. J. Edler, S. Mann, C. J. Murphy, Liquid crystalline assemblies of ordered gold nanorods. *J. Mater. Chem*, **2002**. *12* (10), 2909–2912.

47 Kovtyukhova, N. I., T. E. Mallouk, Nanowires as building blocks for self-assembling logic and memory circuits. *Chem. Eur. J.* **2002**. *8* (19), 4354–4363.

48 Park, S., J.-H. Lim, S.-W. Chung, C. A. Mirkin, Self-assembly of mesoscopic metal-polymer amphiphiles. *Science* **2004**. *303* (5656), 348–351.

49 Love, J. C., A. R. Urbach, M. G. Prentiss, G. M. Whitesides, Three-dimensional self-assembly of metallic rods with submicron diameters using magnetic interactions. *J. Am. Chem. Soc.* **2003**. *125* (42), 12696–12697.

50 Flynn, C. E., S.-W. Lee, B. R. Peelle, A. M. Belcher, Viruses as vehicles for growth, organization and assembly of materials. *Acta Mater.* **2003**. *51* (19), 5867–5880.

51 Mao, C., D. J. Solis, B. D. Reiss, S. T. Kottmann, R. Y. Sweeney, A. Hayhurst, G. Georgiou, B. Iverson, A. M. Belcher, Virus-based toolkit for the directed synthesis of magnetic and semiconducting nanowires. *Science* **2004**. *303* (5655), 213–217.

52 Mucic, R. C., J. J. Storhoff, C. A. Mirkin, R. L. Letsinger, DNA-directed synthesis of binary nanoparticle network materials. *J. Am. Chem. Soc.* **1998**. *120* (48), 12674–12675.

53 Mbindyo, J. K. N., B. D. Reiss, B. R. Martin, C. D. Keating, M. J. Natan, T. E. Mallouk, DNA-directed assembly of gold nanowires on complementary surfaces. *Adv. Mater.* **2001**. *13* (4), 249–254.

54 Braun, E., Y. Eichen, U. Sivan, G. Ben-Yoseph, DNA-templated assembly and electrode attachment of a conducting silver wire. *Nature* **1998**. *391* (6669), 775–778.

55 Li, H., S. H. Park, J. H. Reif, T. H. LaBean, H. Yan, DNA-templated self-assembly of protein and nanoparticle linear arrays. *J. Am. Chem. Soc.* **2004**. *126* (2), 418–419.

56 Liu, D., S. H. Park, J. H. Reif, T. H. LaBean, DNA nanotubes self-assembled from triple-crossover tiles as templates for conductive nanowires. *Proc. Natl. Adac. Sci. U.S.A.* **2004**. *101* (3), 717–722.

57 Shenton, W., S. A. Davis, S. Mann, Directed self-assembly of nanoparticles into macroscopic materials using antibody-antigen recognition. *Adv. Mater.* **1999**. *11* (6), 449–452.

58 Caswell, K. K., J. N. Wilson, U.H. F. Bunz, C. J. Murphy, Preferential end-to-end assembly of gold nanorods by biotin-streptavidin connectors. *J. Am. Chem. Soc.* **2003**. *125* (46), 13914–13915.

59 Douglas, T., M. Young, Virus particles as templates for materials synthesis. *Adv. Mater.* **1999**. *11* (8), 679–681.

60 Douglas, T., E. Strable, D. Willits, A. Aitouchen, M. Libera, M. Young, Protein engineering of a viral cage for constrained nanomaterials synthesis. *Adv. Mater.* **2002**. *14* (6), 415–418.

61 Qi, L., H. Colfen, M. Antonietti, M. Li, J. D. Hopwood, A. J. Ashley, S. Mann, Formation of BaSO4 fibres with morphological complexity in aqueous polymer solutions. *Chem. Eur. J.* **2001**. *7* (16), 3526–3532.

62 Yu, S.-H., M. Antonietti, H. Coelfen, J. Hartmann, Growth and self-assembly of BaCrO4 and BaSO4 nanofibers toward hierarchical and repetitive superstructures by polymer-controlled mineralization reactions. *Nano Lett.* **2003**. *3* (3), 379–382.

63 Tian, Z. R., J. A. Voigt, J. Liu, B. McKenzie, M. J. McDermott, Biomimetic arrays of oriented helical ZnO nanorods and columns. *J. Am. Chem. Soc.* **2002**. *124* (44), 12954–12955.

64 Tian, Z. R., J. A. Voigt, J. Liu, B. McKenzie, M. J. McDermott, M. A. Rodriguez, H. Konishi, H. Xu, Complex and oriented ZnO nanostructures. *Nat. Mater.* **2003**. *2* (12), 821–826.

65 Kim, E., Y. Xia, G. M. Whiteside, Two- and three-dimensional crystallization of polymeric microspheres by micromolding in capillaries. *Adv. Mater.* **1996**. *8* (3), 245–247.

66 Yang, S. M., G. A. Ozin, Opal chips. Vectorial growth of colloidal crystal patterns inside silicon wafers. *Chem. Comm.* **2000**, *24*, 2507–2508.

67 Messer, B., J. H. Song, P. Yang, Microchannel networks for nanowire patterning. *J. Am. Chem. Soc.* **2000**. 10232–10233.

68 Ozin, G. A., S. M. Yang, The race for the photonic chip: colloidal crystal assembly in silicon wafers. *Adv. Functional Mater.* **2001**. 95–104.

69 Yin, Y., Y. Lu, B. Gates, Y. Xia, Template-assisted self-assembly: a practical route to complex aggregates of monodispersed colloids with well-defined sizes, shapes, and structures. *J. Am. Chem. Soc.* **2001**. 8718–8729.

70 Lu, G., X. Chen, J. Yao, W. Li, G. Zhang, D. Zhao, B. Yang, J. Shen, Fabricating ordered voids in a colloidal crystal film-substrate system by using organic liquid patterns as templates. *Adv. Mater.* **2002**. 1799–1802.

71 van Blaaderen, A., R. Rue, P. Wiltzius, Template-directed colloidal crystallization. *Nature* **1997**. 321–324.

72 Yin, Y., Y. Xia, Growth of large colloidal crystals with their (100) planes oriented parallel to the surfaces of supporting substrates. *Adv. Mater.* **2002**. 605–608.

73 Aizenberg, J., P. V. Braun, P. Wiltzius, Patterned colloidal deposition controlled by electrostatic and capillary forces. *Phys. Rev. Lett.* **2000**. *84* (13), 2997–3000.

74 Lee, I., H. Zheng, M. F. Rubner, P. T. Hammond, Controlled cluster size in patterned particle arrays via directed adsorption on confined surfaces. *Adv. Mater.* **2002**. *14* (8), 572–577.

75 Demers, L. M., C. A. Mirkin, Combinatorial templates generated by dip-pen nanolithography for the formation of two-dimensional particle arrays. *Angew. Chem. Int. Ed.* **2001**. *40* (16), 3069–3071.

76 Hayward, R. C., A. Saville, A. Aksay, Electrophoretic assembly of colloidal crystals with optically tunable micropatterns. *Nature* **2000**. *404* (6773), 56–59.

77 Werts, M. H. V., M. Lambert, J.-P. Bourgoin, M. Brust, Nanometer scale patterning of Langmuir-Blodgett films of gold nanoparticles by electron beam lithography. *Nano Lett.* **2002**. *2* (1), 43–47.

78 Hamann, H. F., S. I. Woods, S. Sun, Direct thermal patterning of self-assembled nanoparticles. *Nano Lett.* **2003**. *3* (12), 1643–1645.

79 Guo, Q., X. Teng, S. Rahman, H. Yang, Patterned Langmuir-Blodgett films of monodisperse nanoparticles of iron oxide using soft lithography. *J. Am. Chem. Soc.* **2003**. *125* (3), 630–631.

80 Yao, J., X. Yan, G. Lu, K. Zhang, X. Chen, L. Jiang, B. Yang, Patterning colloidal crystals by lift-up soft lithography. *Adv. Mater.* **2004**. *16* (1), 81–84.

81 Duan, X., Y. Huang, Y. Cui, J. Wang, C. M. Lieber, Indium phosphide nanowires as building blocks for nanoscale electronic and optoelectronic devices. *Nature* 2001. *409* (6816), 66–69.

82 Zhang, H., S. Boussaad, L. Nguyen, N. J. Tao, Magnetic-field-assisted assembly of metal/polymer/metal junction sensors. *Appl. Phys. Lett.*, 2004. *84* (1), 133–135.

83 Giersig, M., P. Mulvaney, Preparation of ordered colloid monolayers by electrophoretic deposition. *Langmuir*, 1993. *9* (12), 3408–3413.

84 Giersig, M., P. Mulvaney, Formation of ordered two-dimensional gold colloid lattices by electrophoretic deposition. *J. Phys. Chem.* 1993. *97* (24), 6334–6336.

85 Trau, M., N. Yao, E. Kim, Y. Xia, G. M. Whitesides, I.A. Aksay, Microscopic patterning of oriented mesoscopic silica through guided growth. *Nature* 1997. *390* (6661), 674–676.

86 Huang, Y., X. Duan, Q. Wei, C. M. Lieber, One-dimensional nanostructures into functional networks. *Science* 2001. *291* (5504), 630–633.

87 Ural, A., Y. Li, H. Dai, Electric-field-aligned growth of single-walled carbon nanotubes on surfaces. *Appl. Phys. Lett.* 2002. *81* (18), 3464–3466.

88 Morkved, T. L., M. Lu, A. M. Urbas, E. E. Ehrichs, H. M. Jaeger, P. Mansky, T. P. Russell, Local control of microdomain orientation in diblock copolymer thin films with electric fields. *Science* 1996. *273* (5277), 931–933.

89 Zhang, Y., A. Chang, J. Cao, Q. Wang, W. Kim, Y. Li, N. Morris, E. Yenilmez, J. Kong, H. Dai, Electric-field-directed growth of aligned single-walled carbon nanotubes. *Appl. Phys. Lett.* 2001. *79* (19), 3155–3157.

90 Jin, L., C. Bower, O. Zhou, Alignment of carbon nanotubes in a polymer matrix by mechanical stretching. *Appl. Phys. Lett.* 1998. *73* (9), 1197–1199.

91 Haggenmueller, R., H. H. Gommans, A. G. Rinzler, J. E. Fischer, K. I. Winey, Aligned single-wall carbon nanotubes in composites by melt processing methods. *Chem. Phys. Lett.* 2000. *330* (3,4), 219–225.

92 Vigolo, B., A. Penicaud, C. Coulon, C. Sauder, R. Pailler, C. Journet, P. Bernier, P. Poulin, Macroscopic fibers and ribbons of oriented carbon nanotubes. *Science* 2000. *290* (5495), 1331–1334.

93 Launois, P., A. Marucci, B. Vigolo, P. Bernier, A. Derre, P. Poulin, Structural characterization of nanotube fibers by x-ray scattering. *J. Nanosci. Nanotechnol.* 2001. *1* (2), 125–128.

94 Fischer, J. E., W. Zhou, J. Vavro, M. C. Llaguno, C. Guthy, R. Haggenmueller, M. J. Casavant, D. E. Walters, R. E. Smalley, Magnetically aligned single wall carbon nanotube films: Preferred orientation and anisotropic transport properties. *J. Appl. Phys.* 2003. *93* (4), 2157–2163.

95 Trau, M., D. A. Saville, I. A. Aksay, Assembly of colloidal crystals at electrode interfaces. *Langmuir*, 1997. *13* (24), 6375–6381.

96 Holgado, M., F. Garcia-Santamaria, A. Blanco, M. Ibisate, A. Cintas, H. Miguez, C. J. Serna, C. Molpeceres, J. Requena, A. Mifsud, F. Meseguer, C. Lopez, Electrophoretic deposition to control artificial opal growth. *Langmuir*, 1999. *15* (14), 4701–4704.

97 Kloepper, K. D., T.-D. Onuta, D. Amarie, B. Dragnea, Field-induced interfacial properties of gold nanoparticles in AC microelectrophoretic experiments. *J. Phys. Chem. B* 2004. *108* (8), 2547–2553.

98 Chandrasekharan, N., P. V. Kamat, Assembling gold nanoparticles as nanostructured films using an electrophoretic approach. *Nano Lett.* 2001. *1* (2), 67–70.

99 Teranishi, T., M. Hosoe, T. Tanaka, M. Miyake, Size control of monodispersed Pt nanoparticles and Their 2D organization by electrophoretic deposition. *J. Phys. Chem. B* 1999. *103* (19), 3818–3827.

100 Affoune, A. M., B. L. V. Prasad, H. Sato, T. Enoki, Electrophoretic deposition of nanosized diamond particles. *Langmuir*, 2001. *17* (2), 547–551.

101 Ke, C., Z. Ni, Y. J. Wang, Y. Tang, Y. Gu, Z. Gao, W. L. Yang, Electrophoretic assembly of nanozeolites: zeolite coated fibers and hollow zeolite fibers. *Chem. Commun.* 2001(8), 783–784.

102 Collier, C. P., R. J. Saykally, J. J. Shiang, S. E. Henrichs, J. R. Heath, Reversible tuning of silver quantum dot monolayers through the metal-insulator transition. *Science* 1997. *277* (5334), 1978–1981.

103 Chung, S. W., G. Markovich, J. R. Heath, Fabrication and alignment of wires in two dimensions. *J. Phys. Chem. B* **1998**. *102* (35), 6685–6687.

104 Tao, A., F. Kim, C. Hess, J. Goldberger, R. He, Y. Sun, Y. Xia, P. Yang, Langmuir-Blodgett silver nanowire monolayers for molecular sensing using surface-enhanced Raman spectroscopy. *Nano Lett.* **2003**. *3* (9), 1229–1233.

105 Yang, P., F. Kim, Langmuir-Blodgett assembly of one-dimensional nanostructures. *Chemphyschem* **2002**. *3* (6), 503–506.

106 Whang, D., S. Jin, Y. Wu, C. M. Lieber, Large-scale hierarchical organization of nanowire arrays for integrated nanosystems. *Nano Lett.* **2003**. *3* (9), 1255–1259.

107 Joo, S. H., S. J. Choi, I. Oh, J. Kwak, Z. Liu, O. Terasaki, R. Ryoo, Ordered nanoporous arrays of carbon supporting high dispersions of platinum nanoparticles. *Nature* **2001**. *412* (6843), 169–172.

108 Sakamoto, Y., M. Kaneda, O. Terasaki, D. Y. Zhao, J. M. Kim, G. Stucky, H. J. Shin, R. Ryoo, Direct imaging of the pores and cages of three-dimensional mesoporous materials. *Nature* **2000**. *408* (6811), 449–453.

109 Wang, D., W. L. Zhou, B. F. McCaughy, J. E. Hampsey, X. Ji, Y.-B. Jiang, H. Xu, J. Tang, R. H. Schmehl, C. O'Connor, C. J. Brinker, Y. Lu, Electrodeposition of metallic nanowire thin films using mesoporous silica templates. *Adv. Mater.* **2003**. *15* (2), 130–133.

110 Martin, C. R., Nanomaterials: a membrane-based synthetic approach. *Science* **1994**. *266* (5193), 1961–1966.

111 Martin, B. R., D. J. Dermody, B. D. Reiss, M. Fang, L. A. Lyon, M. J. Natan, T. E. Mallouk, Orthogonal self-assembly on colloidal gold-platinum nanorods. *Adv. Mater.* **1999**. *11* (12), 1021–1025.

112 Thurn-Albrecht, T., J. Schotter, G. A. Kastle, N. Emley, T. Shibauchi, L. Krusin-Elbaum, K. Guarini, C. T. Black, M. T. Tuominen, T. P. Russell, Ultrahigh-density nanowire arrays grown in self-assembled diblock copolymer templates. *Science* **2000**. *290* (5499), 2126–2129.

113 Attard, G. S., P. N. Bartlett, N. R. B. Coleman, J. M. Elliott, J. R. Owen, J. H. Wang, Mesoporous platinum films from lyotropic liquid crystalline phases. *Science* **1997**. *278* (5339), 838–840.

114 Huang, L., H. Wang, Z. Wang, A. Mitra, K. N. Bozhilov, Y. Yan, Nanowire arrays electrodeposited from liquid crystalline phases. *Adv. Mater.* **2002**. *14* (1), 61–64.

115 Lopes, W. A., H. M. Jaeger, Hierarchical self-assembly of metal nanostructures on diblock copolymer scaffolds. *Nature*, **2001**. *414* (6865), 735–738.

116 Kresge, C., M. Leonowicz, W. Roth, C. Vartuli, J. Beck, Ordered mesoporous molecular sieves synthesized by a liquid-crystal template mechanism. *Nature* **1992**. *359*, 710–712.

117 Lu, Y., R. Ganguli, C. A. Drewien, M. T. Anderson, C. J. Brinker, W. Gong, Y. Guo, H. Soyez, B. Dunn, M. H. Huang, J. I. Zink, Continuous formation of supported cubic and hexagonal mesoporous films by sol-gel dip-coating. *Nature* **1997**. *389* (6649), 364–368.

118 Lu, Y., H. Fan, A. Stump, T. L. Ward, T. Rieker, C. J. Brinker, Aerosol-assisted self-assembly of mesostructured spherical nanoparticles. *Nature* **1999**. *398* (6724), 223–226.

119 Lu, Y., B. F. MaCaughey, D. Wang, J. E. Hampsey, D. Nilesh, C. J. Brinker, Aerosol-assisted formation of mesostructured thin film. *Adv. Mater.* **2003**, accepted.

120 Yang, P., D. Zhao, B. F. Chmelka, G. D. Stucky, Triblock-copolymer-directed syntheses of large-pore mesoporous silica fibers. *Chem. Mater.* **1998**. *10* (8), 2033–2036.

121 Feng, P., X. Bu, G. D. Stucky, D. J. Pine, Monolithic mesoporous silica templated by microemulsion liquid crystals. *J. Am. Chem. Soc.* **2000**. *122* (5), 994–995.

122 Huang, M. H., A. Choudrey, P. Yang, Ag nanowire formation within mesoporous silica. *Chem. Commun.* **2000**. (12), 1063–1064.

123 Coleman, N. R. B., N. O'Sullivan, K. M. Ryan, T. A. Crowley, M. A. Morris, T. R. Spalding, D. C. Steytler, J. D. Holmes, Synthesis and characterization of dimensionally ordered semiconductor nanowires within mesoporous silica. *J. Am. Chem. Soc.* **2001**. *123* (29), 7010–7016.

124 Lee, K.-B., S.-M. Lee, J. Cheon, Size-controlled synthesis of Pd nanowire using a mesoporous silica template via chemical vapor infiltration. *Adv. Mater.* **2001**. *13* (7), 517–520.

125 Zhang, Z., S. Dai, D. A. Blom, J. Shen, Synthesis of ordered metallic nanowires inside ordered mesoporous materials through electroless deposition. *Chem. Mater.* **2002**. *14* (3), 965–968.

126 Kang, H., Y.-w. Jun, J.-I. Park, K.-B. Lee, J. Cheon, Synthesis of porous palladium superlattice nanoballs and nanowires. *Chem. Mater.* **2000**. *12* (12), 3530–3532.

127 Whitehead, A. H., J. M. Elliott, J. R. Owen, G. S. Attard, Electrodeposition of mesoporous tin films. *Chem. Commun.* **1999**(4), 331–332.

128 Nelson, P. A., J. M. Elliott, G. S. Attard, J. R. Owen, Mesoporous nickel/nickel oxide – a nanoarchitectured electrode. *Chem. Mater.* **2002**. *14* (2), 524–529.

129 Liu, L., M. Singh, V. T. John, G. L. McPherson, J. He, V. Agarwal, A. Bose, Shear-induced alignment and nanowire silica synthesis in a rigid crystalline surfactant mesophase. *J. Am. Chem. Soc.* **2004**. *126* (8), 2276–2277.

130 Walsh, D., L. Arcelli, T. Ikoma, J. Tanaka, S. Mann, Dextran templating for the synthesis of metallic and metal oxide sponges. *Nature Mater.* **2003**. *2* (6), 386–390.

131 Huang, J., T. Kunitake, Nano-Precision Replication of Natural Cellulosic Substances by Metal Oxides. *J. Am. Chem. Soc.* **2003**. *125* (39), 11834–11835.

132 Yue, J., V. Sankaran, R. E. Cohen, R. R. Schrock, Interconversion of ZnF2 and ZnS nanoclusters within spherical microdomains in block copolymer films. *J. Am. Chem. Soc.* **1993**. *115* (10), 4409–4410.

133 Sankaran, V., J. Yue, R. E. Cohen, R. R. Schrock, R.J. Silbey, Synthesis of zinc sulfide clusters and zinc particles within microphase-separated domains of organometallic block copolymers. *Chem. Mater.* **1993**. *5* (8), 1133–1142.

134 Horiuchi, S., M. I. Sarwar, Y. Nakao, Nanoscale assembly of metal clusters in block copolymer films with vapor of a metal-acetylacetonato complex using a dry process. *Adv. Mater.* **2000**. *12* (20), 1507–1511.

135 Horiuchi, S., T. Fujita, T. Hayakawa, Y. Nakao, Three-dimensional nanoscale alignment of metal nanoparticles using block copolymer films as nanoreactors. *Langmuir*, **2003**. *19* (7), 2963–2973.

136 Li, D., Y. Xia, Direct fabrication of composite and ceramic hollow nanofibers by electrospinning. *Nano Lett.* **2004**. *4* (5), 933–938.

137 Reneker, D. H., A. L. Yarin, H. Fong, S. Koombhongse, Bending instability of electrically charged liquid jets of polymer solutions in electrospinning. *J. Appl. Phys.* **2000**. *87* (9, Pt. 1), 4531–4547.

138 Greene, L. E., M. Law, J. Goldberger, F. Kim, J. C. Johnson, Y. Zhang, R. J. Saykally, P. Yang, Low-temperature wafer-scale production of ZnO nanowire arrays. *Angew. Chem. Int. Ed.* **2003**. *42* (26), 3031–3034.

139 Huang, M. H., S. Mao, H. Feick, H. Yan, Y. Wu, H. Kind, E. Weber, R. Russo, P. Yang, Room-temperature ultraviolet nanowire nanolasers. *Science* **2001**. *292* (5523), 1897–1899.

140 Melosh, N. A., A. Boukai, F. Diana, B. Gerardot, A. Badolato, P. M. Petroff, J. R. Heath, Ultrahigh-density nanowire lattices and circuits. *Science* **2003**. *300* (5616), 112–115.

141 Li, D., Y. Wang, Y. Xia, Electrospinning nanofibers as uniaxially aligned arrays and layer-by-layer stacked films. *Adv. Mater.* **2004**. *16* (4), 361–366.

142 Li, W. Z., S. S. Xie, L. X. Qian, B. H. Chang, B. S. Zou, W. Y. Zhou, R. A. Zhao, G. Wang, Large-scale synthesis of aligned carbon nanotubes. *Science* **1996**. *274* (5293), 1701–1703.

143 Ren, Z. F., Z. P. Huang, J. W. Xu, J. H. Wang, P. Bush, M. P. Siegel, P. N. Provencio, Synthesis of large arrays of well-aligned carbon nanotubes on glass. *Science* **1998**. *282* (5391), 1105–1107.

144 Tian, Z. R., J. A. Voigt, J. Liu, B. McKenzie, H. Xu, Large oriented arrays and continuous films of TiO2-based nanotubes. *J. Am. Chem. Soc.* **2003**. *125* (41), 12384–12385.

145 Zhou, J., N.-S. Xu, S.-Z. Deng, J. Chen, J.-C. She, Z.-L. Wang, Large-area nanowire arrays of molybdenum and molybdenum oxides: synthesis and field emission properties. *Adv. Mater.* **2003**. *15* (21), 1835–1840.

146 Chen, L. C., S. W. Chang, C. S. Chang, C. Y. Wen, J. J. Wu, Y. F. Chen, Y. S. Huang, K. H. Chen, Catalyst-free and controllable growth of SiCxNy nanorods. *J. Phys. Chem. Solids* **2001**. *62* (9–10), 1567–1576.

147 Liang, L., J. Liu, C. F. Windisch, Jr., G. J. Exarhos, Y. Lin, Direct assembly of large arrays of oriented conducting polymer nanowires. *Angew. Chem. Int. Ed.* **2002**. *41* (19), 3665–3668.

148 Tian, Z. R., J. Liu, J. A. Voigt, B. McKenzie, H. Xu, Hierarchichal and self-similar growth of self-assembled crystals. *Angew. Chem. Int. Ed.* **2003**. *42* (4), 414–417.

149 Liu, J., Y. Lin, L. Liang, J. A. Voigt, D. L. Huber, Z. R. Tian, E. Coker, B. McKenzie, M. J. McDermott, Templateless assembly of molecularly aligned conductive polymer nanowires: A new approach for oriented nanostructures. *Chem. Eur. J.* **2003**. *9* (3), 604–611.

150 Chan, W. C., S. Nie, Quantum dot bioconjugates for ultrasensitive nonisotopic detection. *Science* **1998**. *281* (5385), 2016–2018.

151 Bruchez, M., Jr., M. Moronne, P. Gin, S. Weiss, A. P. Alivisatos, Semiconductor nanocrystals as fluorescent biological labels. *Science* **1998**. *281* (5385), 2013–2016.

152 Taton, T. A., C. A. Mirkin, R. L. Letsinger, Scanometric DNA array detection with nanoparticle probes. *Science* **2000**. *289* (5485), 1757–1760.

153 Elghanian, R., J. J. Storhoff, R. C. Mucic, R. L. Letsinger, C. A. Mirkin, Selective colorimetric detection of polynucleotides based on the distance-dependent optical properties of gold nanoparticles. *Science* **1997**. *277* (5329), 1078–1080.

154 Tiefenauer, L. X., G. Kuehne, R. Y. Andres, Antibody-magnetite nanoparticles: In vitro characterization of a potential tumor-specific contrast agent for magnetic resonance imaging. *Bioconjug. Chem.* **1993**. *4* (5), 347–352.

155 Murthy, S. N., Magnetophoresis: an approach to enhance transdermal drug diffusion. *Pharmazie* **1999**. *54* (5), 377–379.

156 Kong, J., N. R. Franklin, C. Zhou, M. G. Chapline, S. Peng, K. Cho, H. Dailt, Nanotube molecular wires as chemical sensors. *Science* **2000**. *287* (5453), 622–625.

157 Collins, P. G., K. Bradley, M. Ishigami, A. Zettl, Extreme oxygen sensitivity of electronic properties of carbon nanotubes. *Science* **2000**. *287* (5459), 1801–1804.

158 Cui, Y., Q. Wei, H. Park, C. M. Lieber, Nanowire nanosensors for highly sensitive and selective detection of biological and chemical species. *Science* **2001**. *293* (5533), 1289–1292.

159 Huang, J., S. Virji, B. H. Weiller, R. B. Kaner, Polyaniline nanofibers: facile synthesis and chemical sensors. *J. Am. Chem. Soc.* **2003**. *125* (2), 314–315.

160 Favier, F., E. C. Walter, M. P. Zach, T. Benter, R. M. Penner, Hydrogen sensors and switches from electrodeposited palladium mesowire arrays. *Science* **2001**. *293* (5538), 2227–2231.

161 Li, C. Z., H. X. He, A. Bogozi, J. S. Bunch, N. J. Tao, Molecular detection based on conductance quantization of nanowires. *Appl. Phys. Lett.* **2000**. *76* (10), 1333–1335.

162 Bradley, K., J.-C.P. Gabriel, M. Briman, A. Star, G. Gruner, Charge transfer from ammonia physisorbed on nanotubes. *Phy. Re. Lett.* **2003**. *91* (21), 218301/1–218301/4.

163 Star, A., T.-R. Han, J.-C.P. Gabriel, K. Bradley, G. Gruener, Interaction of aromatic compounds with carbon nanotubes: correlation to the Hammett parameter of the substituent and measured carbon nanotube FET response. *Nano Lett.* **2003**. *3* (10), 1421–1423.

164 Someya, T., J. Small, P. Kim, C. Nuckolls, J. T. Yardley, Alcohol Vapor Sensors Based on Single-Walled Carbon Nanotube Field Effect Transistors. *Nano Lett.* **2003**. *3* (7), 877–881.

165 Star, A., J.-C.P. Gabriel, K. Bradley, G. Gruener, Electronic detection of specific protein binding using nanotube FET devices. *Nano Lett.* **2003**. *3* (4), 459–463.

166 Hahm, J.-I., C. M. Lieber, Direct ultrasensitive electrical detection of DNA and DNA sequence variations using nanowire nanosensors. *Nano Lett.* **2004**. *4* (1), 51–54.

167 Li, Z., Y. Chen, X. Li, T.I. Kamins, K. Nauka, R. S. Williams, Sequence-specific label-free DNA sensors based on silicon nanowires. *Nano Lett.* **2004**. *4* (2), 245–247.

168 Kolmakov, A., Y. Zhang, G. Cheng, M. Moskovits, Detection of CO and O2 using tin oxide nanowire sensors. *Adv. Mater.* **2003**. *15* (12), 997–1000.

169 Wang, Y., X. Jiang, Y. Xia, A solution-phase, precursor route to polycrystalline SnO2 nanowires that can be used for gas sensing under ambient conditions. *J. Am. Chem. Soc.* **2003**. *125* (52), 16176–16177.

170 Liu, H., J. Kameoka, D. A. Czaplewski, H. G. Craighead, Polymeric nanowire chemical sensor. *Nano Lett.* **2004**, *4* (4), 671–675.

171 He, H., N. Tao, Interactions of molecules with metallic quantum wires. *Adv. Mater.* **2002**. *14* (2), 161–164.

172 Xu, B., N. J. Tao, Measurement of single-molecule resistance by repeated formation of molecular junctions. *Science* **2003**. *301* (5637), 1221–1223.

173 Bogozi, A., O. Lam, H. He, C. Li, N. J. Tao, L. A. Nagahara, I. Amlani, R. Tsui, Molecular adsorption onto metallic quantum wires. *J. Am. Chem. Soc.* **2001**. *123* (19), 4585–4590.

174 Rajagopalan, V., S. Boussaad, N. J. Tao, Detection of heavy metal ions based on quantum point contacts. *Nano Lett.* **2003**. *3* (6), 851–855.

175 Murray, B. J., E. C. Walter, R. M. Penner, Amine vapor sensing with silver mesowires. *Nano Lett.* **2004**. *4* (4), 665–670.

176 Cui, Y., Z. Zhong, D. Wang, W. U. Wang, C. M. Lieber, High performance silicon nanowire field effect transistors. *Nano Lett.* **2003**. *3* (2), 149–152.

177 Tans, S. J., A. R. M. Verschueren, C. Dekker, Room-temperature transistor based on a single carbon nanotube. *Nature* **1998**. *393* (6680), 49–52.

178 Duan, X., C. Niu, V. Sahi, J. Chen, J. W. Parce, S. Empedocles, J. L. Goldman, High-performance thin-film transistors using semiconductor nanowires and nanoribbons. *Nature* **2003**. *425* (6955), 274–278.

179 Sun, S., C. B. Murray, Synthesis of monodisperse cobalt nanocrystals and their assembly into magnetic superlattices. *J. Appl. Phys.* **1999**. *85* (8, Pt. 2A), 4325–4330.

180 Black, C. T., C. B. Murray, R. L. Sandstrom, S. Sun, Spin-dependent tunneling in self-assembled cobalt-nanocrystal superlattices. *Science* **2000**. *290* (5494), 1131–1134.

181 Skumryev, V., S. Stoyanov, Y. Zhang, G. Hadjipanayis, D. Givord, J. Nogues, Beating the superparamagnetic limit with exchange bias. *Nature* **2003**. *423* (6942), 850–853.

182 Whitney, T. M., J. S. Jiang, P. C. Searson, C. L. Chien, Fabrication and magnetic properties of arrays and metallic nanowires. *Science* **1993**. *261* (5126), 1316–1319.

4
Nanostructured Collagen Mimics in Tissue Engineering

Sergey E. Paramonov and Jeffrey D. Hartgerink

4.1
Introduction

A human body consists of approximately 10^{14} cells. This huge number of cells is organized and maintained at several levels of scale. First, cells must maintain their proper phenotype and not dedifferentiate into more generic cell types or undergo programmed cell death. Next, groups of cells are organized into tissues, which make up organs, and these organs are brought together to make a complete individual. This organization is maintained in large part by the extracellular matrix (ECM). ECM along with the cells it surrounds are the two necessary components of any multicellular tissue. ECM is a very complex biological environment that provides cells not only with a surface on which to attach to move, but also with chemical and biological information stored in the form of bioactive molecules, enzymes, and growth factors. Proper tissue functioning depends strongly on the correct ECM structure and composition, and several severe diseases (for example, *osteogenesis imperfecta*) may arise when the ECM is corrupted. ECM is a diverse material but its primary components are a collagen scaffold that provides the support for cells and a highly hydrated polysaccharide gel that allows the rapid diffusion of nutrients and resists compressive forces.

Recent advances in medicine have dramatically advanced reconstructive surgery and given rise to a whole new area of research: tissue engineering. Tissue engineering attempts to replace or rebuild damaged, diseased, or aged human tissues with those grown outside the body. The tissue growth requires cells, and the recipient can often provide his or her own cells, thus eliminating problems of finding a suitable donor and immune rejection of those donated organs. One key aspect of successful tissue engineering is the proper reconstruction of ECM of a particular tissue. This is a very difficult task, since the ECM is an exceptionally complex biomaterial. In order to tackle this problem it needs to be broken in several more simple tasks. Much research in tissue engineering currently concentrates on scaffold design. The choice of a scaffold is crucial, since it provides a surface on which cells can grow and a shape to which to conform. The variety of currently used scaffolds can be divided into two categories, those with a synthetic origin and those with a biological origin. Synthetic scaffolds include simple polymers such as poly-lactic acid (PLA), poly-gly-

Nanofabrication Towards Biomedical Applications. C. S. S. R. Kumar, J. Hormes, C. Leuschner (Eds.)
Copyright © 2005 WILEY-VCH Verlag GmbH & Co. KGaA, Weinheim
ISBN 3-527-31115-7

colic acid (PGA), PLA–PGA blends, and also inorganic ceramics such as hydroxyapatite (HA).The main design principle in this case is the biodegradability and non-toxicity of the polymer and its degradation products. Traditional advantages of synthetic polymers include precise control over structure and function, good mechanical qualities, and relatively low production costs. The large class of biocompatible synthetic polymers includes poly(acrylic acid) and its derivatives, poly(ethylene oxide) and its copolymers, poly(lactic acid), poly(vinyl alcohol), polyphosphazene, and polypeptides [1].

Although these scaffolds show very exciting properties they lack chemical functionality that can be used for final tuning of these materials upon the needs of a particular tissue; typically they act only to provide macroscopic structure on which cells may grow. Biologic scaffolds typically contain collagen that has been harvested and chemically purified. Various types of naturally occurring and chemically synthesized polymers have been studied and tested as scaffolds for tissue engineering. Among the naturally occurring polymers are the scaffolds based on collagen, hyaluronate, fibrin, alginate, and several others. Harvested collagen is a widely used tissue engineering scaffold. Although it is potentially immunogenic, collagen meets many of the requirements for an artificial matrix, such as biodegradability and cell recognition. Collagen has been successfully utilized for reconstruction of skin [2] and as a tissue scaffold for many different cell types [3]. We believe that ECM reconstruction should begin with a collagen mimic which can be prepared in a synthetic way. These synthetic collagen fibers should be able to be modified to display cell signaling peptides (such as RGD, IKVAV) or entire proteins such as growth factors. The artificially produced collagen can include different proteins, responsible for cell differentiation and proliferation. If this can be done it will be an important step toward replicating the complete ECM and the preparation of designer tissues for medical applications.

This article will discuss the structure and properties of natural collagen and then focus on how this knowledge has been used in attempts at creating synthetic collagen. Next we will discuss tissue engineering approaches which have attempted to mimic one or more properties of the extracellular matrix, and we will conclude with a discussion of our approach to the preparation of collagen mimics.

4.2
Collagen Structural Hierarchy

Collagen is the most abundant protein in the human body and constitutes about 25% of the total protein mass. Collagen has a complicated structure with many levels of hierarchy. At the lowest level is the amino acid sequence, which is found to have a highly conserved three amino acid repeat of Gly-X-Y where positions X and Y (in this case Y refers to a general amino acid, not to tyrosine) contain a high percentage of the imino acids proline and hydroxyproline respectively (Fig. 4.1). Proteins with this sequence are found to preferentially adopt a proline type II helical conformation that allows three of these peptides to assemble into the next level of hierarchy, the triple helix. These helices are bundled into fibrils and are often covalently cross-linked through modified lysine residues. Finally, these fibrils are bundled into macroscopic fibers (Fig. 4.2).

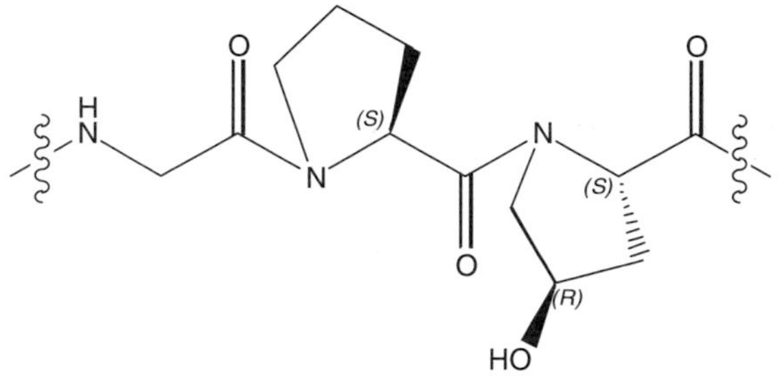

Figure 4.1. The chemical structure of the most common collagen triplet: glycine-proline-hydroxyproline.

When produced *in vivo* collagen is first synthesized as a propeptide that consists of a long triple-helical central region flanked by N and C terminal regions which assist in folding and prevent high-order assembly. This procollagen is excreted into the extracellular space, where it is enzymatically cleaved, leaving only the central triple helix, which then undergoes further assembly into larger structures called fibrils [4]. The dimensions of fibrils vary greatly from 10 to 300 nm in diameter and can be many hundreds of micrometers in length. These fibrils gain additional strength by covalent cross-linking of modified lysine residues. The fibril structure has a characteristic banding pattern revealed in electron microscopy images that results from the packing of triple helices that are offset from one another. The final stage of assembly involves packing of fibrils into macroscopic fibers that are large enough to be observed in a light microscope.

4.3
Amino Acid Sequence and Secondary Structure

Collagen has a unique secondary structure which consists of three poly(proline-II)-like chains. These three chains form a triple helix in which the peptide chains are wound around one another forming a right-handed superhelix [5, 6]. The geometrical parameters of poly(proline-II)-like conformation and the triple-helical structure are slightly different. The triple helix is characterized by 2.9 Å rise per residue and 3.33 residues per turn with dihedral angles $\phi = -75°$ and $\varphi = 145°$, whereas poly(proline-II)-like conformation is described by the same angles but a looser helix with 3.1 Å per residue and 3 residues per turn [5, 7, 8]. The collagen structure varies from tissue to tissue and about 15 types of collagen molecules have been found. A single collagen chain consists of a sequence of a trimeric amino acid repeat Gly-X-Y. Typically X positions are occupied by Pro and Y positions are occupied by 4-hydroxyproline; together these amino acids comprise about 20–25% of a collagen single chain.

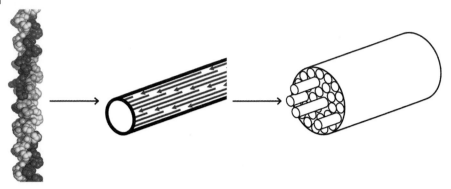

Figure 4.2. From left to right the collagen triple helix, collagen fibril, and collagen fiber.

The presence of Gly at every third position is required to allow tight packing of three single chains to form a triple helix. Proline and hydroxyproline prevent the formation of other secondary structures and provide necessary backbone conformation for triplex formation. The three single polypeptide chains are held together by hydrogen bonds that reside inside the triple helix. Hydrogen bonds are formed between the carbonyl oxygen of the amino acid in the X position and the amido hydrogen of glycine from an adjacent chain, while the carbonyl oxygen of glycine and amino acids in the Y position are oriented toward the exterior of the helix (Fig. 4.3). Supercoiled structure formation requires the three chains to be shifted by one residue relative to each other. This allows one of the three chains to have a Gly residue facing the interior of the triple helix at all times. Glycine, the smallest of the amino

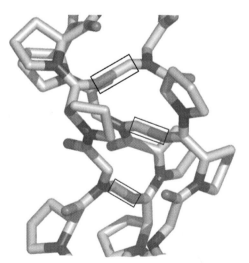

Figure 4.3. View of the collagen triple helix with the interchain hydrogen bonds identified.

acids, is required in this position due to the very tight packing of the helices against one another. The structural environment of Gly and residues in the X and Y positions is very different: side chains of amino acids in the X and Y positions point away from triple helix while Gly is completely buried inside the supercoiled structure.

4.4
Experimental Observation of the Collagen Triple Helix

Studying collagen and collagen model peptides is a relatively complex task. Direct observation of collagen triple helical structure is difficult and restricted to a few experimental methods. Among the most used techniques are circular dichroism spectroscopy (CD) and nuclear magnetic resonance (NMR). X-ray structure analysis, although very powerful for direct structure determination, is limited due to the difficulty of preparing quality crystals, and only a few structures have been reported.

CD is a very robust tool that has been used to observe protein secondary structure for decades. A review of this technique was published by Sreerama and Woody in 2000 [9]. It features a large negative peak around 200 nm and a small positive peak between 215 and 227 nm [Fig. 4.4(a)] [10]. Unfortunately, it is hard to distinguish between true triple helix, present in a solution, and a polyproline-II-like single chain. The latter exhibits a very similar CD spectrum with a small positive CD band at about 228 nm and a large negative band at about 205 nm [11]. Nevertheless, both the collagen spectrum and the spectrum of the polyproline-II-like conformation differ significantly from the CD spectrum of unordered polypeptides, which have a dynamic structure that accepts a wide range of conformations. Their characteristic spectra have a small negative band near 225 nm and a somewhat more intense negative band near 200 nm [11, 12].

The ambiguity of collagen CD data can be addressed by exploring temperature dependencies of the optical rotation and the CD molar ellipticity. During the temperature transition the triple-helical structure falls apart, releasing the three polypeptide chains that were held together by hydrogen bonds. This transition can be easily seen by monitoring the nonlinear, cooperative change in magnitude of either the positive or the negative band in the CD spectra with increasing temperature [Fig. 4.4(b)]. In contrast, polypeptide chains that are in a polyproline-II-like or "random coil" conformation undergo a linear thermal transition and their temperature-dependent optical rotation curves appear as straight lines.

The presence of a triple helical structure in solution can also be observed by NMR spectroscopy. Folding of three polypeptide chains results in a unique spin system with interchain nuclear Overhauser effects (NOEs) that cannot be observed for unfolded chains [13]. Studies by Brodsky et al. [14] indicate that the spin system of a model (Pro-Hyp-Gly)$_{10}$ peptide can be identified by means of double quantum-filtered correlation spectroscopy (DQF-COSY) and total correlation spectroscopy (TOCSY) spectra. Glycines can be distinguished from Pro and Hyp by the amide proton and by a pair of $C_\alpha H$ resonances. Although it has the same connectivity

a)

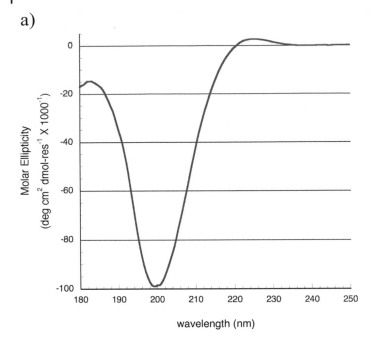

b)

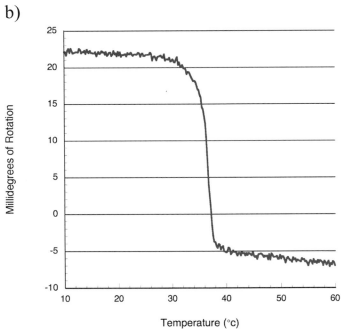

Figure 4.4. (a) CD spectrum of typical triple helix. (b) Temperature dependence of natural collagen unfolding as observed by CD at 220 nm.

pattern as Pro, Hyp is easily distinguished by the $C_\gamma H$ which is shifted downfield by the OH group. Triple helix formation can be proven by analyzing NOE spectra of peptides. NOEs occur between two atoms at distances smaller than 5 Å. The spin–spin coupling constants depend on dihedral angle and along with distances derived from NOEs can be used to determine peptide structure. Unfortunately it is very difficult to distinguish between interchain and intrachain NOEs in triple helix since both can lie within 5 Å distance. One way to solve this problem has been to compare observed NOEs of backbone with those calculated from peptide models based on X-ray structural data [14]. The other approach includes insertion of labeled Gly (^{15}N-Gly, ^{13}C-Gly) in the middle range of a sequence and observation of heteronuclear ^{1}H-^{15}N correlation spectra as well as ^{15}N relaxation and heteronuclear NOEs [13, 15].

4.5
Folding Kinetics

Collagen folding kinetics attract much attention since they provide an approach for understanding the complex process of triple helix formation. Collagen contains a large number of imino acids, which participate in the folding process. It is known that an imide bond of Pro can adopt cis conformation more easily than an amide bond and a great number of Pro may exist in cis conformation in an unfolded poly-Pro chain [16]. Thus cis-trans isomerization of peptide bonds, involving proline and hydroxyproline, can limit the rate of triple helix propagation in collagen. During the formation of a triple helix all cis imino bonds must be converted to the trans conformation. This type of folding is referred to as a "zipper-like propagation" [17]. The folding of natural collagen has been characterized for type I and III collagen [18]. Collagen folding starts with the self-association of three collagen propeptides. The C-terminus of these propeptides is rich in Gly-Pro-Hyp repeats and these repeat units initiate triple helix formation. The propagation of the triple-helix conformation, limited by cis-trans isomerization, from the C to the N terminus of the protein finishes the formation of a collagen triple-helical structure. Studies on collagen model peptides reveal that the rate-limiting step depends on concentration. At high concentration the rate-limiting step is cis-trans isomerization that follows after the nucleation reaction where three monomeric peptide chains combine to initiate triple helix formation. At low concentration the rate-limiting step is the nucleation reaction while the propagation step comparatively is fast. Xu et al. [19] studied the early steps of folding of a collagen-like peptide with a C-terminal (Gly-Pro-Hyp)$_4$ and N-terminal 18-residue sequence from α1(I) chain of type I collagen. They found that the peptide adopts a rigid triple helical conformation, with a melting temperature near 26 °C [20]. The kinetic scheme they derived suggests that 40–50% of the starting monomer population needs to have the required trans bonds in order to begin helix formation. They argue that little accumulation of intermediates, such as partially folded dimers, is possible since third-order reaction is sufficient to describe the folding from the isomerized monomer to a triple helix. A negative activation energy was observed for this third-order folding reaction, which is similar to the folding of globular proteins or α-helical coiled coils. The negative activation energy suggests that an entropic barrier plays greater role than the enthalpic barrier and suggests the existence of a rapid pre-equilibration between two or three collagen chains before the rate-limiting nucleation step. Boudko et al. [21] studied the kinetics of collagen formation in a broad range of concentrations in order to elucidate the nucleation mechanism. They noticed a shift from third to first order with increasing concentration and attributed this to a transition of the rate-limiting step from nucleation to propagation. They argued that the observed third-order rate constant for (Pro-Pro-Gly)$_{10}$ at low concentration would include the simultaneous collision of three single peptide chains, which is an extremely infrequent event. That led them to assume that the formation of a very unstable dimer, which is in fast pre-equilibrium with monomeric chain, exists prior to the nucleation of a triple helix. The authors stress that the dimers are very unstable and they appear as two-stranded short nuclei of a suitable conformation for addition of the third chain. By fitting the concentration dependence of initial rates of single chains the rate of triple helix propagation was found to be $k = 0.0007$ s^{-1} at 7 °C. They found this propagation rate to be in good agreement with the rate of cis-trans isomerization obtained by Bächinger et al. The extreme slowness of the collagen folding process, compared to other biological molecules such as α-helical coiled coils or DNA [22], may be explained by taking into account the great instability of collagen dimers. These dimers were never detected, whereas DNA or α-helical coiled coils have well-known double-stranded structures. The instability of collagen dimers may also explain the necessity of nucleation domains in collagen.

4.6
Stabilization Through Sequence Selection

Recently many research efforts have been focused on stabilizing collagen and on how the different variations in primary collagen structure affect the overall stability of a triple helix. It is well known that Gly has to be present as every third residue to allow three collagen chains to pack into triplex. However, the X and Y positions may contain various amino acids, which are not restricted to imino acids. Nevertheless, stable triple-helical structure requires about 20% of imino acids to be present in the X and Y positions. There are more than 400 possible Gly-X-Y combinations, but only about 25 possible triplets occur with more than 1% frequency. Some residues appear more frequently in the X (e.g., Phe, Leu, Glu) or Y positions (e.g., Arg, Lys), and some are rarely or never found in collagen (e.g. ,Trp, Tyr). Among imino acids, Pro is usually found in the X position while Hyp, which results from posttranslational modifications of Pro, is usually in the Y position. Pro-Hyp-Gly is the most stabilizing triplet and frequently occurs in the collagen sequence. Any substitution of Pro or Hyp drops the stability of the triple helix, yet many other residues are found therein. The reason for their occurrence may therefore be sought in their ability to induce the self-assembly of other levels of the collagen hierarchy, such as fibril and fiber formation. For example, the electrostatic charges or hydrophobicity of amino acids can help to mediate the staggered association of collagen triple helices [23]. Non Gly-Pro-Hyp triplets also bear biological functions, such as providing recognizable regions for collagenases, providing binding sites for integrins, proteoglycans, and fibronectin [23], and allowing covalent cross-linking through lysine side chains.

Substitution of Gly in the Gly-X-Y triplet with any other residue provokes a dramatic decrease of collagen stability. It was found that the genetic disease *osteogenesis imperfecta* results from such substitution. In this disease the defective mineralization of type I collagen either in $\alpha1$ or $\alpha2$ chain leads to the formation of brittle, fragile bones. The severity of *osteogenesis imperfecta* depends on the actual mutations: for some substituting amino acids (Ser, Cys) it can be nonlethal, while for others (Asp and Val) it is lethal in all cases, unless the substituting amino acids are located near the N-terminus. Studies by Long et al. [24] indicate that a single Gly to Ala substitution in the model peptide (Pro-Hyp-Gly)$_{10}$ still allows the formation of a triple helix but lowers its thermal stability by 30 °C. X-ray data obtained by Bella et al. [8] for this peptide show that the peptide is slightly distorted at the site of substitution and water bridges replace the usual interchain hydrogen bonds. Persikov et al. [25] determined the order of relative thermal stability of different peptide sequences X-Pro-Hyp. In the case of X = Gly the melting temperature was 44.5 °C, and it decreased dramatically for any other residue. In some cases the authors were not able to detect any formation of a triple-helical structure and the corresponding melting temperatures were estimated by linear extrapolation of the data obtained in trimethylamine N-oxide. Based on their data the following order of relative thermal stability can be derived: Gly > Ser > Ala > Cys > Arg > Val > Glu > Asp, with Asp possessing an estimated melting temperature of −17 °C.

A propensity scale for different amino acids in the X and Y positions were studied using host peptides with a stabilizing peptide sequence [26]. Variable guest Gly-X-Y units were inserted in the middle of a (Gly-Pro-Hyp)$_3$-Gly-X-Hyp-(Gly-Pro-Hyp)$_4$ or a (Gly-Pro-Hyp)$_3$-Gly-Pro-Y-(Gly-Pro-Hyp)$_4$ sequence and their thermal stabilities were studied by CD spectroscopy. Without doubt Pro in the X position gives the most stable triple-helical structure, with a melting temperature (T_m) of 47.3 °C. The presence of any other residue in the X position drops the stability and the lowest T_m = 31.9 °C is observed in the case of X = Trp. Next to Pro the most stabilizing set of amino acids includes all charged amino acids (Arg, Glu, Asp and Lys) as well as Gln and Ala. The stabilization effect may arise from side-chain interactions with accessible backbone carbonyl groups. Nonpolar residues Met, Val, Ile, and Leu together with Ser and Asn form less stable triple helices, and the presence of Thr, His, and Cys in the X position lowers the stability even more. Gly and the aromatic residues are the most destabilizing residues, which is most likely due to unfavorable conformations and steric interactions.

The variations of amino acid residues in the Y position show that the most stabilizing residues are Hyp (T_m = 47.3 °C) and Arg (T_m = 47.2 °C), while aromatic residues are again the most destabilizing (Trp, T_m = 26.1 °C). The remainder of the amino acids falls in a continuous range of transition temperatures with no special preferences for charged or nonpolar residues.

Comparison of the effects of different amino acids in the X and Y positions reveals that the greatest stabilization occurs when X = Pro and Y = Hyp, while the presence of Gly or aromatic residues significantly destabilizes the triple helix. The lack of correlation between charged and nonpolar residues suggests that the interchain interactions and solvent exposure are different for the X and Y sites of a triple-helical structure. The relationship between frequency of amino acid occurrence in fibrillar collagen sequences and transition temperature of corresponding host–guest peptides allows one to assume that this frequency is related to the triple helical stability. The presence of destabilizing residues may be necessary for fibril formation and binding of various extracellular matrix components, and at the same time it provides the appropriate denaturation temperature of the triple helical structure close to the upper limit temperature of the body.

4.7
Stabilization via Hydroxyproline: The Pyrrolidine Ring Pucker

As was noted earlier, the posttranslational hydroxylation of Pro occurs in the Y position, and studies of model peptides reveal that the presence of Hyp in the X position actually destabilizes triple helical structure [27]. To understand this observation one must understand the mechanism by which Hyp in the Y position leads to stabilization. Recent findings suggest two alternative explanations. The first explanation states that the stabilization arises from the hydrogen bonds formed between the hydroxyl group of Hyp and carbonyl oxygen of Gly of an adjacent chain [8, 28]. The second explanation suggests that the electronegativity of hydroxyl group plays major role by favoring the appropriate pyrrolidine ring conformation of Hyp, while hydrogen bonds do not contribute as much.

The pyrrolidine ring of Pro and Hyp can exist in two distinct conformations or puckerings (Fig. 4.5). These two different puckered forms are usually referred to as C^γ-exo (up) and C^γ-endo (down). The differences between the two arise from the displacement of the C^β and the C^γ atoms from the mean plane of the ring. A down-puckering of prolines occurs in the X position while those in the Y position adopt up-puckering. Such alternation of puckering X(endo)-Y(exo) is observed in X-ray crystal structures of collagen model peptides [29]. The puckering pattern strongly correlates with the backbone ϕ dihedral angles and together they suggest that the proper ring puckering with Pro down in the X position and Pro or Hyp up in the Y position is very important for the triple helix formation.

The strongest evidence for this mechanism was obtained by Raines and coworkers [30]. They synthesized a polypeptide containing 4(R)-fluoroproline (fPro^R), (Pro-fPro^R-Gly)$_{10}$, and found that the substitution of hydroxyl group with fluorine greatly increases the melting temperature to about 90 °C, which is the highest among known collagen-like polypeptides. The fact that fluorine is more electronegative *and* cannot act as a hydrogen bond donor supports the idea that stereoelectronic factors are the primary reason for helix stabilization, as opposed to hydrogen bonding. Furthermore, an analogous study by the same group on a sequence containing 4(S)-fluoro-

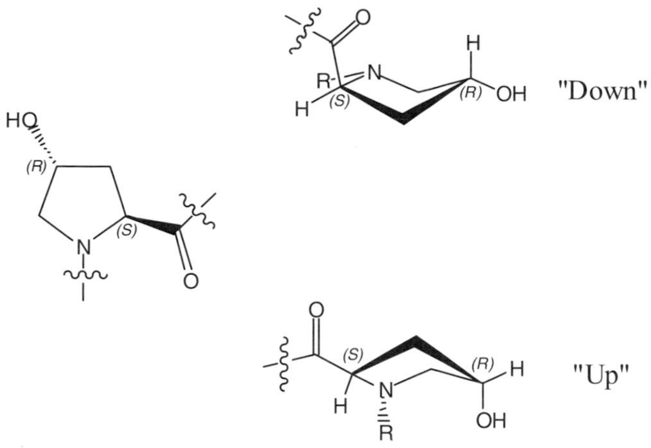

Figure 4.5. "Up" and "down" conformation of hydroxyproline.

proline (fPro^S), (Pro-fPro^S-Gly)$_7$, showed that the transition, if any, occurs at a temperature below 2 °C.

Barone and coworkers [31] performed quantum mechanical and molecular mechanical computations on model polypeptides, (Pro-Pro-Gly)$_n$, in order to investigate the importance of vicinal and long-range interresidue effects on triple helical stability. They found that the inclusion of interresidue interactions does not stabilize the X-down/Y-up conformation. Moreover, the X-down/Y-down isomer was always more stable than the X-down/Y-up isomer, which is found in triple helix. Thus, the preference for X-down/Y-up in collagen cannot be an intrinsic property of the Pro-Pro-Gly sequence. This adds weight to the stabilizing role of Hyp or fPro^R in the Y position. It was shown by the same authors [32] that the presence of electronegative 4(R) substituent favors the up-puckering and moreover the relative stability of the up isomer increases with an increase of substituent electronegativity. Proper preorganization of the peptide backbone, which has Hyp in the Y position, decreases the entropic penalty for triple helix formation. This finding agrees with the experimentally observed enhancement in triple helix stability in the case of fPro^R containing sequences.

The effects of fPro^R and fPro^S in the X position have been studied by several groups [33]. As one would expect these two residues preferentially adopt two different puckering with fPro^R-up and fPro^S-down and their effect on the triple-helix stability should be opposite. Indeed, it was shown that (fPro^S-Pro-Gly)$_7$ has a transition temperature around 33°C, whereas (fPro^R-Pro-Gly)$_7$ does not form a stable triple-helix. With the increase of the peptide length the transition temperature also increases and (fPro^S-Pro-Gly)$_{10}$ undergoes such a transition at 58°C while the triple-helix formation for (fPro^R-Pro-Gly)$_{10}$ is still undetectable. Overall it seems more likely that the stabilization impact of Hyp originates in the stereoelectronic preference of hydroxylated pyrrolidine ring to adopt the proper up ring puckering.

4.8
Triple Helix Stabilization Through Forced Aggregation

Several attempts have been made to force or enhance triple helix formation using preorganized molecular scaffolds. In this approach the stabilization effect is achieved by covalently fixing three polypeptide chains together in close proximity, thus reducing the entropic costs for nucleation and increasing local peptide concentration.

For example 1,2,3-propanetricarboxylic acid was prepared by Germann and Heidmann [34] to bridge three polypeptide chains, which were found to form a triple helix. A more conformationally constrained organic molecule was used as a template by Goodman et al. [35]. They synthesized a derivative of the Kemp triacid (KTA) [36] (Fig. 4.6). This acid contained three axial carboxylic acid functionalities parallel to each other, which were further modified to include Gly spacers in order to make the resulting structure more flexible and reduce steric hindrance. The authors showed that the KTA-based template stabilized the triple helical structure. It was found that the KTA-[Gly-(Gly-Pro-Hyp)$_5$-NH$_2$]$_3$ transition temperature was 70 °C, which was 52 °C higher than that of a simple Ac-(Gly-Pro-Hyp)$_5$-NH$_2$. Moreover, this scaffold was found to induce triple-helical folding of the very short peptide, containing only three Gly-Pro-Hyp repeats. Another chiral triacid scaffold, based upon a cone-shaped cyclotriveratrylene (CTV), was used by Liskamp et al. [37] to enforce triple helical folding of both model and native collagen sequences. As was stated in the preceding paragraphs, the substitution of imino acids in a Gly-Pro-Hyp polypeptide by non-imino acids greatly destabilizes triple helix. Therefore the authors were interested in studying the influence of a scaffold on the stability of native collagen sequences containing a large percentage of non-imino residues. The data obtained showed that (+)-CTV was more potent than (–)-CTV in enhancing the correct folding. In the case of the native collagen sequence, no folding was measurable for (–)-CTV, while the (+) stereoisomer was shown to be suitable for inducing triple helix formation.

Recently, tris(2-aminoethyl)amine (TREN) coupled with the succinic acid spacers (suc) was incorporated to assemble triple helices containing Gly-Nleu-Pro sequences (Nleu: *N*-isobutylglycine) [38]. CD and NMR studies along with thermal denaturation and molecular modeling revealed that the TREN-[suc-(Gly-Nleu-Pro)$_n$-NH$_2$]$_3$ where $n=5$ and 6 formed a stable triple helix in water. Comparison between the TREN and KTA scaffolds indicates that the TREN induces triple helix formation more effectively than the KTA scaffold. Most likely the observed enhance in stability is due to the greater flexibility of the TREN molecule.

Another approach to stabilizing the triple helix used by Tirrell et al. [39] is to attach long-chain fatty acids to the N or C terminal end of a collagen model peptide, creating a peptide amphiphile. The fatty acid tails have a strong propensity to aggregate in water, which brings together the attached peptide chains, resulting in an effective increase in local peptide concentration and thus helps to stabilize the triple helix. Their peptide amphiphiles consisted of two parts: a "head group" composed of a peptide with a collagen-like sequence, and a "tail group" made of a saturated

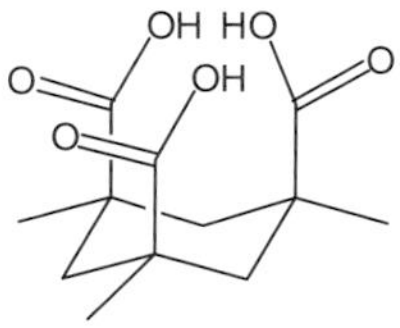

KTA 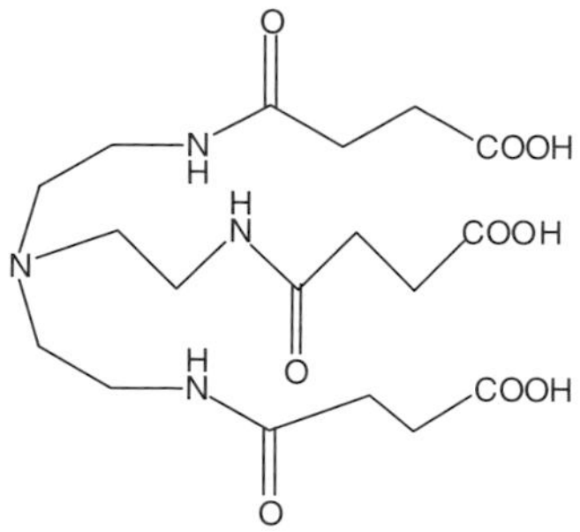   TREN-(suc-OH)$_3$

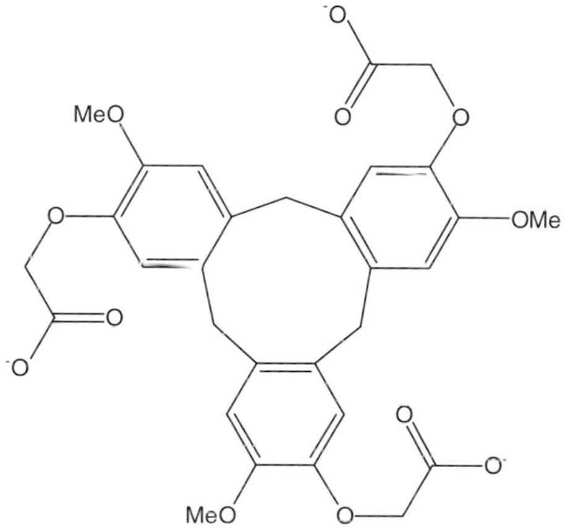

(+/−)CTV[CH$_2$CO$_2^-$Bu$_4$N$^+$]$_3$

Figure 4.6. Chemical structures of KTA, TREN, and CTV, which have been used to tether peptide strands together and enhance the triple helix character.

carbon chain. The "head group" has a propensity to adopt triple-helical conformation, thus providing a structural element that can be recognized by cells and other biomolecules. The hydrophobic "tail group" provides a proper register of peptide strands and can induce triple-helical folding due to the self-association in polar solvents. Furthermore, tail hydrophobicity can be beneficial for interaction with other biological surfaces (e.g., cell membrane).

According to their monolayer observations, CD and NMR data peptide amphiphiles were indeed able to fold into highly ordered polyproline-II-like triple-helical structures. CD melting curves confirmed that the peptide amphiphiles experience

4.10
"Sticky Ends" and Supramolecular Polymerization

Our strategy for the preparation of collagen mimics falls into the category of ECM mimics which attempt to copy the structure of the ECM. Our approach is to design short peptides with a general sequence of $(XYG)_7$ which can self-assemble into triple helices and, depending on the selected sequence, carry out further assembly into much longer triple-helical strands. If successful, this will address one of the significant problems in chemical collagen synthesis, which is that natural collagen peptides are typically on the order of 1000 amino acids in length while conventional solid-phase peptide synthesis is limited to approximately 100 amino acids.

In general a peptide with sequence $(XYG)_n$ will assemble into a triple helix, assuming that n is large enough and X and Y are selected to have a high proportion of proline and hydroxyproline respectively. This provides a desirable conformation and also satisfies the hydrogen bonding between peptides. Because of the requirement that a glycine residue must be in the interior of the triple helix at all times and that this glycine residue is contributed from alternating peptide strands, a triple helix composed of three identical peptides is naturally offset by one amino acid (Scheme 4.1).

```
XYGXYGXYGXYGXYGXYGXYG
  XYGXYGXYGXYGXYGXYGXYG
    XYGXYGXYGXYGXYGXYGXYG
```

Scheme 4.1. A 21-amino-acid triple helix. The requirement for glycine to pack in the interior of the triple helix causes a single amino acid offset between peptides. X and Y refer to any amino acid.

Table 4.1. Five peptides examined. N-termini and C-termini were prepared as free amine and carboxylic acid respectively. O, hydroxyproline; P, proline; G, glycine; E, glutamic acid; K, lysine.

Peptide	Sequence	Net charge
1	PPGPPGPPGPPGPPGPPGPPG	0
2	PPGPOGPPGPOGPPGPOGPPG	0
3	POGPPG**E**PGPOGP**K**GPOGPPG	0
4	POGPPG**E**PGPOGP**E**GPOGPPG	−2
5	POGPPG**K**PGPOGP**K**GPOGPPG	+2

This means that any given section perpendicular to the long axis of the triple helix must have one glycine, one amino acid from position X, and one amino acid from position Y. It has been demonstrated that proper selection of charged amino acids in a-helical-coiled coils can force the two helices out of register with one another and create "sticky ends" [51]. These sticky ends can then be used for further assembly of peptides which results in a non-covalent or supramolecular polymerization. The same idea may be applied with respect to a collagen triple helix. Proper selection of charged amino acids such as lysine and glutamic acid may lead to preferred pairings that create a larger amino acid offset than a single amino acid. To test this idea we have prepared five peptides with an XYG repeat as shown in Tab. 4.1.

Peptides 1 and 2 were designed as control peptides to examine the baseline level of triple helix formation when no charged residues were employed. Peptide 1 contains no hydroxyproline while peptide 2 contains three hydroxyproline residues in the Y position. Peptides 3–5 all contain three hydroxyproline residues in Y positions and also have two charged residues, one in an X position and one in a Y position. Peptide 3 has both glutamic acid and lysine while peptide 4 has only glutamic acid and peptide 5 has only lysine. If peptide 4 is dissolved at a pH of 7, the charges on its glutamate side chains would be expected to discourage assembly through electrostatic repulsion, as shown in Scheme 4.2.

POGPPG**E**PGPOGP**E**GPOGPPG
  POGPPG**E**PGPOGP**E**GPOGPPG
    POGPPG**E**PGPOGP**E**GPOGPPG

POGPPG**E**PGPOGP**E**GPOGPPG
      POGPPG**E**PGPOGP**E**GPOGPPG
        POGPPG**E**PGPOGP**E**GPOGPPG

Scheme 4.2. Alternate assembly modes for peptide 4 in which multiple glutamate residues are unfavorably aligned with one another while glycine alignment is enforced.

However, if peptide 5 is added to the solution in a 1:1 ratio it may be expected that this electrostatic repulsion could be converted to a favorable attraction between oppositely charged glutamate and lysine, as shown in Scheme 4.3. This leads to large overhanging, "sticky" ends which can lead to further assembly of the peptides. This self-assembly can continue indefinitely and is the basis for the preparation of very long triple-helical strands.

POGPPG**E**PGPOGP**E**GPOGPPG
      POGPPG**K**PGPOGP**K**GPOGPPG
        POGPPG**E**PGPOGP**E**GPOGPPG

POGPPG**E**PGPOGP**E**GPOGPPG**POGPPGKPGPOGPKGPOGPPG**
      POGPPG**K**PGPOGP**K**GPOGPPG**POGPPGEPGPOGPEGPOGPPG**
        POGPPG**E**PGPOGP**E**GPOGPPG**POGPPGKPGPOGPKGPOGPPG**

Scheme 4.3. Proposed mechanism of assembly of peptides **4** and **5**. Assembly of staggered triple helices is promoted by charge pairing between the oppositely charged glutamic acid and lysine residues. The first three peptides assemble, forming "sticky ends". Additional peptides can then assemble, propagating the sticky ends.

fibril-like super structures. However, without more detailed analysis this theory cannot yet be confirmed.

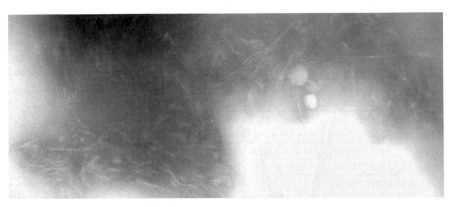

Figure 4.8. Negatively stained TEM image of a mixture of peptides 4 and 5. This image shows what appears to be an assortment of nanofibers approximately 100 nm in length and 10 nm in diameter.

4.11
Conclusion

The extracellular matrix is a rich and complicated environment. Mimicking its nanostructure and function remains a significant challenge despite a great deal of knowledge regarding its chief constituent, collagen. Much of the challenge in recreating collagen will be in unraveling how the higher levels of association – the fibril and fiber – are controlled. This understanding is likely to be obtained not from small peptide studies but with larger systems. We hope that our work will help to provide access to the mechanism of fibril and fiber assembly and ultimately to synthetic materials that can be used in sophisticated tissue engineering approaches.

Abbreviations

The usual single and three letter codes for the amino acid abbreviations were used.
CD – circular dichroism spectroscopy
CTV – cyclotriveratrylene
DNA – deoxyribonucleic acid
DQF-COSY – double quantum-filtered correlation spectroscopy
ECM – extracellular matrix
FMOC – 9-fluorenylmethoxycarbonyl
fPro^R – 4(R)-fluoroproline
fPro^S – 4(S)-fluoroproline
HA – hydroxyapatite
KTA – Kemp triacid

Nleu – isobutylglycine

NMR – nuclear magnetic resonance

NOE – nuclear Overhauser effect

PC12 – pheochromocytoma cells

PGA – poly-glycolic acid

PLA – poly-lactic acid

suc – succinic acid

TE – tissue engineering

TEM – transmission electron microscopy

TOCSY – total correlation spectroscopy

TREN – tris(2-aminoethyl)amine

References

1 K. Y. Lee, D. J. Mooney, Hydrogels for tissue engineering, *Chem. Rev.* **2001**, *101*, 1869.

2 F. A. Auger, M. Rouabhia, F. Goulet, F. Berthod, V. Moulin, L. Germain, Tissue-engineering human skin substitutes developed from collagen-populated hydrated gels: clinical and fundamental applications, *Med. Biol. Engin. Comput.* **1998**, *36*, 801.

3 P. M. Kaufmann, S. Heimrath, B. S. Kim, D. J. Mooney, Highly porous polymer matrices as a three-dimensional culture system for hepatocytes, *Cell Transplant.* **1997**, *6*, 463; D. Seliktar, R. A. Black, R. P. Vito, R. M. Nerem, Dynamic mechanical conditioning of collagen-gel blood vessel constructs induces remodeling in vitro, *Ann. Biomed. Eng.* **2000**, *28*, 351.

4 K. E. Kadler, D. F. Holmes, J. A. Trotter, J. A. Chapman, Collagen fibril formation, *Biochem. J.* **1996**, *316*, 1.

5 A. Rich, F. H. C. Crick, The molecular structure of collagen, *J. Mol. Biol.* **1961**, *3*, 483.

6 G. N. Ramachandran, *Treatise on Collagen*, Academic, New York, **1964**.

7 P. M. Cowan, S. McGavin, The structure of poly-L-proline, *Nature* **1955**, *176*, 501.

8 J. Bella, M. Eaton, B. Brodsky, H. M. Berman, Crystal and molecular structure of a collagen-like peptide at 1.9 A resolution, *Science* **1994**, *266*, 75.

9 N. Sreerama, R. W. Woody, in *Circular dichroism: principles and applications*, 2nd ed. (Eds.: N. Berova, K. Nakanishi, R. W. Woody), John Wiley & Sons, Inc., New York, **2000**, pp. 601.

10 M. G. Venugopal, J. A. M. Ramshaw, E. Braswell, D. Zhu, B. Brodsky, Electrostatic interactions in collagen-like triple-helical peptides, *Biochemistry* **1994**, *33*, 7948; F. R. r. Brown, J. P. Carver, E. R. Blout, Low temperature circular dichroism of poly (glycyl-L-prolyl-L-alanine), *J. Mol. Biol.* **1969**, *39*, 307.

11 D. D. Jenness, C. Sprecher, W. C. J. Johnson, Circular dichroism of collagen, gelatin, and poly(proline) II in the vacuum ultraviolet, *Biopolymers* **1976**, *15*, 513.

12 A. A. Adzhubei, M. J. E. Sternberg, Left-handed polyproline II helices commonly occur in globular proteins, *J. Mol. Biol.* **1993**, *229*, 472; D. J. Russell, G. Pearce, C. A. Ryan, J. D. Satterlee, Proton NMR assignments of systemin, *J. Protein Chem.* **1992**, *11*, 265.

13 B. Brodsky, M. Li, C. G. Long, J. Apigo, J. Baum, NMR and CD studies of triple-helical peptides, *Biopolymers* **1992**, *32*, 447.

14 M. Li, P. Fan, B. Brodsky, J. Baum, Two-dimensional NMR assignments and conformation of (Pro-Hyp-Gly)10 and a designed collagen triple-helical peptide, *Biochemistry* **1993**, *32*, 7377.

15 J. Baum, B. Brodsky, Real-time NMR investigations of triple-helix folding and collagen folding diseases, *Folding Des.* **1997**, *2*, R53; K. H. Mayo, NMR and x-ray studies of collagen model peptides, *Biopolymers (Pept. Sci.)* **1996**, *40*, 359; C. G. Long, M. H. Li, J. Baum, B. Brodsky, Nuclear magnetic resonance and circular dichroism studies of a triple-helical peptide with a glycine substitution, *J. Mol. Biol.* **1992**, *225*, 1.

16 J. F. Brandts, H. R. Halvorson, M. Brennan, Consideration of the possibility that the slow step in protein denaturation reactions is due to cis-trans isomerism of proline residues, *Biochemistry* **1975**, *14*, 4953; T. E. Creighton, Possible implications of many proline residues for the kinetics of protein unfolding and refolding, *J. Mol. Biol.* **1978**, *125*, 401.

17 J. Engel, D. J. Prockop, The zipper-like folding of collagen triple helices and the effects of mutations that disrupt the zipper, *Annu. Rev. Biophys. Biophys. Chem.* **1991**, *20*, 137.

18 P. Bruckner, E. Eikenberry, Procollagen is more stable in cellulo than in vitro, *Eur. J. Biochem.* **1984**, *140*, 397; H. P. Bachinger, P. Bruckner, R. Timpl, D. J. Prockop, J. Engel, Folding mechanism of the triple helix in type-III collagen and type-III pN-collagen. Role of disulfide bridges and peptide bond isomerization, *Eur. J. Biochem.* **1980**, *106*, 619.

19 Y. Xu, M. Bhate, B. Brodsky, Characterization of the nucleation step and folding of a collagen triple-helix peptide, *Biochemistry* **2002**, *41*, 8143.

20 W. Yang, M. L. Battineni, B. Brodsky, Amino acid sequence environment modulates the disruption by osteogenesis imperfecta glycine substitutions in collagen-like peptides, *Biochemistry* **1997**, *36*, 6930; X. Liu, S. Kim, Q.-H. Dai, B. Brodsky, J. Baum, Nuclear magnetic resonance shows asymmetric loss of triple helix in peptides modeling a collagen mutation in brittle bone disease, *Biochemistry* **1998**, *37*, 15528.

21 S. Boudko, S. Frank, R. A. Kammerer, J. Stetefeld, T. Schulthess, R. Landwehr, A. Lustig, H. P. Bachinger, J. Engel, Nucleation and propagation of the collagen triple helix in single-chain and trimerized peptides: translation from third to first order kinetics, *J. Mol. Biol.* **2002**, *317*, 459.

22 E. Durr, H. R. Bosshard, Folding of a three-stranded coiled coil, *Protein Sci.* **2000**, *9*, 1410; D. Porschke, M. Eigen, Co-operative non-enzymic base recognition. 3. Kinetics of the helix-coil transition of the oligoribouridylic–oligoriboadenylic acid system and of aligoriboadenylic acid alone at acidic pH, *J. Mol. Biol.* **1971**, *62*, 361.

23 D. J. S. Hulmes, A. Miller, D. A. D. Parry, K. A. Piez, Analysis of the primary structure of collagen for the origins of molecular packing, *J. Mol. Biol.* **1973**, *79*, 137.

24 C. G. Long, E. Braswell, D. Zhu, J. Apigo, J. Baum, B. Brodsky, Characterization of collagen-like peptides containing interruptions in the repeating Gly-X-Y sequence, *Biochemistry* **1993**, *32*, 11688.

25 A. V. Persikov, J. A. M. Ramshaw, B. Brodsky, Collagen model peptides: sequence dependence of triple-helix stability, *Biopolymers* **2000**, *55*, 436.

26 A. V. Persikov, J. A. M. Ramshaw, A. Kirkpatrick, B. Brodsky, Amino acid propensities for the collagen triple-helix, *Biochemistry* **2000**, *39*, 14960.

27 J. Engel, D. J. Prockop, Does bound water contribute to the stability of collagen? *Matrix Biol.* **1998**, *17*, 679.

28 J. Bella, B. Brodsky, H. M. Berman, Hydration structure of a collagen peptide, *Structure* **1995**, *3*, 893.

29 R. Berisio, L. Vitagliano, L. Mazzarella, A. Zagari, Crystal structure of a collagen-like polypeptide with repeating sequence Pro-Hyp-Gly at 1.4 Å resolution: implications for collagen hydration, *Biopolymers* **2000**, *56*, 8.

30 S. K. Holmgren, K. M. Taylor, L. E. Bretscher, R. T. Raines, Code for collagen's stability deciphered, *Nature* **1998**, *392*, 666.

31 R. Importa, F. Mele, O. Crescenzi, C. Benzi, V. Barone, Understanding the role of stereoelectronic effects in determining collagen stability. 2. A quantum mechanical/molecular mechanical study of (proline-proline-glycine)n polypeptides, *J. Am. Chem. Soc.* **2002**, *124*, 7857.

32 R. Importa, C. Benzi, V. Barone, Understanding the role of stereoelectronic effects in determining collagen stability. 1. A quantum mechanical study of proline, hydroxyproline, and fluoroproline dipeptide analogues in aqueous solution, *J. Am. Chem. Soc.* **2001**, *123*, 12568.

33 M. Doi, Y. Nishi, S. Uchiyama, Y. Nishiuchi, T. Nakazawa, T. Ohkubo, Y. Kobayashi, Characterization of collagen model peptides containing 4-fluoroproline; (4(S)-fluoro-proline-Pro-Gly)10 forms a triple helix, but (4(R)-fluoroproline-Pro-Gly)10 does not, *J. Am. Chem. Soc.* **2003**, *125*, 9922; J. A. Hodges, R. T. Raines, Stereoelectronic effects on collagen stability: the dichotomy of 4-fluoroproline diastereomers, *J. Am. Chem. Soc.* **2003**, *125*, 9262.

34 H. P. Germann, E. Heidemann, A synthetic model of collagen: an experimental investigation of the triple-helix stability, *Biopolymers* **1988**, *27*, 157.

35 Y. Feng, G. Melacini, J. P. Taulane, M. Goodman, Acetyl-terminated and template-assembled collagen-based polypeptides composed of Gly-Pro-Hyp sequences. 2. Synthesis and conformational analysis by circular dichroism, ultraviolet absorbance, and optical rotation, *J. Am. Chem. Soc.* **1996**, *118*, 10351.

36 D. S. Kemp, K. S. Petrakis, Synthesis and conformational analysis of *cis,cis*-1,3,5-trimethyl-cyclohexane-1,3,5-tricarboxylic acid, *J. Org. Chem.* **1981**, *46*, 5140.

37 E. T. Rump, T. S. Dirk, D. T. S. Rijkers, H. W. Hilbers, P. G. De Groot, R. M. J. Liskamp, Cyclotriveratrylene (CTV) as a new chiral triacid scaffold capable of inducing triple helix formation of collagen peptides containing either a native sequence or Pro-Hyp-Gly repeats, *Chem. Eur. J.* **2002**, *8*, 4613.

38 J. Kwak, A. De Capua, E. Locardi, M. Goodman, TREN (tris(2-aminoethyl) amine): an effective scaffold for the assembly of triple-helical collagen mimetic structures, *J. Am. Chem. Soc.* **2002**, *124*, 14085.

39 Y.-C. Yu, P. Berndt, M. Tirrell, G. B. Fields, Self-assembling amphiphiles for construction of protein molecular architecture, *J. Am. Chem. Soc.* **1996**, *118*, 12515.

40 T. Gore, Y. Dori, Y. Talmon, M. Tirrell, H. Bianco-Peled, Self-assembly of model collagen peptide amphiphiles, *Langmuir* **2001**, *17*, 5352.

41 H. Shin, S. Jo, A. G. Mikos, Biomimetic materials for tissue engineering, *Biomaterials* **2003**, *24*, 4353.

42 M. J. Humphries, S. K. Akiyama, A. Komoriya, K. Olden, K. M. Yamada, Identification of an alternatively spliced site in human plasma fibronectin that mediates cell type-specific adhesion, *J. Cell Biol.* **1986**, *103*, 2637.

43 W. Dai, J. Belt, W. M. Saltzman, Cell-binding peptides conjugated to poly(ethylene glycol) promote neural cell aggregation, *Bio/Technology* **1994**, *12*, 797.

44 J. A. Hubbell, S. P. Massia, N. P. Desai, P. D. Drumheller, Endothelial cell-selective materials for tissue engineering in the vascular graft via a new receptor, *Biotechnology* **1991**, *9*, 568.

45 J. P. Ranieri, R. Bellamkonda, E. J. Bekos, J. T. G. Vargo, J. A. G. Aebischer, Neuronal cell attachment to fluorinated ethylene propylene films with covalently immobilized laminin oligopeptides YIGSR and IKVAV. II. *J. Biomed. Mater. Res.* **1995**, *29*, 779.

46 J. L. West, J. A. Hubbell, Polymeric biomaterials with degradation sites for proteases involved in cell migration, *Macromolecules* **1999**, *32*, 241.

47 J. D. Hartgerink, E. Beniash, S. I. Stupp, Self-assembly and mineralization of peptide-amphiphiles nanofibers, *Science* **2001**, *294*, 1684; G. A. Silva, C. Czeisler, K. L. Niece, E. Beniash, D. A. Harrington, J. A. Kessler, S. I. Stupp, Selective differentiation of neural progenitor cells by high-epitope density nanofibers, *Science* **2004**, *303(5662)*, 1352.

48 K. L. Niece, J. D. Hartgerink, J. J. J. M. Donners, S. I. Stupp, Self-assembly combining two bioactive peptide-amphiphile molecules into nanofibers by electrostatic attraction, *J. Am. Chem. Soc.* **2003**, *125*, 7146.

49 M. D. Pierschbacher, E. Ruoslahti, Cell attachment activity of fibronectin can be duplicated by small synthetic fragments of the molecule, *Nature* **1984**, *309*, 30.

50 S. Weiner, H. D. Wagner, The material bone: structure-mechanical function relations, *Annu. Rev. Mater. Sci.* **1998**, *28*, 271; W. Traub, T. Arad, S. Weiner, Three-dimensional ordered distribution of crystals in turkey tendon collagen fibers, *Proc. Natl. Acad. Sci. U.S.A.* **1989**, *86*, 9822.

51 M. J. Pandya, G. M. Spooner, M. Sunde, J. R. Thorpe, A. Rodger, D. N. Woolfson, Sticky-end assembly of a designed peptide fiber provides insight into protein fibrillogenesis, *Biochemistry* **2000**, *39*, 8728.

5

Molecular Biomimetics: Building Materials Nature's Way, One Molecule at a Time[1]

Candan Tamerler and Mehmet Sarikaya

Physical and chemical functions of organisms are carried out by a very large number (billions) of proteins, of differing variety ($\sim 10^5$ in humans), through predictable and self-sustaining interactions, developed through evolution. Using biology as a guide, in the molecular biomimetics approach we select, design, genetically tailor, synthesize, and utilize short polypeptides as molecular erectors in self-assembly, ordered organization, and biofabrication of nanoinorganic materials and molecularly hybrid systems in nanotechnology (molecular electronics, magnetics, and photonics) and nanobiotechnology (biosensors, bioassays, and biomaterials). These polypeptides are usually 7–15 amino acids long, and are obtained via combinatorial biology using cell surface or phage display libraries. Once selected, the inorganic binding polypeptides can be further engineered using genetic engineering to tailor their properties for specific material surfaces, morphologies, and crystal chemistries, and for designed applications. The potential of engineered polypeptides in nanotechnology is enormous due to molecular recognition, self- and co-assembly, and manipulation via DNA technologies.

5.1
Introduction

Future functional materials systems, developed either for nanobiotechnology or nanotechnology, could include protein(s) in the assembly, formation, and, perhaps, in the final structure as an integral component leading to specific and controllable functions similar to those seen in biological soft and hard tissues (Fig. 5.1). In the new field of *molecular biomimetics* – a true marriage of traditional physical and biological fields – hybrid materials could potentially be assembled from the molecular level using the recognition properties of proteins that specifically bind to inorganics [1, 2].

Molecular biomimetics offer three simultaneous solutions to the problem of the control and fabrication of large-scale nanostructures and ordered assemblies of materials in two and three dimensions (Fig. 5.2) [3]. The first is that inorganic-binding peptides and proteins are selected and designed at the molecular level and

---

1) This article is adapted from a recent review paper prepared for *Annual Review of Materials Research* (Ref. [3]).

Nanofabrication Towards Biomedical Applications. C. S. S. R. Kumar, J. Hormes, C. Leuschner (Eds.)
Copyright © 2005 WILEY-VCH Verlag GmbH & Co. KGaA, Weinheim
ISBN 3-527-31115-7

Only a few polypeptides have been identified that specifically bind to inorganics. These are mostly biomineralizing proteins extracted from hard tissues followed by isolation, purification, and cloning. Although this approach is difficult, time-consuming, and has major limitations, several proteins isolated in this fashion have been used as nucleators, growth modifiers, or enzymes in the synthesis of certain inorganics [4–9]. Some examples include amelogenins in mammalian enamel synthesis [5], silicatein, effective in sponge spicular formation [9], and calcite- and aragonite-forming polypeptides in mollusk shells [4, 6, 7]. The preferred approach to

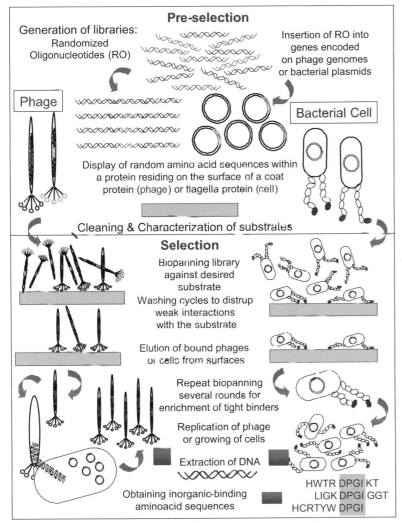

Figure 5.3. Principles of phage display (left) and cell surface display (right) protocols adapted for selecting polypeptide sequences with binding affinity to a given inorganic substrate [3].

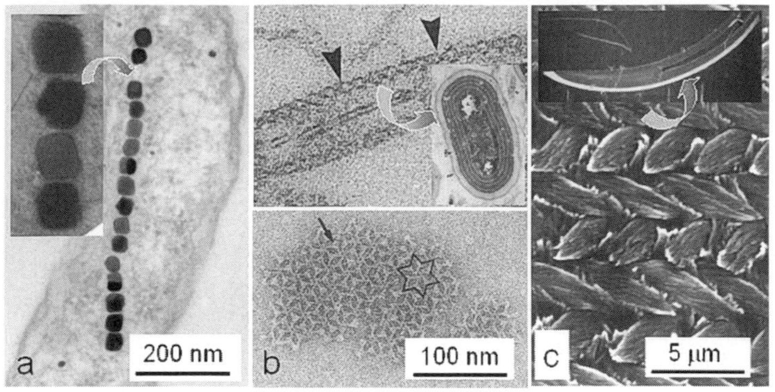

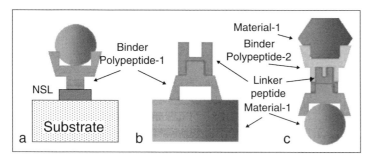

200 nm   a   b   100 nm   c   5 μm

Figure 5.1. Examples of biologically synthesized hybrid materials with a variety of physical properties. (a) Magnetic nanoparticles formed by a magnetotactic bacterium (*Aquaspirillum magnetotacticum*) are single-crystalline, single-domained magnetite (Fe$_3$O$_4$) particles (inset: higher magnification image revealing cubo-octahedral particle shape) (M. Sarikaya, unpublished). (b) Nanostructurally ordered thin-film calcite on the outer layer of an S-layer bacterium, *Synechococcus* strain GL24, that serves as a protective coating. (c) Mouse enamel is a hard, wear-resistant material with a highly ordered micro/nano architecture consisting of hydroxyapatite crystallites that assemble into a woven rod structure (SEM image). Each rod is composed of thousands of hydroxyapatite particles (inset: cross-sectional image of a mouse incisor tooth; white region is enamel, backed by grayish dentine) [3].

through genetics. This allows control at the lowest dimensional scale possible. The second is that such proteins can be used as *linkers* or *molecular erector sets* to join synthetic entities, including nanoparticles, functional polymers, or other nanostructures on molecular templates. The third solution is that the biological molecules self- and co-assemble into ordered nanostructures. This ensures a robust assembly process for the construction of complex nanostructures, and possibly hierarchical structures, similar to those found in nature (self-assembly).

NSL   Binder Polypeptide-1   Material-1 Binder Polypeptide-2   Linker peptide   Material-1   Substrate   a   b   c

Figure 5.2. Potential utility of inorganic-binding proteins as: (a) linkers for nanoparticle immobilization, (b) functional molecules that assemble on specific substrates, and (c) heterofunctional linkers involving two (or more) binding proteins adjoining several nanoinorganic units. NSL, nonspecific linker [1, 3].

obtaining inorganic-binding polypeptides is to use combinatorial biological techniques [1, 10–13]. In this approach, a large, random library of peptides with the same number of amino acids, but varying compositions, are screened to identify specific sequences that strongly bind to an inorganic material of practical use. In molecular biomimetics, the ultimate goal is to generate a "molecular erector set" in which different proteins, each engineered to bind to a specific surface, size, or morphology of an inorganic compound, promote the assembly of intricate, hybrid structures composed of inorganics, proteins, and even functional polymers [1–3]. Achieving this would be a giant leap towards realizing nanoscale building blocks in which the protein and its binding characteristics are tailored using DNA technologies [14] while the inorganic component is synthesized for its specific functions (e.g., electronic, optical, or magnetic) [15]. These short polypeptides (or small proteins) are called genetically engineered proteins for inorganics (GEPIs) [1, 3]. In the following section, we provide an overview of the display technologies that can be used to select polypeptides that recognize inorganic compounds and highlight unique aspects of using these systems. We next give examples of achievements involving the use of inorganic-binding polypeptides. Finally, we present future prospects in bio- and nanotechnologies.

5.2
Inorganic Binding Peptides via Combinatorial Biology

Since the invention of phage display nearly two decades ago [16], display technologies have proven an extraordinarily powerful tool for a myriad of biotechnological and biological applications. These include the characterization of receptor and antibody binding sites, the study of protein–ligand interactions, and the isolation and evolution of proteins or enzymes exhibiting improved or otherwise altered binding characteristics for their ligands. The three most common approaches, phage display (PD), cell surface display (CSD), and ribosome display (RD), have recently been reviewed [17–21]. All technologies are based on the common theme of linking phenotype and genotype. Both PD and CSD rely on the use of chimeric proteins that consist of a target sequence fused within (or to) a protein that naturally localizes on the surface of a bacteriophage (a bacterial virus) or a cell to achieve display (Fig. 5.3). Using standard molecular biology techniques [17], the DNA sequence of the target region (for instance, the active site of an enzyme or the complete sequence of a small polypeptide) can be randomized to create a library of phages or cells, each of which will synthesize a different version of the chimera on its surface. By contacting the library with an immobilized ligand, washing out weak or non-binders and repeating the process to enrich for tight binders, a subset can be selected from the original library exhibiting the ability to interact tightly with the desired ligand, a process is known as biopanning. Because the chimera is encoded within the phage genome or on a plasmid carried by the cell, the identity of the selected sequences (e.g., their amino acid compositions) can be deduced by DNA sequencing (Fig. 5.3).

The growing interest in hybrid materials incorporating both inorganic components and peptides or proteins for nanotechnology or nanobiotechnology applications has made PD and CSD very appealing for the isolation of polypeptides capable of binding inorganic materials with high affinity. Others and we adapted both technologies in our laboratory for the selection of inorganic binders for metals, semiconductors, and dielectrics. To date, CSD has been used to identify peptides recognizing iron oxide [22], gold [10], zinc oxide [23], and zeolites [24], whereas PD has been employed to isolate sequences binding to gallium arsenide [11], silica [13], silver [25], zinc sulfide [26], calcite [27], cadmium sulfide [28], and noble metals such as platinum and palladium [S. Dincer et al. 2003, unpublished]. Some of these peptides have been used to assemble inorganic particles [11, 26, 28] and some others for formation (control the nucleation and growth) of the compounds they were selected for [13, 25, 27, 29].

Inorganic materials are very different substrates from proteinaceous ligands and surprisingly little attention has been paid to adapting display technologies developed with biology in mind to the realm of materials science. For instance, many materials rapidly develop an oxide layer on their surface, expose different crystallographic faces to the solvent, and may become chemically or physically modified when incubated in the biological media used during the panning process. To avoid becoming a victim of the first law of directed evolution ("you get what you screen for") [30], it is therefore imperative to characterize inorganic surfaces before and after panning using spectroscopic and imaging techniques [31]. It may also be useful to monitor wash or elution buffers (e.g., using atomic adsorption spectroscopy to detect metals and metalloids). If evidence of surface modification or deterioration is obtained, buffer conditions should be optimized to guarantee compatibility with the target inorganic material.

Inorganic compounds come in a variety of forms, from polydisperse and morphologically uncharacterized powders to single crystals. The nature of the inorganic substrate may disqualify a particular display technology. For instance, PD is suitable for work with powders even if a gradient centrifugation step is used to harvest complexes between binding phages and particles. On the other hand the CSD system would not be amenable to such an enrichment process since centrifugal forces would shear off the flagella from the cell. Similarly, while both PD and CSD are theoretically suitable for panning on single crystals, tightly bound cells or phages may be very difficult to elute from the material, thereby leading to the loss of high-affinity clones. In such cases, the use of the bacterial system may be advantageous since all binders have an equal likelihood to be recovered following flagellar breakage.

In traditional biological applications of peptide libraries (e.g., antibody epitope characterization, mapping of protein–protein contacts, and the identification of peptide mimics of nonpeptide ligands), three to four biopanning cycles are usually performed in PD while four to five are carried out in CSD. After these cycles of enrichment, the selected sequences typically converge towards a consensus consisting of identical or conservatively replaced residues (e.g., an isoleucine for a leucine). Such consensus sequences reflect precise interactions between the side chains of the protein under study and those of the selected polypeptides. However, all available evidence indicates that this rule does not hold true in the case of inorganic binding sequences where similarities rather than a strict consensus are generally observed. This presumably reflects the heterogeneity of the inorganic substrate at the atomic and crystallographic levels and the fact that there are multiple solutions to the problem of inorganic binding. One could, for example, envision binding strategies relying on shape complementarities, electrostatic interactions, van der Waal's interactions, or various combinations of these mechanisms. Clearly, a better understanding of the rules that govern the binding of polypeptides to inorganic compounds is needed to understand the nature of specificity, predict cross-specificity and affinity, and, ultimately, for the design of hybrid materials exhibiting controlled topology and composition.

5.3
Physical Specificity and Molecular Modeling

The specificity of a polypeptide for an inorganic surface is likely to be rooted in its molecular structure and the atomic (nanoscale) surface structure of the inorganic. Structural information is, therefore, essential not only to elucidate the fundamentals of the recognition process, but also for practical applications. Such knowledge would allow genetic or chemical modifications to create additional functionalities (e.g., attachment of conducting or light-emitting polymers to create hybrid, hetero-functional molecules, ability to bind DNA or proteins), thereby yielding a molecular "tinker-toy". Ideally, using molecular dynamics/simulated annealing protocols and solution and/or solid-state NMR constraints, one could obtain an averaged lowest-energy structure for as many GEPIs as is feasible, and utilize these structures along with a simulation program in modeling the orientation and binding energetics at

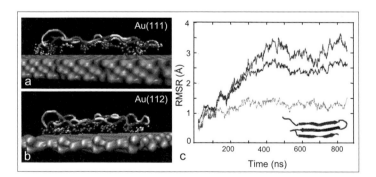

Figure 5.4. (a) and (b) are molecular dynamic rendering of gold binding protein (3-repeat GBP-1) on Au{111} and {112} surfaces viewed edge-on, respectively. Coloring corresponds to residue type: polar residues are highlighted in green, charged in blue, and hydrophobic in white. (c) Root-mean-square displacements (RMSD) of C_α atoms on Au{111} were a result of an equilibration relative to the predicted starting structure (black). The protein is stable after 500 ps. Polar residues (green) exhibit a small RMSD compared with the fluctuations observed in the hydrophobic residues (red) [32]

specific interfaces. These data could then be used to rank peptides by interfacial interaction energies, allowing the identification of important side chains and preferred chain alignments for each polypeptide with specific interfaces. Experimental findings together with structural information from simulations should add up to give a coherent understanding of GEPI–inorganic surface interactions.

In a collaborative work, we have recently performed computer structure modeling studies to predict the shape of a GEPI (*viz.* gold-binding protein, GBP-1) in solution, and present a summary here as an illustration [Fig. 5.4(a)] [32]. This preliminary work was carried out in the hope that any coincidence between the amino acid residues of the peptide and the spacing of the inorganic atomic lattice would shed light on how a GEPI binds to an inorganic surface. Raw amino acid sequence data for three repeats of the 14 amino acid GBP-1 was compared to all known protein structures using FASTA searches and various first-order secondary-structure prediction algorithms (Chou-Fasman and Holley/Karplus) [32]. Figure 5.4(a) shows the three-repeat GBP-1 above a {111} Au atomic lattice that highlights the correspondence of OH⁻ groups to gold atom positions. These initial results suggest that binding repeats of GBP-1 form an anti-parallel β-pleated sheet conformation, which places OH⁻ groups from serine and threonine residues into a regular lattice based on energy minimized *in vacuo* using X-PLOR [Fig. 54(b)]. We also showed in these preliminary studies that GBP-1 does not bind to Au{112} surface as tightly [Fig. 4(a)], because of the migration of water molecules through the atomic grooves on this crystallographic surface, which decouples the protein from the surface.

5.4
Applications of Engineered Polypeptides as Molecular Erectors

Controlled binding and assembly of proteins onto inorganic substrates is at the core of biological materials science and engineering with wide ranging applications [33, 34] (Fig. 5.5). Protein adsorption and macromolecular interactions at solid surfaces

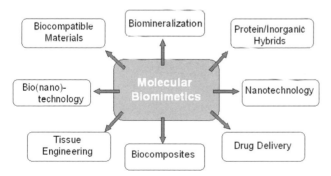

Figure 5.5 Potential applications of molecular biomimetics in nanotechnology and nanobiotechnology using combinatorially selected and genetically engineered inorganic-binding polypeptides [3].

play key roles in the performance of implants and hard-tissue engineering. DNA and proteins adsorbed specifically onto probe substrates are used to build microarrays suitable for modern genomics [35], pharmogenetics [36], and proteomics [37, 38]. Engineered polypeptides hybridized with functional synthetic molecules could be used as heterofunctional building blocks in molecular electronics and photonics [39, 40]. The unique advantages of engineered polypeptides, created through molecular biomimetics discussed here, include highly specific molecular surface recognition of inorganics, self-assembly into ordered structures, and tailoring of their molecular structures and functions through molecular biology and genetics protocols. Using inorganic binding polypeptides, one can create molecular erectors sets for potential nanotechnological and nanobiotechnological applications.

5.4.1
Self-Assembly of Inorganic-Binding Polypeptides as Monolayers

One of the central questions related to the genetically engineered proteins is whether they can assemble with a long-range order on a given crystallographic surface of a material in addition to chemically recognizing it. While this aspect of molecular biomimetics is in its nascency, the atomic-force microscopy (AFM) image of Fig. 5.6 shows that it is possible to assemble a one-monolayer-thick gold-binding protein layer on the {111} surface of gold (H.M. Zareie, M. Sarikaya, 2003, unpub-

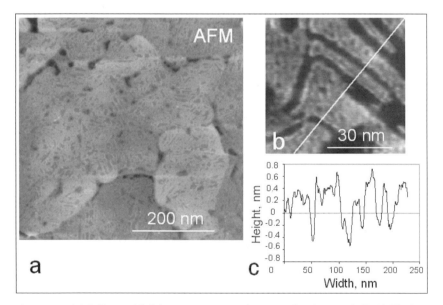

Figure 5.6. (a) Self-assembled three-repeat GBP-1 on Au{111} (AFM image) showing ordered-domained structure. The angle among the domain boundaries is either 120° or 60°, implying recognition of the symmetry of the top surface layer on Au{111}. The line across several domains in the AFM image (b) produces a profile with 0.5-nm-high platforms, revealing monolayer thickness of the GEPI film (H.M. Zareie, M. Sarikaya, 2003, unpublished].

lished results). In fact, close inspection of this and other images reveal that GBP-1 assembles into domains with clear and straight boundaries that actually make a 60°/120° angle, suggesting that the polypeptide actually recognizes the six-fold lattice symmetry on the Au{111} surface. What is also significant in these results is that the assembly process progresses until the whole surface is completely covered. Instead of using the traditional thiol or silane linkers in self-assembled monolayers [41, 42], this result demonstrates self-assembly of GEPI monolayer that may be used as functional linkages – a central premise in this research.

5.4.2
Morphogenesis of Inorganic Nanoparticles via GEPIs

In biomineralization, a significant aspect of biological control over materials formation is via protein/inorganic interactions that control the growth morphology in tissue, such as bone, dentin, mollusk shells, and bacterial and algal particle formation [43–46]. Studies aiming at finding how such proteins affect biomineralization have traditionally focused mostly on templating [4–9], nucleation [47, 48], and enzymatic reactions [8, 9]. With the emergence of combinatorially selected peptides that strongly bind inorganics, a natural step is to examine how these polypeptides affect inorganic formation, and investigate various effects in mineralization (including nucleation, growth, morphogenesis, and enzymatic effects). These studies have been carried out under aqueous environments necessary for biological functions of selected GEPIs, most notably those binding to Au [29] and Ag [25]. The morphology

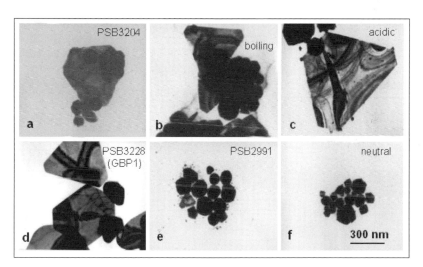

Figure 5.7. Effect of GEPI on nanocrystal morphology. Two mutants out of more than 50 from a library of gold-binding GEPIs were tested to result in flat gold particle formation as shown in (a) and (d) similar to those formed under acidic (b) or boiling conditions (c). Particles formed in the presence of vector-encoded alkaline phosphatase and neutral conditions did not result in morphological change of gold particles, as shown in (e) and (f), respectively [29].

of solution-grown gold particles was controlled by the presence of gold-binding GEPIs selected by CSD [29]. More than 50 mutants were tested for their influence on the rate of crystallization of nanogold particles formed by the reduction of AuCl₃ with Na citrate. Two mutants accelerated the rate of crystal growth and changed the particle morphology from cubo-octahedral (the usual shape of gold particles grown under equilibrium conditions) to flat, triangular, or hexagonal (Fig. 5.7). Similar crystals usually form under acidic conditions in the absence of gold-binding polypeptides [49]. It may be concluded, therefore, that slightly basic polypeptides may have acted as an acid in flat gold crystallization.

5.4.3
Assembly of Inorganic Nanoparticles via GEPIs

Organization and immobilization of inorganic nanoparticles in two- or three-dimensional geometries are fundamental in the utilization of the nanoscale effects [50]. For example, quantum dots could be produced using vacuum techniques (e.g., molecular beam epitaxy, MBE), and such an organization is shown in Fig. 5.8(a) in the GaInAs/GaAs system (T. Pearsall, 1996, unpublished). However, this can only be accomplished under stringent conditions of high temperature, very low pressures, and a toxic environment. A desirable approach would not only synthesize the inorganic particles under less stringent conditions but also assemble/immobilize them via self-assembly using functional molecules as coupling agents. Inorganic particles have been coupled and functionalized using synthetic molecules (e.g., thiols, lipids, and biological molecules, including amino acids, polypeptides, and ligand-functionalized DNA) and assembled to generate novel materials using the recognition properties of these molecules [1, 2, 40, 51]. Nanoparticles synthesized in wet-chemical conditions in the presence of these molecules (e.g., citrate, thiol, silane, lipid, and amino acids) not only cap the particle, resulting in controlled growth, but also prevent their uncontrolled aggregation [40, 52, 53].

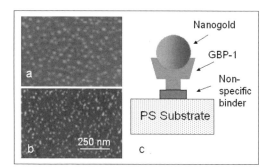

Figure 5.8 The AFM images show quantum (GaInAs) dots assembled on GaAs substrate (a) via high-vacuum (molecular beam epitaxy, MBE) strain-induced self-assembly and (b) via seven-repeat GBP-1. (c) Schematic illustration of (b), where PS is polystyrene substrate, and the nonspecific binder is glutaraldehyde.

Synthetic and biological molecules used as coupling agents, however, are nonspecific to a given material. For example, thiols couple gold as well as silver nanoparticles in similar ways [40]. Likewise, citrate ions cap noble metals indiscriminately [40, 52, 53]. A desirable next step in molecular recognition and assembly via molecules would be to use polypeptide sequences that recognize inorganics specifically. In this context, GEPIs could potentially be used for assembly of nanoparticles. In addition to inorganic surface recognition, a further advantage of the GEPI is that it can be genetically fused to other functional biomolecular units or ligands to produce heterobifunctional (or multifunctional) molecular entities [54, 55]. In Fig. 5.8(b), we show the assembly of nanogold particles on GBP-1-coated flat mica surfaces [56]. When seven-repeat GBP-1 was used as the linker, high-density gold particles formed on the surface, resembling a distribution not unlike in that in Fig. 5.8(a). Clearly, in this case, unlike the conditions of the MBE, the assembly was accomplished under ambient conditions of temperature or pressure and in an aqueous solution. Therefore, the homogeneous decoration of the surface with nanogold suggests that proteins may be useful in the production of tailored nanostructures as quantum dots. The recognition activity of the protein would provide the ability to control the particle distribution, while particle preparation conditions would allow size control.

5.5
Future Prospects and Potential Applications in Nanotechnology

Proteins hold great promise for the creation of architectures at the molecular or nanoscale levels [1–3, 40]; the ultimate aim is to use proteins for controlled binding and assembly onto inorganics. Genetically engineered proteins for inorganics (GEPIs) represent a new class of biological molecules that are combinatorially selected to bind to specific inorganic surfaces [1, 3]. The ordered assembly of GEPI on inorganic surfaces could have a significant impact on molecular technology applications, offering several novel practical advantages. Our results described above are the first demonstrations that combinatorially selected polypeptides can self-assemble specifically on a selected inorganic single crystal surface, and that a GEPI may be molecularly recognizing an inorganic surface. Realizing the fact that thiol and silane linkages are the other two major molecular linkers for noble metal and oxide (silica) surfaces that have constituted the field of self-assembled molecules up until now, it is naturally expected that self-assembled GEPI monolayers – SA(GEPI)M – as "molecular erector sets" will open up new avenues for designing and engineering new and novel functional surfaces. We have already demonstrated that inorganic materials can be assembled at the nanoscale by proteins that have been genetically engineered to bind to selected materials surfaces. It is also the first time that engineered proteins were shown to affect crystal morphology [29]. The combinatorial genetic approach is a general one, which should be applicable to numerous surfaces [10–13, 22–27]. The modularity of binding motifs should allow genetic fusion of peptide segments recognizing two different materials. The resulting heterobifunctional molecules could be used to attach different materials to each other and may permit assembly of complex nanocomposite and hybrid materials. These results could lead to new avenues in nanotechnology, biomimetics, biotechnology, and crystal and tissue engineering such as in the formation, shape modification, and assembly of materials, and development of surface-specific protein coatings.

One particular potential utility of GEPI as a molecular linker is in nanotechnology. Both nanostructured inorganics and functional molecules are becoming fundamental building blocks for future functional materials such as nanoelectronics, nanophotonics, and nanomagnetics [1–3, 39, 40, 57]. Before nanoscience could be implemented in practical and working systems, however, there are numerous challenges that must be addressed. Some of these challenges include molecular and nanoscale ordering and scale up into larger architectures. A nanotechnological system, for example, could require several components made up of materials of different physical and chemical characteristics. These different materials have to be connected and assembled without an external manipulation. As schematically shown in Fig. 5.9, the components may include two or more inorganic nanoparticles, several functional molecules, and nanopatterned multimaterial substrates, all assembled through the specific interaction with the appropriate GEPI molecules.

Although significant advances have been made in developing protocols for surface-binding polypeptides, many questions need to be answered before their robust design and practical applications are effectively realized. These questions include: (i) What are the physical and chemical bases for recognition of inorganic surfaces by the genetically engineered polypeptides? (ii) What are the long-range assembly characteristics, kinetics, and stability of the binding? (iii) What are the molecular mechanisms of engineered polypeptide binding onto (noble) metals compared to those on

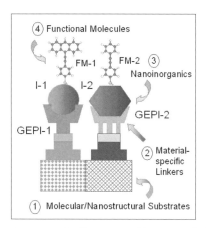

Figure 5.9 An illustration of the potential of utility of GEPIs as molecular erectors in functional nanotechnology. Two different GEPIs (GEPI-1 and GEPI-2) assemble on a patterned substrate. One could use either a designer protein, followed by genetic fusion of the respective GEPIs, or directed-assemble GEPIs on a nanopatterned substrate. Two different inorganic particles I-1 and I-2 are immobilized selectively on GEPI-1 and GEPI-2, respectively. Synthetic molecules (i.e., conducting or photonic) via functionalized side groups or hybridized GEPIs are assembled on the nanoparticles [3].

nonmetals? (iv) How do surface characteristics affect binding? (v) Based on the insights achieved, can we develop a "road map" in using GEPIs as molecular linkers and open new avenues in self-assembled molecular systems in nanotechnology based on biology? Some of these questions can be answered using the existing nano-technological and molecular biological tools, and some require adaptation of biology methodologies into molecular biomimetics. For example, new algorithms need to be developed to adapt current protein-folding models into molecular recognition, reconformation, and binding of engineered proteins on inorganic compounds with various crystallographical, surface-topological, and chemical characteristics. Modeling studies coupled with rapid development of the experimental tools – such as high-resolution scanned probe microscopies and spectroscopies, local surface spectroscopies (such as surface-enhanced RAMAN and surface plasmon resonance), x-ray photoelectron spectroscopy, small- and grazing-angle x-ray diffraction, and liquid- and solid-state nuclear magnetic resonance spectroscopy – is expected to lead into new predictive protocols in designing novel polypeptide sequences that specifically bind to desired inorganics and self-assemble on them and can be used as molecular erectors. Considering the present rapid developments in the inorganic-binding polypeptide selection protocols, the increased variety of materials utilized as substrates, and novel modeling adaptations, it is expected that many of the above questions will find answers in the near future with subsequent significant impact in broad multidisciplinary fields with potentially wide-ranging nanotechnology and nanomedicine applications.

Acknowledgments

We thank our colleagues A. Jen, F. Ohuchi, D. Schwartz, F. Baneyx, and B. Traxler (University of Washington, Seattle), J. Evans (New York University), K. Schulten (University of Illinois, Urbana), and S. Brown (University of Copenhagen) for invaluable discussions, and our students and researchers S. Dinçer, D. Heidel, R. Braun, M. H. Zareie, M. Duman, E. Ventkatasubramanian, V. Bulmus, E. Ören, D. Sahin, and H. Fong for technical help. This research was supported by USA-ARO through the DURINT program (PM Dr. R. Campbell) under grant no. DAAD19–01–1-04999 (Defense University Research Initiative on Nanotechnology).

Abbreviations
AFM – atomic force microscopy
CSD – cell surface display
GBP – gold-binding protein
GEPI – genetically engineered polypeptides for inorganics
MBE – molecular beam epitaxy
PD – phage display
PS – polystyrene
RD – ribosomal display

References

1 M. Sarikaya, C. Tamerler, A. Y. Jen, K. Schulten, F. Baneyx, Molecular biomimetics: nanotechnology through biology. *Nature Mater.* **2003**, *2*, 577–585.

2 P. Ball, Life's lessons in design. *Nature* **2001**, *409*, 413–416.

3 M. Sarikaya, C. Tamerler, D. T. Schwartz, F. Baneyx, Materials assembly and formation using engineering polypeptides, *Ann. Rev. Mater. Res.* **2004**, *34*, 373–408.

4 M. A. Cariolou, D. E. Morse, Purification and characterization of calcium-binding conchiolin shell peptides from the mollusk, *Haliotisrufescens*, as a function of development. *J. Comp. Physiol. B.* **1988**, *157*, 717–729.

5 M. L. Paine, M. L. Snead, Protein interactions during assembly of the enamel organic extracellular matrix. *J. Bone Miner. Res.* **1996**, *12*, 221–226.

6 A. Berman, L. Addadi, S. Weiner, Interactions of sea-urchin skeleton macromolecules with growing calcite crystals: A study of intracrystalline proteins. *Nature* **1988**, *331*, 546–548.

7 A. Weizbicki, C. S. Sikes, J. D. Madura, B. Darke, Atomic force microscopy and molecular modeling of proteins and peptide binding to calcite. *Calcif. Tissue Intl.* **1994**, *54*, 133–141.

8 N. Kröger, R. Deutzman, M. Sumper, Polycationic peptides from diatom biosilica that direct silica nanosphere formation. *Science* **1999** *286*, 1129–1132.

9 J. N. Cha, K. Shimizu, Y. Zhou, S. C. Christiansen, B. F. Chmelka, et al., Silicatein filaments and subunits from a marine sponge direct the polymerization of silica and silicones *in vitro*. *Proc. Natl. Acad. Sci. USA* **1999**, *96*, 361–365.

10 S. Brown, Metal recognition by repeating polypeptides. *Nature Biotechnol.* **1997**, *15*, 269–272.

11 S. R. Whaley, D. S. English, E. L. Hu, P. F. Barbara, M. A. Belcher, Selection of peptides with semiconducting binding specificity for directed nanocrystal assembly. *Nature* **2000**, *405*, 665–668.

12 D. J. H. Gaskin, K. Strack, E. N. Vulfson, Identification of inorganic crystal-specific sequences using phage display combinatorial library of short peptides: A feasibility study. *Biotech. Lett.* **2000**, *22*, 1211–1216.

13 R. R. Naik, L. Brott, S. J. Carlson, M. O. Stone, Silica precipitating peptides isolated from a combinatorial phage display libraries. *J. Nanosci. Nanotechnol.* **2002**, *2*, 95–100.

14 D. Voet, J. G. Voet, *Biochemistry*. **2003**, Wiley, New York

15 B. Muller, Natural formation of nanostructures: from fundamentals in metal heteroepitaxy to applications in optics and biomaterials science. *Surf. Rev. Lett.* **2001**, *8*, 169–228.

16 G. P. Smith, Filamentous fusion phage: Novel expression vectors that display cloned antigens on the virion surface. *Science* **1985**, *228*, 1315–1317.

17 M. Dani, Peptide display libraries: design and construction. *J Recept. Signal Transduct. Res.* **2001**, *21*, 469–488.

18 K. D. Wittrup, Protein engineering by cell surface display. *Curr. Opin. Biotechnol.* **2001**, *12*, 395–399.

19 H. M. E. Azzazy, W. E. Jr. Highsmith, Phage display technology: clinical applications and recent innovations. *Clin. Biochem.* **2002**, *35*, 425–445.

20 W. J. Dower, L. C. Mattheakis, In vitro selection as a powerful tool for the applied evolution of proteins and peptides. *Curr. Opin. Chem. Biol.* **2002**, *6*, 390–398.

21 P. Samuelson, E. Gunneriusson, P. Å. Nygren, S. Ståhl, Display of proteins on bacteria. *J. Biotechnol.* **2002**, *96*, 129–154.

22 S. Brown, Engineering iron oxide-adhesion mutants of *Escherichia coli* phage λ-receptor. *Proc. Natl. Acad. Sci. USA* **1992**, *89*, 8651–8655.

23 K. Kiargaard, J. K. Sorensen, M. A. Schembri, P. Klemm, Sequestration of zinc oxide by fimbrial designer chelators. *Appl. & Env. Microbiol.* **2000**, *66*, 10–14.

24 Nygaard S, Wandelbo R, Brown S. Surface-specific zeolite-binding proteins. *Adv. Mater.* **2002**, *14*, 1853–1856.

25 R. R. Naik, S. J. Stringer, G. Agarwal, S. E. Jones, M. O. Stone, Biomimetic synthesis and patterning of silver nanoparticles. *Nature Mater.* **2002**, *1*, 169–172.

26 W. S. Lee, C. Mao, C. E. Flynn, A. M. Belcher, Ordering quantum dots using genetically engineered viruses. *Science* **2002**, *296*, 892–895.

27 C. M. Li, G. D. Botsaris, D. L. Kaplan, Selective *in vitro* effects of peptides on calcium carbonate crystallization. *Crystal Growth & Design* **2002**, *2*, 387–393.

28 C. Mao, C. E. Flynn, A. Hayhurst, R. Sweeney, J. Qi, G. Georgiou, B. Iverson, A. M. Belcher, Viral assembly of oriented quantum dot nanowires. *Proc. Natl. Acad. Sci. USA* **2003**, *100*, 6946–6951.

29 S. Brown, M. Sarikaya, E. Johnson, Genetic analysis of crystal growth. *J. Mol. Biol.* **2000**, *299*, 725–732.

30 C. Schmidt-Dannert, F. H. Arnold, Directed evolution of industrial enzymes. *Trends. Biotechnol.* **1999**, *17*, 135–136.

31 H. Dai, C. K. Thai, M. Sarikaya, F. Baneyx, D. T. Schwartz. Through-mask anodic patterning of copper surfaces and film stability in biological media. *Langmuir*, **2004**, *20*, 3483–3486.

32 R. Braun, M. Sarikaya, K. S. Schulten, Genetically engineered gold-binding polypeptides: Structure prediction and molecular dynamics. *J. Biomater. Sci.* **2002**, *13*, 747–758.

33 S. E. Sakiyama-Elbert, J. A. Hubbell, Design of novel biomaterials. *Ann. Rev. Mater. Res.* **2001**, *31*, 183–201.

34 B. Ratner, F. Schoen, A. Hoffman, J. Lemone (eds.), *Biomaterials Science: Introduction to Materials in Medicine*. Academic Press, San Diego, **1996**.

35 J. R. Epstein, I. Biran, D. R. Walt, Fluorescence based nucleic acid detection and microarrays *Anal. Chim. Acta* **2002**, *469*, 3–36.

36 M. L. Yarmush, A. Yarayam, Advances in proteomics technologies. *Ann. Rev. Biomed. Eng.* **2002**, *4*, 349–373.

37 P. Cutler, Protein Arrays: The current state-of-the-art. *Proteomics* **2003**, *3*, 3–18.

38 M. E. Chicurel, D. D. Dalma-Weiszhausz, Microarrays in pharmacogenomics – advances and future promise. *Pharmacogenomics* **2002**, *3*, 589–601.

39 D. I. Gittins, D. Bethell, D. J. Schiffrin, R. J. Michols, A nanometer-scale electronic switch consisting of a metal cluster and redox-addressable groups. *Nature* **2000**, *408*, 67–69.

40 C. M. Niemeyer, Nanoparticles, proteins, and nucleic acids: Biotechnology meets materials science. *Angew. Chem. Int. Ed.* **2001**, *40*, 4128–4158.

41 G. M. Whitesides, J. P. Mathias, C. T. Seto, Molecular Self-assembly and nanochemistry-A chemical strategy for the synthesis of nanostructures. *Science* **1991**, *254*, 1312–1319.

42 A. Ulman (ed.), *Self-assembled Monolayers of Thiols*. Academic Press, San Diego, **1998**.

43 R. B. Frankel, R. P. Blakemore (ed.), *Iron Biominerals*. Plenum, New York, **1991**.

44 M. Glimcher, M. Nimni, Collagen cross-linking and biomineralization. *Connect. Tissue Res.* **1992**, *27*, 73–83.

45 S. Schultze, G. Harauz, T. I. Beveridge, Participation of a cyanobacterial S layer in fine-grain mineral formation. *J. Bacteriol.* **1992**, *174*, 7971–7981.

46 M. Sarikaya, Biomimetics. Biomimetics: Materials fabrication through biology *Proc. Natl. Acad. Sci. USA*, **1999**, *96*, 14183–14185.

47 A. M. Belcher et al., Control of crystal phase switching and orientation by soluble mollusk-shell proteins. *Nature* **1996**, *381*, 56–58.

48 P. Mukharjee, S. Senapati, D. Mandal, A. Ahmad, M. I. Khan, et al., Extracellular synthesis of gold nanoparticles by the fungus *Fusarium oxysporum*. *Chem. Biol. Chem.* **2002**, *3*, 461–463.

49 J. Türkevich, P. C. Stevenson, J. Hillier, A study of the nucleation and growth of processes in the synthesis of colloidal gold. *Trans. Faraday Soc. Disc.* **1951**, *11*, 55–75.

50 B. Muller, Natural formation of nanostructures: from fundamentals in metal heteroepitaxy to applications in optics and biomaterials science. *Surf. Rev. Lett.* **2001**, *8*, 169–228.

51 O. Yamauchi, A. Odani, S. Hirota, Metal ion assisted weak interactions involving biological molecules. From small complexes to metalloproteins. *Bull. Chem. Soc. Jpn.* **2001**, *74*, 1525–1545.

52 R. Djulali, Y. Chen, H. Matsui, Au nanowires from sequenced histidine-rich peptide. *J. Am. Chem. Soc.* **2002**, *124*, 13660–13661.

53 Y. Xia, Y. Sun, Shape-Controlled Synthesis of Gold and Silver Nanoparticles. *Science* **2002**, *298*, 2176–2179.

54 A. R. McMillan, C. D. Pauola, J. Howard, S. L. Chen, N. J. Zaluzec, et al., Ordered nanoparticle arrays formed on engineered chaperonin protein templates. *Nature Mater.* **2002**, *1*, 247–52.

55 D. Moll et al. S-layer-streptavidin fusion proteins as template for nanopatterned molecular arrays. *Proc. Natl. Acad. Sci. USA* **2002**, *99*, 14646–14651.

56 M. Sarikaya, H. Fong, D. Heidel, R. Humbert, Biomimetic assembly of nanostructured materials. *Mater. Sci. Forum* **1999**, *293*, 83–87.

57 National Nanobiotechnology Initiative. **2003**, *http://www.nano.gov.*

II
Characterization Tools for Nanomaterials and Nanosystems

Nanofabrication Towards Biomedical Applications. C. S. S. R. Kumar, J. Hormes, C. Leuschner (Eds.)
Copyright © 2005 WILEY-VCH Verlag GmbH & Co. KGaA, Weinheim
ISBN 3-527-31115-7

magnifications of > 100K and resolution that is capable of resolving crystal lattices has become one of the major techniques since its development in the early 1980s, the development of high-resolution electron diffraction, at the convergence of several microscopy technologies, is relatively new. The development of field emission guns (FEG) in the 1970s and their adoption in conventional transmission electron microscopes (TEM) brought high source brightness, small probe size, and coherence to electron diffraction. The significant impact is the ability to record diffraction patterns to obtain crystallographic information from very small (nano) structures. The electron energy filter, such as the in-column energy filter with four magnets that bend the electron beam into an Ω shape (Ω-filter), allows inelastic background from plasmon and higher electron energy losses to be removed with an energy resolution of a few electron volts. The development of array detectors of charge-coupled device (CCD) cameras or imaging plates enables parallel recording of diffraction patterns and quantification of diffraction intensities over a large dynamic range that was not available to electron microscopy before. The post specimen lenses of the TEM give the flexibility of recording electron diffraction patterns at different magnifications. Last, but just as important, the development of efficient and accurate algorithms to simulate electron diffraction and modeling structures on a first-principle basis using fast modern computers has significantly improved our ability to interpret experimental diffraction patterns.

For readers new to electron microscopy, there are a number of introductory books [1–4]. Kinematical approximation for electron diffraction and the diffraction geometry for qualitative analysis of electron diffraction patterns are covered in Ref. [1]. The book by Williams and Carter is an excellent introductory textbook for students [3]. More specialized books on electron imaging and diffraction can be found in Refs. [5–8].

6.2
Electron Diffraction and Geometry

Electron optics in a microscope can be configured for different diffraction modes, from parallel-beam illumination to convergent beams. Figure 6.1 illustrates three modes of electron diffraction: selected area electron diffraction (SAED), nanoarea electron diffraction (NED) and convergent-beam electron diffraction (CBED). Variations from these three techniques include large-angle CBED [9], convergent-beam imaging [10], electron nanodiffraction [11], and their modifications [12]. For nanostructure characterization, the electron nanodiffraction technique developed by Cowley [11] and others in the late 1970s using a scanning TEM (STEM) is particularly relevant. In this technique, a small electron probe of a size from a few angstroms to a few nanometers is placed directly onto the sample. A diffraction pattern can thus be obtained from a localized area as small as a single atomic column, which is very sensitive to local structure and probe positions. Readers interested in these techniques can find their description and applications in above references.

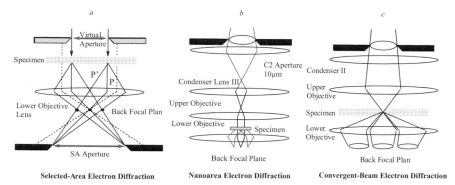

Figure 6.1. Three modes of electron diffraction. Both (a) selected area electron diffraction (SAED) and (b) nanoarea electron diffraction (NED) use parallel illumination. SAED limits the sample volume contributing to electron diffraction by using an aperture in the image plane of the image-forming lens (objective). NED achieves a very small probe by imaging the condenser aperture on the sample using a third condenser lens. Convergent-beam electron diffraction (CBED) uses a focused probe.

6.2.1
Selected-Area Electron Diffraction

SAED is formed by placing an aperture in the imaging plane of the objective lens [see Fig. 6.1(a)]. Only rays passing through this aperture contribute to the diffraction pattern at the far field. For a perfect lens without aberration, these rays come from an area defined by the back-projected aperture image. The aperture image is typically a factor of 20 smaller because of the demagnification of the objective lens. In a conventional electron microscope, the difference in the focus for rays at different angles to the optic axis due to the objective lens aberration results a displaced aperture image for each diffracted beams. Take the rays marked P and P' for an example. While ray P being parallel to the optical axis defines the ideal back-projected aperture image, ray P' at an angle of α will move by aberration a distance $y=C_s\alpha^3$. For a microscope with $C_s=1$ mm and $\alpha=50$ mrad, this gives a displacement of 125 nm. The smallest area that can be selected in SAED is thus limited by the objective lens aberration.

The combination of imaging and diffraction in SAED mode makes it particularly useful for setting diffraction conditions for electron imaging in a TEM, such as lattice images or diffraction contrast. It is also one of the major electron microscopy techniques for materials phase identification and orientation determination. The interpretation of SAED for materials phase identification, orientation relationships, and defects is described by Edington [1].

6.2.2
Nano-Area Electron Diffraction

Figure 6.1(b) shows a schematic diagram of the principle of parallel-beam electron diffraction from a nanometer-sized area in a TEM. The electron beam is focused to the focal plane of the objective prefield, which then forms a parallel-beam illumination on the sample. For a condenser aperture of 10 μm in diameter, the probe diameter is ~50 nm, which is much smaller than the smallest area that can be achieved in conventional SAED, and does not suffer from aberration-induced image shift (see above). The diffraction pattern recorded in this mode is similar to that in SAED, e.g., the diffraction pattern consists of sharp diffraction spots for perfect crystals.

NED in a FEG microscope also provides higher beam intensity than SAED (the probe current intensity is ~10^5 e s^{-1} nm^{-2}), since all electrons illuminating the sample are recorded in the diffraction pattern. The small probe size enables us to select an individual nanostructure for electron diffraction.

The application of nanoarea electron diffraction for electron nanocrystallography is demonstrated in Fig. 6.2, which shows an experimental diffraction pattern recorded from a single Au nanocrystal close to the [110] zone axis. The small illumination enables the isolation of a single nanocrystal for diffraction. The parallel beam gives the high angular resolution for the recording of details of diffuse scattering that comes from finite size and deviations from a ideal crystal structure.

The NED technique described here is different from electron nanodiffraction in a STEM [9]. In electron nanodiffraction, the beam is focused on, or near, the sample. A small condenser aperture is used to reduce the beam convergence angle. The beam convergence limits the angular resolution in recorded diffraction patterns. Many nanostructures are nonperiodic and lack the perfect crystal order resulted from either the small size or atomic arrangements. Their diffraction patterns are diffuse as shown in Fig. 6.2. High angular resolution is thus required for recording diffuse scattering.

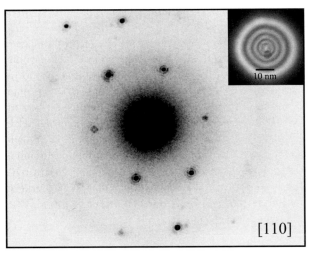

Figure 6.2. An example of nanoarea electron diffraction. The diffraction pattern was recorded from a single Au nanocrystal of ~4 nm near the zone axis of [110]. Around each diffraction spot, two rings of oscillation are clearly visible. The rings are not continuous because of the shape of the crystal. The electron probe and an image of the nanocrystal are shown top right.

A third condenser lens, or a minilens, provides the flexibility and demagnification for the formation of a nanometer-sized parallel beam. In the design of electron microscopes with a two condenser lenses illumination system, the first lens is used to demagnify the electron source and the second lens transfers the demagnified source image to the sample at focus (for probe formation) or underfocused to illuminate a large area. The condenser aperture is placed after the second lens.

6.2.3
Convergent-Beam Electron Diffraction

CBED is formed by focusing the electron probe at the specimen [see Fig. 1(c)]. Compared to SAED, CBED has two main advantages for studying *perfect crystals* or the local structure of polycrystalline materials or crystals with defects:

1. The pattern is taken from a much smaller area with a focused probe; The smallest electron probe currently available in a high-resolution FEG-STEM is close to 1 Å. Thus, in principle and in practice, CBED can be recorded from individual atomic columns. For crystallographic applications, CBED patterns are typically recorded with a probe of a few to tens of nanometers.
2. CBED patterns record diffraction intensities as a function of incident-beam directions. Such information is very useful for symmetry determination and quantitative analysis of electron diffraction patterns.

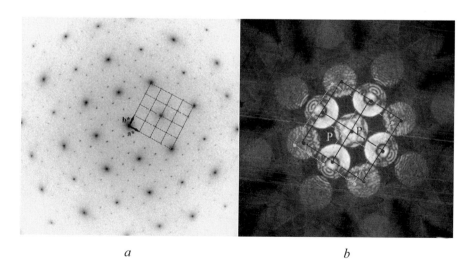

a   *b*

Figure 6.3. A comparison between CBED and SAED. (a) Recorded diffraction pattern along [001] from magnetite cooled to liquid nitrogen temperature. There are two types of diffraction spots, strong and weak ones. The weak ones come from the low-temperature structural transformation. All diffraction spots in this pattern can be indexed based on two reciprocal lattice vectors (a* and b*). (b) Recorded CBED pattern from spinel $MgAl_2O_4$ along [100] at 120 kV.

A comparison between SAED and CBED is given in Fig. 6.3. CBED patterns consist of disks. Each disk can be divided into many pixels, and each pixel approximately represents one incident beam direction. For an example, let us take the beam P in Fig. 6.3. This particular beam gives one set of diffraction pattern shown as the full lines. The diffraction pattern by the incident beam P is the same as the selected area diffraction pattern with a single parallel incident beam. For a second beam P′, which comes at different angle compared to P, the diffraction pattern in this case is displaced from that of P by α/λ with α as the angle between the two incident beams.

Experimentally, the size of the CBED disk is determined by the condenser aperture and the focal length of the probe-forming lenses. In modern microscopes with an additional minilens placed in the objective prefield, it is also possible to vary the convergence angle by changing the strength of the minilens. Underfocusing the electron beam also gives a smaller convergence angle, but it leads to a bigger probe, which can be an issue for specimens with a large wedge angle.

The advantage of being able to record diffraction intensities over a range of incident beam angles makes CBED readily accessible for comparison with simulations. Because of this, CBED is a quantitative diffraction technique. In the past 15 years, CBED has evolved from a tool primarily for crystal symmetry analysis to the most accurate technique for structure refinement and strain and structure factor measurement [13]. For defects, large-angle CBED technique can characterize individual dislocations, stacking faults, and interfaces. For applications to defect structures and structure without three-dimensional periodicity, parallel-beam illumination with a very small beam convergence is required.

6.3
Theory of Electron Diffraction

Electron diffraction from a nanostructure can be alternatively described by electrons interacting with an assembly of atoms (ions) or from a crystal of finite size and shape. Which description is more appropriate depends on which is a better approximation of the structure. Both cases are covered here. We start with kinematical electron diffraction from a single atom, then move on to the assembly of atoms and to crystals. Electron multiple scattering, or electron dynamic diffraction, is strong in perfect crystals, but can be neglected to the first-order approximation for very small nanostructures or macromolecules such as carbon nanotubes made of mostly light atoms.

6.3.1
Kinematic Electron Diffraction and Electron Atomic Scattering

Electron interacts with an atom by the Coulomb potential of the positive nucleus and the electrons surrounding the nucleus. The relationship between the potential and the atomic charge is given by the Poisson equation:

$$\nabla^2 V(\vec{r}) = -\frac{e[Z\delta(\vec{r}) - \rho(\vec{r})]}{\varepsilon_o} \tag{1}$$

Here V is the potential, ρ the electron density, e the electron charge, ε_o the vacuum permitivity, and Z the atomic number. If we take a small volume, $d\vec{r} = dxdydz$, of the atomic potential at position $\vec{r}$, the exit electron wave ϕ_e from this small volume is approximately given by:

$$\phi_e \approx (1 + i\pi\lambda U dxdydz)\phi_o \tag{2}$$

Here we take $U = 2meV(\vec{r})/h^2$, where m is relativistic mass of high-energy electrons and h is the Planck constant. λ is the electron wavelength. The interaction potential U is treated as constant within the small volume. Equation (2) is known as the weak-phase-object approximation. For a parallel beam of incident electrons, the incident wave is described by the plane wave $\exp(2\pi i \vec{k}_o \cdot \vec{r})$. For high-energy electrons with $E \gg V$, the scattering by the atom is weak and we have approximately:

$$\phi_s = \frac{2\pi me}{h^2} \int \frac{V(\vec{r})}{|\vec{r} - \vec{r}'|} e^{2\pi ik|\vec{r} - \vec{r}'|} e^{2\pi i\vec{k}_o \cdot \vec{r}''} d\vec{r}'' \tag{3}$$

Far away from the atom, we have $|\vec{r}| \gg |\vec{r}'|$. By replacing $|\vec{r} - \vec{r}'|$ with $|\vec{r}|$ in the denominator and

$$|\vec{r} - \vec{r}'| \approx r - \frac{\vec{r}' \cdot \vec{r}}{r} \tag{4}$$

for the exponential, we obtain

$$\phi_s = \frac{2\pi me}{h^2} \int \frac{V(\vec{r}'')}{|\vec{r} - \vec{r}'|} e^{2\pi ik|\vec{r} - \vec{r}'|} e^{2\pi i\vec{k}_o \cdot \vec{r}''} d\vec{r}'' \approx \frac{2\pi me}{h^2} \frac{e^{2\pi i\vec{k}_o \cdot \vec{r}}}{r} \int V(\vec{r}'') e^{2\pi i(\vec{k} - \vec{k}_o) \cdot \vec{r}''} d\vec{r}'' \tag{5}$$

Here $\vec{k}$ is the scattered wave vector and the direction is taken along $\vec{r}$. The half of the difference between the scattered wave and incident wave, $\vec{s} = (\vec{k} - \vec{k}_o)/2$, is defined as the scattering vector. As we will see later, this vector is half the reciprocal lattice vector (g) for the Bragg diffraction of a crystal. From Eq. (5), we define electron atomic scattering factor

$$f(s) = \frac{2\pi me}{h^2} \int V(\vec{r}'') e^{4\pi i\vec{s} \cdot \vec{r}''} d\vec{r}'' \tag{6}$$

The potential is related to the charge density; Fourier transform of electron charge density is commonly known as the x-ray scattering factor. It can be shown that the relationship between the electron and x-ray scattering factors is given by

$$f(s) = \frac{me^2}{8\pi\varepsilon_o h^2} \frac{(Z - f^x)}{s^2} = 0.023934 \frac{(Z - f^x)}{s^2} \text{ (Å)} \tag{7}$$

The x-ray scattering factor in the same unit is given by

$$f^x(s) = \left(\frac{e^2}{mc^2}\right) f^x = 2.82 \times 10^{-5} f^x \text{ (Å)} \tag{8}$$

For typical value of $s\sim0.2\,\text{Å}^{-1}$, the ratio $f/(e^2/mc^2)f^x\sim10^4$. Thus electrons are scattered by an atom much more strongly than x-ray. There are two consequences as the result of this: one is that electrons are much more sensitive to a small volume of materials, such as nanostructures, and the other is multiple scattering. For a thick crystal, electron multiple scattering is a serious effect, while multiple scattering is generally weak in most x-ray structural analyses.

The electron distribution in an atom depends on the atomic electronic structure and bonding with neighboring atoms. At sufficiently large scattering angles, we can approximate the atoms in a crystal by spherical free atoms or ions. Atomic charge density and the Fourier transform of the charge density can be calculated with high accuracy. The results of these calculations are published in the literature and tabulated in the international table for crystallography. Tables optimized for electron diffraction applications are also available [14].

6.3.2
Kinematical Electron Diffraction from an Assembly of Atoms

Here, we extend our treatment of kinematical electron scattering from a single atom to an assembly of atoms. For nanostructures of a few nanometers, the treatment outlined here forms the basis for electron diffraction pattern analysis and interpretation. For a large assembly of atoms in a crystal with well-defined 3-D periodicity, we use this section to introduce the concepts of lattice and reciprocal space, which are the foundation for the treatment of crystal diffraction.

Kinematic scattering from an assembly of atoms follows the same treatment as for a single atom:

$$\phi_s \approx \frac{2\pi me\,e^{2\pi i\vec{k}_o\cdot\vec{r}}}{h^2}\frac{}{r}\int V(\vec{r}')e^{2\pi i(\vec{k}-\vec{k}_o)\cdot\vec{r}'}\,d\vec{r}'' = \frac{2\pi me\,e^{2\pi i\vec{k}_o\cdot\vec{r}}}{h^2}\frac{}{r}FT(V(\vec{r}')) \qquad (9)$$

Here FT denotes Fourier transform. The potential of an assembly of atoms can be expressed as a sum of potentials from individual atoms

$$V(\vec{r}) = \sum_i\sum_j V_i\left(\vec{r}-\vec{r}_j\right) \qquad (10)$$

The summations over i and j are for the type of atoms and atoms of each type respectively.

To see how an atomic assembly diffracts differently from a single atom, we first look at a row of atoms that are separated periodically by an equal spacing of a. Each atom contributes to the potential at point $\vec{r}$. The total potential is obtained by summing up the potential of each atom. If we take the x-direction along the row, then

$$V\left(\vec{r}\right) = \sum_{n=1}^{N} V_A(\vec{r}-na\hat{x}) \qquad (11)$$

The sum can be considered as placing an atom on each of a collection of points; these points describe the geometrical arrangement of the atoms. In a crystal, the periodic arrangement of points defines the lattice. In both cases, the potential of an assembly of atoms is a convolution of the atomic potential and the lattice:

$$V\left(\vec{r}\right) = V_A(\vec{r}) * \sum_{n=1}^{N} \delta(\vec{r} - na\hat{x}) \tag{12}$$

The Fourier transform of this potential [see Eq. (9)] is the product of Fourier transform of the atomic potential and the Fourier transform of the lattice:

$$FT[V(\vec{r})] = FT[V_A(\vec{r})] \cdot FT\left[\sum_{n=1}^{N} \delta(\vec{r} - na\hat{x})\right] \tag{13}$$

The Fourier transform of atomic potential gives the atomic scattering factor. The Fourier transform of an array of delta functions gives

$$FT\left[\sum_{n=1}^{N} \delta(\vec{r} - na\hat{x})\right] = \sum_{n=1}^{N} e^{-2\pi i(\vec{k}-\vec{k}_o)\cdot\vec{x}na} = \frac{\sin\left[\pi\left(\vec{k}-\vec{k}_o\right)\cdot\vec{x}Na\right]}{\sin\left[\pi\left(\vec{k}-\vec{k}_o\right)\cdot\vec{x}a\right]} \tag{14}$$

The function of Eq. (14) has an infinite number of maxima at the condition:

$$\left(\vec{k} - \vec{k}_o\right) \cdot \hat{x} = h/a \tag{15}$$

Here h is an integer from $-\infty$ to ∞. The maxima are progressively more pronounced with increasing N. For sufficiently large N, Eq. (14) reduces to a periodic array of delta functions with the spacing of $1/a$.

For a three-dimension periodic array of atoms, where the position of atoms is given by the integer displacement of the unit cell **a**, **b**, and **c**, the potential is given by

$$V\left(\vec{r}\right) = V_A\left(\vec{r}\right) * \sum_{n,m,l} \delta\left(\vec{r} - n\vec{a} - m\vec{b} - l\vec{c}\right) \tag{16}$$

The Fourier transform of the three-dimensional lattice of a cube N×N×N gives

$$FT\left[\sum_{n=1}^{N}\sum_{m=1}^{N}\sum_{l=1}^{N} \delta\left(\vec{r} - n\vec{a} - m\vec{b} - l\vec{c}\right)\right] =$$

$$\frac{\sin\left[\pi\left(\vec{k}-\vec{k}_o\right)\cdot\vec{a}N\right]}{\sin\left[\pi\left(\vec{k}-\vec{k}_o\right)\cdot\vec{a}\right]} \frac{\sin\left[\pi\left(\vec{k}-\vec{k}_o\right)\cdot\vec{b}N\right]}{\sin\left[\pi\left(\vec{k}-\vec{k}_o\right)\cdot\vec{b}\right]} \frac{\sin\left[\pi\left(\vec{k}-\vec{k}_o\right)\cdot\vec{c}N\right]}{\sin\left[\pi\left(\vec{k}-\vec{k}_o\right)\cdot\vec{c}\right]} \tag{17}$$

Similar to the one-dimensional case, Eq. (17) defines an array of peaks. The position of the peaks is placed where

$$\left(\vec{k} - \vec{k}_o\right) \cdot \vec{a} = h$$
$$\left(\vec{k} - \vec{k}_o\right) \cdot \vec{b} = k \tag{18}$$
$$\left(\vec{k} - \vec{k}_o\right) \cdot \vec{c} = l$$

with h, k, and l as integers. It can be shown that

$$\Delta \vec{k} = \vec{k} - \vec{k}_o = h\vec{a}* + k\vec{b}* + l\vec{c}* \tag{19}$$

and

$$\vec{a}* = \frac{\left(\vec{b} \times \vec{c}\right)}{\vec{a} \cdot \left(\vec{b} \times \vec{c}\right)}, \; \vec{b}* = \frac{\left(\vec{c} \times \vec{a}\right)}{\vec{a} \cdot \left(\vec{b} \times \vec{c}\right)}, \; \vec{c}* = \frac{\left(\vec{a} \times \vec{b}\right)}{\vec{a} \cdot \left(\vec{b} \times \vec{c}\right)} \tag{20}$$

The vectors $a*$, $b*$, and $c*$ together define the three-dimensional reciprocal lattice.

For nanostructures that have the topology of a periodic lattice, we can be describe the structure by the lattice plus a lattice-dependent displacement:

$$V\left(\vec{r}\right) = V_A\left(\vec{r}\right) * \sum_{n,m,l} \delta\left(\vec{r} - \vec{R} - \vec{u}(\vec{R})\right), \text{ with } \vec{R} = n\vec{a} - m\vec{b} - l\vec{c} \tag{21}$$

For simplicity, we will restrict the treatment to monoatomic primitive lattices. Generalization to complex cases with nonprimitive lattices follows the same principle and will lead to similar qualitative conclusions, but more complex expressions. The Fourier transform of the lattice gives a sum of two terms:

$$FT\left[\sum_{n,m,l} \delta\left(\vec{r} - \vec{R} - \vec{u}(\vec{R})\right)\right] = \sum_{n,m,l} \exp\left(2\pi i \Delta\vec{k} \cdot \vec{R}\right)\exp\left[2\pi i \Delta\vec{k} \cdot \vec{u}\left(\vec{R}\right)\right] \tag{22}$$

For small displacement, Eq. (22) can be expanded to the first order

$$FT\left[\sum_{n,m,l} \delta\left(\vec{r} - \vec{R} - \vec{u}(\vec{R})\right)\right] =$$

$$\sum_{n,m,l} \exp\left(2\pi i \Delta\vec{k} \cdot \vec{R}\right) + \sum_{n,m,l} 2\pi i \Delta\vec{k} \cdot \vec{u}\left(\vec{R}\right)\exp\left(2\pi i \Delta\vec{k} \cdot \vec{R}\right) \tag{23}$$

The first term is the same as Eq. (17), which defines an array of diffraction peaks; the position of each peak is defined by the reciprocal lattice of the averaged crystal structure. Compared to infinite crystals, the diffraction peak of a finite crystal has a broad distribution, which is defined by the shape of the crystal. The second term describes the diffuse scattering around a reflection defined by the crystal reciprocal lattice. If we take the reflection as g and write

$$\Delta\vec{k} = \vec{g} + \vec{q} \text{ and } \vec{g} \cdot \vec{R} = 2n\pi \tag{24}$$

For $|g| >> |q|$, the diffuse scattering term can be rewritten as

$$\sum_{n,m,l} 2\pi i \Delta\vec{k} \cdot \vec{u}\left(\vec{R}\right)\exp\left(2\pi i \Delta\vec{k} \cdot \vec{R}\right) \approx \sum_{n,m,l} 2\pi i \vec{g} \cdot \vec{u}\left(\vec{R}\right)\exp\left(2\pi i \vec{q} \cdot \vec{R}\right) \tag{25}$$

Equation (25) is the Fourier sum of the displacements along the g direction. The intensity predicted by this equation will increase with a g^2-dependence. The atomic scattering contains the Debye–Waller factor, which describes the damping of high-angle scattering because of thermal vibrations. The balance of these two terms results in a maximum contribution to the diffuse scattering from the structural deviation from the ideal crystal lattice [5].

The oscillations from the finite size of the nanocrystals are clearly visible in the diffraction pattern shown in Fig. 6.2. The subtle differences in the intensity oscillations for different reflections can come from several factors, including surface relaxation, the small tilt and curvature of the Ewald sphere, and nonnegligible multiple scattering effects for heavy atoms such as Au. The surface relaxation can be treated using the approximation described above.

In summary, we have described a kinematical theory for an atomic assembly by considering the interplay between ordering among atoms, the finite sizes, and the structural modification. Qualitatively, we expect to find several features in the diffraction pattern of an assembly of atoms, including intensity maxima from ordering of atoms, diffuse scattering from the outline shape of the atomic assembly, and structural modification in the form of atomic displacements. The characteristics of diffuse scattering from the shape and atomic displacements are different. While the shape function is same for all reflections, diffuse scattering from atomic displacement has a characteristic distribution. This difference can be used in experimental studies to distinguish these two effects [5].

6.4
High-Resolution Electron Microscopy

Transmission high-resolution electron images are formed by recombining diffracted beams of the back focal plane of the image formation (objective) lens at the imaging plane (see Fig. 6.4). While electron diffraction records structural information in the reciprocal space, electron imaging gives direct information about local structure and morphology with near-atomic resolution. Information from HREM is complemen-

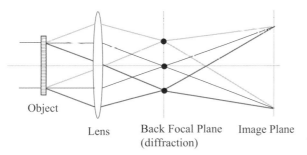

Object

Lens   Back Focal Plane   Image Plane
            (diffraction)

Figure 6.4. High-resolution electron image formation, where the diffraction pattern is formed at the back focal plane of the objective lens and diffracted beams recombine to form image.

tary to electron diffraction. However, interpretation of HREM images requires the knowledge of image formation and the contrast transfer function of the electron image formation lens.

For very small nanostructures or weakly scattering objects, the object potential is weak. Under such condition, following Eq. (2), the exit electron wave function is approximately given by

$$\phi_e(x, y, z) \approx 1 + i\pi\lambda\bar{U}(x, y) \tag{26}$$

Here

$$\bar{U}(x, y) = \int_0^t U(x, y, z)dz \tag{27}$$

is the projected potential along the electron beam direction. The potential $U(x,y,z)$ is measured in ångstrom^{-2} (see the relationship between U and V in Section 3.1). U is electron acceleration voltage dependent because of relativistic effects. Equation (27) is often called the weak phase approximation. For $U \sim 5 \times 10^{-2}$ Å^{-2}, electron wavelength $\lambda \sim 0.02$ Å, Eq. (27) is a reasonable approximation for thickness $t < 100$ Å.

The contrast of electron image depends on focus and the objective lens aberrations. Among many forms of aberrations, the spherical aberration C_s and chromatic aberration C_c are the dominant ones for a conventional TEM. To look at the effect of these two aberrations, the aperture and defocus on imaging, let us start with a sinusoidal weak phase object such that

$$\phi_e(x, y) \approx 1 + i\varepsilon\cos(2\pi qx) \tag{28}$$

The Fourier transform of this wave function is three delta functions:

$$\phi_e(k_x, k_y) \approx \delta(k_x, k_y) + i\varepsilon[\delta(k_x - q, k_y) + \delta(k_x + q, k_y)]/2 \tag{29}$$

For the purpose of discussion now, we omit the aperture function. In this case, the wave function at the back focal plane is given by

$$\phi_f(k_x, k_y) \approx \{\delta(k_x, k_y) + i\varepsilon[\delta(k_x - q, k_y) + \delta(k_x + q, k_y)]/2\}\exp(i\chi(k_x, k_y)) \tag{30}$$

Here

$$\chi(k_x, k_y) = \left[k_x^2 + k_y^2\right]\left[\frac{C_s\lambda^2}{2}\left[k_x^2 + k_y^2\right] + \Delta f\right] \tag{31}$$

is the phase shift introduced by the spherical aberration (C_s) of the image formation lens and defocus (Δf). The electron image is a Fourier transform of Eq. (30)

$$\phi_i(x) = 1 - \varepsilon\cos(2\pi qx)\sin\chi(q) + i\varepsilon\cos(2\pi qx)\sin\chi(q) \tag{32}$$

From this we obtain the image intensity

$$I(x) = \left[1 - \varepsilon\cos(2\pi qx)\sin\chi(q)\right]^2 + \varepsilon^2 \cos^2(2\pi qx)\sin^2\chi(q)$$

$$= 1 - 2\varepsilon\cos(2\pi qx)\sin\chi(q) + \varepsilon^2 \cos^2(2\pi qx) \tag{33}$$

$$\approx 1 - 2\varepsilon\cos(2\pi qx)\sin\chi(q)$$

The approximation is for small ε under the weak phase object assumption.

Equation (33) shows that in addition to the amplitude of the original wave, the image intensity modulation also depends on the sine function of the aberrated phase. The difference between the maximum and minimum intensities gives the image contrast. The maximum contrast is obtained by having $\sin\chi(q) = 1$ or $\chi(q) = \pi/2$, in which case, $I_{Max} - I_{Min} = 4\varepsilon$. The contrast transfer function is the ratio of the difference with $\sin\chi(q) \neq 1$ and 4ε, defined by:

$$CTF = \frac{I_{Max} - I_{Min}}{\left(I_{Max} - I_{Min}\right)_{MAX}} = \frac{|4\varepsilon\sin\chi(q)|}{4\varepsilon} = |\sin\chi(q)| \tag{34}$$

Thus, the objective lens's imaging properties are described by the contrast transfer function (CTF). Since CTF is independent of the object, it provides a convenient mechanism for understanding the imaging process and contrast. For an object with many frequencies, CTF gives the relative contribution of each frequency to the final image in coherent imaging.

The maximum contrast is obtained when the phase shift is $\pi/2$ and the optimum is that this is achieved for as many frequencies as possible. The condition for this, known as the Scherzer focus, is obtained under the following conditions for aperture angle α and defocus Δf:

$$\alpha_{opt} = 1.41\left(\frac{\lambda}{C_s}\right)^{1/4} ; \quad \Delta f_{opt} = -(C_s\lambda)^{1/2} \tag{35}$$

At Scherzer focus the half-width of the image intensity of a point object is given by

$$\delta_h = 0.43\left(C_s\lambda^3\right)^{1/4} \tag{36}$$

This is often used as a measurement of the electron microscope resolution. For C_s=0.5 mm, λ=0.025 Å, δ=1.28 Å.

The focal length of electron lenses depends on electron energy. Chromatic aberration comes from the finite electron energy spread, objective lens current fluctuations, and high voltage instability. All of these introduce an envelope function to the contrast transfer function

$$K_C = \exp\left[-\left(\frac{q}{\Lambda}\right)^2\right] \tag{37}$$

Here, Λ is a function of the chromatic aberration coefficient, the energy spread, lens current, and high voltage fluctuations.

Figure 6.5 shows a high-resolution image of gold nanoparticles recorded at the Scherzer focus condition. The image was recorded using the Lawrence Berkeley Laboratory One-Ångstrom Microscope [15]. The basic instrument is a modified Philips CM300FEG/UT, a TEM with a field-emission electron source and an ultra-twin objective lens with low spherical aberration ($C_s = 0.65$ mm) and a native resolution of 1.7 Å. The size of the particles ranges from 2 to ~3 nm. The particles were supported on a thin carbon film.

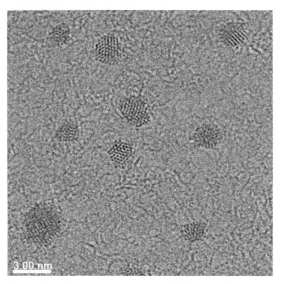

Figure 6.5. High-resolution image of 2~3 nm gold nanoparticles under the Scherzer focus condition. The direct, atomic-resolution image clearly reveals the diversity of structures, from single-crystal to twinned particles.

Another high-resolution electron imaging technique is scanning transmission electron microscopy (STEM). In STEM, electrons are focused into a very small probe and scanned across the sample. Electrons scattered by the sample are detected to form an image. The image intensity is approximately proportional to the square of atomic number, Z^2, if only electrons scattered into high angles are detected using a high-angle annual dark-field (HAADF) detector [16].

6.5
Experimental Analysis

6.5.1
Experimental Diffraction Pattern Recording

The optimum setup for quantitative electron diffraction is the combination of a flexible illumination system, an imaging filter, and an array of 2-D detectors with a large dynamic range. The three diffraction modes described in Section 6.2 can be achieved through a three-lens condenser system. Currently both the JEOL (JEOL, USA) and Zeiss microscopes offer this capability. Two types of energy filter are currently employed: one is the in-column Ω-energy filter [17] and the other is the post-column Gatan imaging filter (GIF) [18]. Each has its own advantages. The in-column Ω-filter takes the full advantage of the postspecimens lens of the electron microscope and can be used in combination with detectors such as film or imaging plates (IP), in addition to the CCD camera. For electron diffraction, geometric distortion, isochromaticity, and angular acceptance are the important characteristics of filters [19]. Geometrical distortion complicates the comparison between experiment and theory and is best corrected by experiment. Isochromaticity defines the range of electron energies for each detector position. Ideally this should be the same across the whole detector area. Angular acceptance defines the maximum range of diffraction angles that can be recorded on the detector without a significant loss of isochromaticity.

Current 2-D electron detectors include CCD cameras and imaging plates (IP). The performance of the CCD camera and IP for electron recording has been measured [20]. Both are linear with large dynamic range. At low dose range, the CCD camera is limited by the readout noise and dark current of CCD. The readout noise places a limit on what information can be recovered from recorded images if the CCD camera has a limited resolution. IP has a better performance at low dose range due to the low dark current and readout noise of the photomultiplier. For medium and high dose, the IP is limited by linear noise due to the granular variation in the phosphor and instability in the readout system. The CCD camera is limited by the linear noise in the gain image, which can be made very small using averaging. There is an uncertainty in the uniformity of CCD because of its dependence on the gain image as prepared in the electron microscope. The performance of CCD also varies from one camera to another, which makes the individual characterization necessary.

The noise in the experimental data can be estimated using the measured detector quantum efficiency (DQE) of the detector

$$\mathrm{var}(I) = mgI/DQE(I) \tag{38}$$

Here I is the estimated experimental intensity, var denotes the variance, m is the mixing factor defined by the point spread function, and g is the gain of the detector [20]. This expression allows an estimation of variance in experimental intensity once DQE is known, which is especially useful in the χ^2-fitting, where the variance is used as the weight.

6.5.2
The Phase Problem and Inversion

In the kinematic approximation, the diffracted wave is proportional to the Fourier transform of the potential [Eq. (9)]. If both the amplitude and phase of the wave are known, then an image can be reconstructed, which would be proportional to the projected object potential. In recording a diffraction pattern, however, what is recorded is the amplitude square of the diffracted wave. The phase is lost. The missing phase is known as the phase problem. In the case of kinematic diffraction, the missing phase prevents reconstruction of the object potential by inverse Fourier synthesis. The phase is preserved in imaging up to the information limit. In electron imaging, the scattered waves recombine to form an image by the transformation of a lens and the intensity of the image is recorded, not the diffraction. The complication in imaging is the lens aberration. Spherical aberration introduces an additional phase to the scattered wave. This phase oscillates rapidly as the scattering angle increases. Additionally, chromatic aberration and the finite energy spread of the electron source impose a damping envelope to the CTF [Section 6.4] and limit the highest-resolution information (information limit) that passes through the lens. As result, the phase of scattered waves with $\sin\theta/\lambda > 1 \text{ Å}^{-1}$ is typically lost in electron images and the resolution of image is ~1 Å for the best microscopes currently available. Phase retrieval is a subject of great interest in both electron and x-ray diffraction. If the phase of a diffraction pattern can be found, then an image can be formed without a lens.

In electron diffraction, the missing phase has not been a major obstacle to its application. The reason is electron imaging and that, for most electron diffraction applications, electrons are multiply scattered. The missing phase is the phase of the exit wave function. Inversion of exit wave function to object potential is not as straightforward as an inverse Fourier transform. A theory developed by two research groups (Spence at Arizona State University and Allen in Australia) shows the principle of inversion using data sets of multiple thicknesses, orientations, and overlapping coherent electron diffraction [21]. The inversion is based on the scattering matrix that relates the scattered wave to the incident wave. This matrix can be derived based on the Bloch wave method, which has a diagonal term of exponentials of the product of eigenvalues and thickness. Electron diffraction intensities determine the moduli of all elements of the scattering matrix. Using the properties of scattering matrix (unitarity and symmetries), an overdetermined set of nonlinear equations can be obtained from these data. Solution of these equations yields the required phase information and allows the determination of a (projected) crystal potential by inversion [22].

For materials' structural characterization, in many cases, the structure of materials is approximately known. What is to be determined is the accurate atomic positions and unit cell sizes. Extraction of these parameters can be done in a more efficient manner using the refinement technique [23]. Another important fact is that the phase of object potential is actually contained in the diffraction pattern through electron interference when electrons are elastically scattered for multiple times [24].

For nanostructures such as carbon nanotubes laid horizontally, the number of atoms is low along the incident electron beam direction. Electron diffraction, to a good approximation, can be treated kinematically. Many nanostructures also have a complicated structure. Modeling, as required in the refinement technique, is difficult because of the lack of knowledge about the structure. The missing phase then becomes an important issue. Fortunately, the missing phase is actually easier to retrieve for nonperiodic objects than for periodic crystals. The principle and technique for phase retrieval are described next.

6.5.3
Electron Diffraction Oversampling and Phase Retrieval for Nanomaterials

Electron phase retrieval uses a coherent electron probe. The formation of a coherent electron probe in NED mode follows the same principle as STEM probe formation, but in a reverse optical geometry. In STEM, a parallel coherent illumination is brought into focus by the electron objective lens. The phase difference from lens aberration, defocus, and convergence angle defines the size and shape of the electron probe. In NED mode, a focused probe (by the condenser lens) at the front focal plane of the pre-objective lens is imaged into a nearly parallel beam. Because of the lens aberration, beams at different angles to the optic axis are imaged at different distances.

For small nanostructures, such as carbon nanotubes, electron diffraction is well described by the kinematical approximation. At a small scattering angle, the scattered electron wave is a sum of the scattered waves over the volume of the structure:

$$\phi_s\left(\vec{k}\right) \approx \int [1 + i\pi\lambda\, U(\vec{r}')] e^{-2\pi i \vec{k}\cdot\vec{r}'} \phi_o(\vec{r}')d\vec{r}' = \phi_o\left(\vec{k}\right) + i\pi\lambda \int U(\vec{r}') e^{-2\pi i \vec{k}\cdot\vec{r}'} \phi_o(\vec{r}')d\vec{r}' \tag{39}$$

Here $U(\vec{r})$ is the interaction potential defined in Section 3.1 and $\vec{k}$ is the scattered wave vector. The illuminating electron wave function $\phi_o(\vec{r})$ is formed by the electron lens as described above. Information about the structure is carried in the second term. For an idealized plane wave illumination, the electron diffraction intensity can be expressed through the Fourier transform of the potential $U(\vec{k})$ as:

$$I\left(\vec{k}\right) \approx \delta\left(\vec{k}\right) + (\pi\lambda)^2 \left| U\left(\vec{k}\right) \right|^2 \tag{40}$$

For a finite nanostructure, the diffraction intensity is continuous in reciprocal space.

Using the combination of coherent NED and phase retrieval, we have demonstrated for the first time that atomic resolution can be achieved from diffraction intensities without an imaging lens [25]. This technique was applied to image the atomic structure of a double-wall carbon nanotube (DWNT). The electron diffraction pattern from a single DWNT was recorded and phased. The resolution is limited by the diffraction intensity. The resolution obtained for the DWNT was 1 Å from a microscope of a nominal resolution of 2.3 Å.

The principle of phase retrieval for a localized object is based on the sampling theory. For a localized object of size S, the minimum sampling frequency (Nyquist frequency) in reciprocal space is $1/S$. Sampling with a smaller frequency (oversampling) increases the field of view. Wavefunctions at these oversampled frequencies are a combination of the wavefunctions sampled at the Nyquist frequency. Because of this, phase information is preserved in oversampled diffraction intensities. Oversampling can be achieved only for a localized object. For a periodic crystal, the smallest sampling frequency is the Nyquist frequency. The iterative phasing procedure works by imposing the amplitude of the diffraction pattern in the reciprocal space and the boundary conditions in real space. The procedure was first developed by Fienup [26] and improved by incorporating other constraints such as symmetry [27]. The approach of diffractive imaging, or imaging from diffraction intensities, appears to solve many technical difficulties in conventional imaging of nonperiodic objects, namely, resolution limit by lens aberration, sample drift, instrument instability, and low contrast in electron images.

The phase retrieval procedure works by starting with the measured amplitude of the Fourier transform and random phases: an estimate of the projected potential is computed and modified to satisfy the real-space constraints. The modified potential is Fourier transformed and the calculated amplitudes are replaced by their measured values. The procedure is repeated until a self-consistent solution of the potential is found. The amplitudes of the Fourier transform of the potential are obtained based on Eq. (40). There are two major constraints that can be used for electron phase retrieval: one is the approximate shape of the object from low-resolution imaging and the other is positivity. Outside the image, we assume a constant project potential, which acts as the support. The projected potential is assumed positive (the same sign).

We use the modified hybrid-input-output (HIO) algorithm outlined by Millane and Stroud [27] for iterative phase retrieval. The working principle of this algorithm is shown in Fig. 6.6. The procedure starts with an estimated image f_n and calculate the constrained image c_n. The support constrain is applied by using

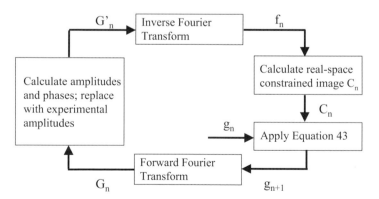

Figure 6.6. The flow chart of hybrid input and output algorithm for iterative phase retrieval. (After Ref. [273].)

$$C_n(\vec{r}) = 0, \vec{r} \in S \tag{41}$$

and

$$C_n(\vec{r}) > 0, \text{ and real } \vec{r} \in O \tag{42}$$

where S and O denote regions of support and the object. From the initial image and the calculated constrained image c_n, a driver function g_{n+1} is derived:

$$g_{n+1}(\vec{r}) = \begin{cases} f_n(\vec{r}) & if\,|C_n(\vec{r}) - f_n(\vec{r})| < \varepsilon \\ g_n(\vec{r}) + \beta[C_n(\vec{r}) - f_n(\vec{r})] & if\,|C_n(\vec{r}) - f_n(\vec{r})| > \varepsilon \end{cases} \tag{43}$$

When the phase of this driver function is combined with the experimental amplitude data they yield an image f_{n+1} that more closely satisfies the real-space constraint than the original image f_{n+1}. The alternative is to take

$$g_{n+1}(\vec{r}) = C_n(\vec{r}) \tag{44}$$

giving the error-reduction (ER) algorithm. It has been found that it is often useful to mix HIO with a few iterations of ER. We found that the ER algorithm is often efficient at the initial stage of iteration, but HIO is always more efficient at late-stage iterations.

Figure 6.7 shows a small DWNT reconstructed from the recorded electron diffraction pattern. The scattering of carbon is generally too weak for direct imaging of the atomic structure in an electron microscope. Diffractive imaging avoids this problem by recording the diffraction pattern, which has a better signal-to-noise ratio than the image because of the highly ordered (helical) structure of carbon nanotubes. Details of the tube structure are clearly visible. The potential profile demonstrates the type of information that can be obtained from the reconstructed image.

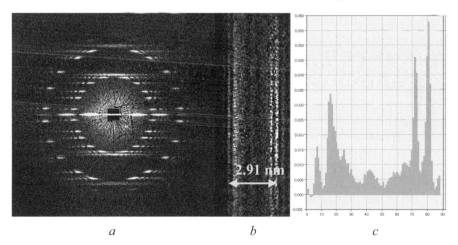

a *b* *c*

Figure 6.7. (a) The recorded electron from a single double-wall carbon nanotube, (b) the reconstructed image using HIO algorithm, (c) the profile of reconstructed potential averaged over the middle section of the image; the profile is consistent with a concentric hollow tube.

6.6
Applications

6.6.1
Structure Determination of Individual Single-Wall Carbon Nanotubes

For simple structures, such as single-wall carbon nanotubes (SWNTs), the structure can be determined uniquely from the diffraction pattern alone. Carbon nanotubes have attracted extraordinary attention due to their unique physical properties, from atomic structure to mechanical and electronic properties, since Iijima showed the first high-resolution TEM image and electron diffraction of multiwall carbon nanotubes [28]. A SWNT can be regarded as a single layer of graphite that has been rolled up into a cylindrical structure. In general, the tube is helical with the chiral vector (n, m) defined by $\vec{c} = n\vec{a} + m\vec{b}$, where $\vec{c}$ is the circumference of the tube and $\vec{a}$ and $\vec{b}$ are the unit vectors of the graphite sheet. A striking feature is that tubes with $n - m = 3l$ (l is a integer) are metallic, while others are semiconductive [29]. This unusual property, plus the apparent stability, has made carbon nanotubes an attractive material for constructing nanoscale electronic devices. As-grown SWNTs have a dispersion of chirality and diameters. Hence, a critical issue in carbon nanotube applications is the determination of individual tube structure and its correlation to the properties of the tube. This requires a structural probe that can be applied to individual nanotubes.

Gao et al. have developed a quantitative structure determination technique for SWNTs using NED [30]. This, coupled with improved electron diffraction pattern interpretation, allows determination of both the diameter and the chiral angle, and thus the chiral vector (n, m), of an individual SWNT. The carbon nanotubes they studied were grown by chemical vapor deposition. TEM observation was carried out in a JEOL2010F TEM with a high voltage of 200 keV.

Figure 6.8 shows the diffraction pattern from a SWNT. The main features of this pattern are: (i) a relatively strong equatorial oscillation which is perpendicular to the tube direction, and (ii) some very weak diffraction lines from the graphite sheet, which are elongated in the direction normal to the tube direction [31]. The intensities of diffraction lines are very weak in this case. In their experimental setup, the strongest intensity of one pixel is about 10 counts, which corresponds to ~12 electrons.

The diameter of the tube is determined from the equatorial oscillation, while the chiral angle is determined by measuring the distances from the diffraction lines to the equatorial line. The details are following. The diffraction of SWNT is well described by kinematic diffraction theory (Section 6.3). The equatorial oscillation in the Fourier transformation of a helical structure like SWNT is a Bessel function with n=0 [32] which gives:

$$I_0(X) \propto J_0^2(X) \propto \left| \int_0^{2\pi} \cos^{X \cos\Omega} d\Omega \right|^2 \tag{45}$$

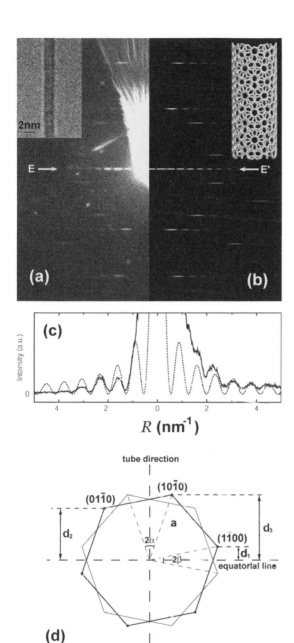

Figure 6.8. (a) A diffraction pattern from an individual SWNT 1.4 nm in diameter. The inset is a TEM image. The radial scattering around the saturated (000) is an artifact from aperture scattering. (b) Simulated diffraction pattern of a (14,6) tube. The inset is the corresponding structure model. (c) Profiles of equatorial oscillation along EE' from (a, b) and simulation for (14,6). (d) Schematic diagram of electron diffraction from an individual SWNT. The two hexagons represent the first order graphite-like {100} diffraction spots from the top and bottom of the tube.

Here $X = 2\pi R r_0 = \pi R D_0$, R is the reciprocal vector which can be measured from the diffraction pattern, and D_0 is the diameter of the SWNT. We use the position of $J_0^2(X)$ maxima (X_n, $n =0, 1, 2, ...$) to determine the tube diameter. With the first several maxima saturated and inaccessible, X_n/X_{n-1} can be used to determine the number N for each maximum in the equatorial oscillation. Thus, by comparing the experimental equatorial oscillation with values of X_n, the tube diameter can be uniquely determined.

To measure chirality from the diffraction pattern, Fig. 6.8(d) is considered, which shows the geometry of the SWNT diffraction pattern based on the diffraction of the top-bottom graphite sheets. The distances d_1, d_2, d_3 relate to the chiral angle α by:

$$d_1 + d_2 = d_3,$$

$$\alpha = \mathrm{atan}(\frac{1}{\sqrt{3}} \cdot \frac{d_2 - d_1}{d_3}) = \mathrm{atan}(\frac{1}{\sqrt{3}} \cdot \frac{2d_2 - d_3}{d_3}), \tag{46}$$

$$\text{or } \beta = \mathrm{atan}(\sqrt{3} \cdot \frac{d_1}{d_2 + d_3}) = \mathrm{atan}(\sqrt{3} \cdot \frac{d_3 - d_2}{d_2 + d_3}).$$

These relationships are not affected by the tilting angle of the tube (see below). Because d_2 and d_3 correspond to the diffraction lines having relatively strong intensities and are further from the equatorial line, they are used in our study instead of d_1 to reduce the error. The distances can be measured precisely from the digitalized patterns. The errors are estimated to be <1% for the diameter determination and <0.2° for the chiral angle.

Using the above methods, the SWNT giving the diffraction pattern shown in Fig. 6.8 was determined to have a diameter of 1.40 nm (±0.02 nm) and a chiral angle of 16.9° (±0.2°). Among the possible chiral vectors, the best match is (14,6), which has a diameter of 1.39 nm and a chiral angle of 17.0°. The closest alternative is (15,6), having a diameter eof 1.46 nm and a chiral angle of 16.1°, which is well beyond the experiment error. Figure 6.8(b) plots the simulated diffraction pattern of (14,6) SWNT from the structure model shown in the inset. Figure 6.8(c) compares the equatorial intensities of experiment and simulation. These results show an excellent agreement.

6.6.2
Structure of Supported Small Nanoclusters and Epitaxy

Nanometer-sized structures in the forms of clusters, dots, and wires have recently received considerable attention for their size-dependent transport, optical, and mechanical properties. The focus is on synthesizing nanostructures of desired shapes with narrow size distributions. For clusters or nanocrystals on crystalline substrates, epitaxy gives lower interface energy and can lead to enhanced stability and better control over the interfacial electronic properties. At the nanometer scale, cluster equilibrium shape is also determined by the surface, interface, and strain energies. A challenge, thus, is how to determine the structure of individual clusters. The case highlighted below on Ag on Si (100) is taken from the experimental work by Li et al.

at University of Illinois, Urbana-Champaign. Over the past three years, they have carried out a systematic study of nanoclusters' structure and interfaces using a combination of electron diffraction and microscopy [33–36].

Figure 6.9 shows the electron diffraction patterns for samples deposited on H-terminated Si (100) surfaces. The diffraction patterns were taken off the [100] zone axis to avoid strong multiple scattering in zone axis orientation. The diffraction pattern of the as-deposited sample consists of a strong and continuous Ag {111} ring, short Ag {200} arcs on a weak Ag {200} ring, short Ag {220} arcs on a weak Ag {220} ring and weak {311} rings. Upon annealing at 400 °C, Ag (020) and (002) reflection intensities increase significantly. Both have diffuse streaks along (011) and (0–11) directions. The Ag (020) and (002) are asymmetrical because of the off-zone axis orientation of the diffraction pattern. Meanwhile, the diffraction intensity in the continuous ring decreases significantly, but remains visible. Figure 6.9 shows high-resolution images of Ag clusters on H-Si(100) after annealing with strong moiré fringe contrast. These images were taken at the Si [100] zone axis. At this orientation, the Si (022) and (0–22) planes are imaged. Most as-grown Ag clusters show no visible moiré fringes, which is consistent with the diffraction pattern that is dominated by a {111} ring. A few clusters with moiré fringes are often defective. In Fig. 9, the clusters of dark contrast with no moiré fringes contribute to the {111} ring in the diffraction pattern. For as deposited clusters (not shown here), at first sight, the orientation of these clusters appears to be random. However, a close inspection of the diffraction pattern shows a much weaker {200} ring than what it would be in a powder diffraction pattern of random polycrystalline Ag. For single crystals oriented with Ag(111)//Si(100) or Ag(100)//Si(100), strong Ag {220} is expected in both cases, while a strong Ag {200} is also expected in the case of Ag(111)//Si(100). Both of these cases can be ruled out. In Fig. 6.9, the square Ag clusters with 2-D moiré fringes (from interference between Ag and Si lattices [2]) perfectly

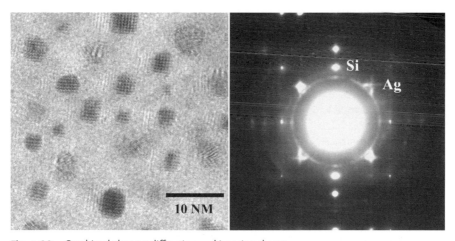

Figure 6.9. Combined electron diffraction and imaging characterization of epitaxial Ag nanoclusters/nanocrystals on Si (100) substrate.

parallel to Si (220) lattice planes, in good agreement with the electron diffraction analysis. At this stage, the transformation from random orientation to epitaxial growth is not finished, because we still see weak-contrast Ag clusters, supposed to be random Ag clusters. The Ag {200} reflections have the shape of a plus-sign, centered at the Ag {200} position, suggestive of perfect cubic Ag nanocrystals, with their edges perfectly aligned to the Si (011) and (01–1) directions.

6.7
Conclusions and Future Perspectives

This chapter has described the practice and theory of electron imaging and diffraction for structural analysis of nanomaterials and has demonstrated that the information obtainable from electron diffraction with a small probe and strong interactions complements other characterization techniques, such as x-ray and neutron diffraction that samples a large volume and real-space imaging by HREM with a limited resolution.

The challenge is to extend the applications of microscopy techniques to soft and biomaterials, where radiation-induced structural damage is more likely to occur at high electron dose levels. While radiation tolerance can be significantly improved with cryo-electron microscopy, the ultimate image resolution will be limited by the low signal-to-noise ratio resulting from the low electron dose that the sample can tolerate [4, 37]. The sensitivity demonstrated here for carbon nanotubes would also be useful for imaging molecular structures.

Abbreviations

CBED – convergent-beam electron diffraction
CCD – charge coupled device
CTF – contrast transfer function
DQE – detector quantum efficiency
DWNT – double-wall carbon nanotube
FEG – field emission guns
GIF – Gatan imaging filter
HAADF – high-angle annual dark-field
HIO – hybrid-input-output
HREM – high-resolution electron microscopy
IP – imaging plates
NED – nanoarea electron diffraction
SAED – selected area electron diffraction
STEM – scanning transmission electron microscope
SWNT – single- wall carbon nanotube
TEM – transmission electron microscopes

Acknowledgments

Work was supported by DOE DEFG02–01ER45923 and DEFG02–91ER45439 and uses the TEM facility of the Center for Microanalysis of Materials at Materials Research Laboratory. The author would like to thank R. Zhang and L. Nagahara (Motorola Labs) for providing the carbon nanotubes, Dr. M. O'Keefe and Chris Nelson for Fig. 6.5, Dr. Min. Gao for Fig. 6.8, and Boquan Li for Fig. 6.9.

References

1 J.W. Edington, Practical Electron Microscopy in Materials Science, Monograph 2, *Electron Diffraction in the Electron Microscope*, MacMillan, Philips Technical Library, **1975**.

2 P. Hirsch et al., *Electron Microscopy of Thin Crystals*, R. E. Krieger, Florida, p. 19, **1977**.

3 D. B. Williams, C. B. Carter, *Transmission Electron Microscopy*, Plenum, New York, **1996**.

4 L. Reimer, *Transmission Electron Microscopy*, 4th edn, Springer, Berlin,**1997**.

5 J. M. Cowley, *Diffraction Physics*, North-Holland, New York, **1981**.

6 J. C. H. Spence, J. M. Zuo, *Electron Microdiffraction*, Plenum, New York, **1992**.

7 Z. L. Wang, *Elastic and Inelastic Scattering in Electron Diffraction and Imaging*, Plenum, New York, **1995**.

8 E. J. Kirkland, *Advanced Computing in Electron Microscopy*, Plenum Press, New York, **1998**.

9 M. Tanaka, M. Terauchi, T. Kaneyama, *Convergent-Beam Electron Diffraction*, JEOL, Tokyo, **1988**.

10 J. P. Morniroli, *Electron Diffraction, Dedicated Software to Interpret LACBED Patterns*, USTL, Lille, France, **1994**.

11 J. M. Cowley, Electron nanodiffraction, *Microsc. Res. Tech.* 46, 75, **1999**.

12 L. J. Wu, Y. M. Zhu, J. Tafto, Picometer accuracy in measuring lattice displacements across planar faults by interferometry in coherent electron diffraction, *Phys. Rev. Lett.* 85, 5126, **2000**.

13 J. M. Zuo, Quantitative convergent-beam electron diffraction, *Materials Trans. JIM* 39, 938–946, **1998**.

14 L. M. Peng, Electron atomic scattering factors and scattering potentials of crystals, *Micron* 30, 625, **1999**.

15 M. A. O'Keefe, C. J. D. Hetherington, Y. C. Wang, E. C. Nelson, J. H. Turner, C. Kisielowski, J. O. Malm, R. Mueller, J. Ringnalda, M. Pan, A. Thust, *Ultramicroscopy*, 89, 215, **2001**.

16 P. D. Nellist, S. J. Pennycook, High angular dark field scanning transmission electron microscopy, in *Advances in Imaging and Electron Physics, vol. 113, ed. P. W. Hawkes, Academic Press, San Diego, p. 147*, **2000**.

17 L. Reimer, ed., *Energy-Filtering Transmission Electron Microscopy*, Springer, Berlin, 1995.

18 O. L. Krivanek, S. L. Friedman, A. J. Gubbens, B. Kraus, An imaging filter for biological applications, *Ultramicroscopy*, 59, 267, **1995**.

19 H. Rose, in *Energy-Filtering Transmission Electron Microscopy*, ed. L. Reimer, Springer, Berlin, **1995**.

20 J. M. Zuo, Electron detection characteristics of a slow-scan CCD camera, imaging plates and film, and electron image restoration, *Microsc. Res. Tech.* 49, 245, **2000**.

21 J. C. H. Spence, Direct inversion of dynamical electron diffraction patterns to structure factors, *Acta Cryst.* A54, 7, **1998**.

22 L. J. Allen, T. W. Josefsson, H. Leeb, Obtaining the crystal potential by inversion from electron scattering intensities, *Acta Cryst.* A54, 388, **1998**.

23 J. M. Zuo, Measurements of g electron densities in Solids, *Rep. Prog. Phys.* 67, 2053, **2004**.

24 J. M. Zuo, J. C. Spence, R. Hoier, Accurate structure-factor phase determination by electron-diffraction in noncentrosymmetric crystals, *Phys. Rev. Lett.* 62, 547, **1989**.

25 J. M. Zuo, I. Vartanyants, M. Gao, R. Zhang, L. A. Nagahara, Atomic resolution imaging of a single double-wall carbon nanotube from diffraction intensities, *Science, 300,* 1419–1421, **2003**.

26 J. Fienup, Phase retrieval algorithms – a comparison, *Appl. Opt., 21, 2758,* **1982.**

27 R. P. Millane, W. J. Stroud, Reconstructing symmetric images from their undersampled Fourier intensities, *J. Opt. Soc. Am. A. 14, 568,* **1997.**

28 S. Iijima, Helical microtubules of graphitic carbon, *Nature 354,* 56, **1991.**

29 J. W. Mintmire, B. I. Dunlap, C. T. White, Are fullerene tubules metallic? *Phys. Rev. Lett. 68,* 631, **1992.**

30 M. Gao, J. M. Zuo, R. D. Twesten, I. Petrov, L. A. Nagahara, R. Zhang, Structure determination of individual single-wall carbon nanotubes by nano-area electron diffraction, *Appl. Phys. Lett. 82,* 2703–2706, **2003.**

31 S. Amelinckx, A. Lucas, P. Lambin, Electron diffraction and microscopy of nanotubes, *Rep. Prof. Phys. 62,* 1471, **1999.**

32 D. Sherwood, *Crystal, X-rays and Proteins.* John Wiley & Sons, New York, **1976.**

33 B. Q. Li, J. M. Zuo, Self-assembly of epitaxial Ag nanoclusters on H-terminated Si (111) surfaces, *J. Appl. Phys. 94,* 743–748, **2003.**

34 J. K. Bording, Y. F. Shi, B. Q. Li, J. M. Zuo, Size- and shape-dependent energetics of nano-crystal interfaces: experiment and simulation, *Phys. Rev. Lett., 90,* 226104, **2003.**

35 B. Q. Li, J. M. Zuo, The development of epitaxy of nanoclusters on lattice-mismatched substrates: Ag on H-Si(111) surfaces, *Surf. Sci. 520,* 7–17, **2002.**

36 J. M. Zuo, B. Q. Li, Nanostructure evolution during cluster growth: Ag on H-terminated Si(111) surfaces, *Phys. Rev. Lett. 88,* 255502, **2002.**

37 R. Henderson, The potential and limitations of neutrons, electrons and x-rays for atomic-resolution microscopy of unstained biological molecules, *Q. Rev. Biophys. 28,* 171, **1995.**

7
X-Ray Methods for the Characterization of Nanoparticles

Hartwig Modrow

7.1
Introduction

Physics and physical chemistry provide a broad range of analytical techniques for the characterization of matter, many of which can be applied for the characterization of nanoparticles, as shown impressively in some of the chapters in this book. Still, in my opinion x-ray based methods play a special role in the characterization of this type of matter. What is special about this class of experimental techniques? It is their penetration strength and element sensitivity, which was evident from the moment of the first discovery of x-rays by W.C. Röntgen in 1899. Why is this property so crucial for studies on nanoparticles in general, but especially if they are to be used for biomedical applications? In order to prevent particles from agglomerating, their surface is usually protected by a surfactant shell and thus not well accessible, e.g., for scanning microscopy techniques. Also, the (long lasting) stability of the particles in a given environment, e.g., in the environment of the human body, is to be guaranteed, and thus experiments which require special environmental conditions, e.g., ultrahigh vacuum, cannot make definite statements about the situation encountered under these conditions. Furthermore, by exploiting the element sensitivity of x-rays, the analysis of complex systems can be broken down into simpler, complementary steps and subsystems.

At the same time, x-ray methods can supply a broad bandwidth of information on a given set of particles, from the arrangement of atoms to the shape and morphology of particles and their chemical composition and electronic structure. In this chapter the three x-ray based techniques which are the respective "champions" in these three disciplines, x-ray diffraction (XRD), (anomalous) small-angle x-ray scattering [(A)SAXS], and x-ray absorption spectroscopy (XAS), will be introduced and discussed before some examples of their application are given. In order to understand what properties of matter can be investigated using these techniques, it is crucial to understand how x-rays – i.e., electromagnetic waves – interact with matter – i.e., electrons and positively charged atomic cores – which can be considered as a charge distribution in space for the description of this interaction. In fact, the different techniques rely on the occurrence of two different physical processes: In XAS, the absorption of a photon is observed; in XRD and SAXS, it is elastic scattering. To

Nanofabrication Towards Biomedical Applications. C. S. S. R. Kumar, J. Hormes, C. Leuschner (Eds.)
Copyright © 2005 WILEY-VCH Verlag GmbH & Co. KGaA, Weinheim
ISBN 3-527-31115-7

make this chapter easier reading for those without an extensive background in physics, this is done in quite a perceptual way, starting from what is measured and showing what information can be derived from it based on selected (partly idealized) examples. For a more rigorous formal description of these processes, which leads to the formulas and statements used in the following sections and any in-depth approach, the interested reader is referred to the Appendix of this chapter. After this introduction to the potential of the respective techniques, their specific application to nanoparticles is discussed from a critical perspective in relation to three examples, which are intended to show not only the strengths but also the weaknesses of the respective methods.

7.2
X-Ray Diffraction: Getting to Know the Arrangement of Atoms

In x-ray diffraction, elastic scattering processes are observed, i.e., the coupled values photon wavelength and photon energy remain constant, and so does the state of the scattering system. At the same time, the scattered x-rays stay coherent. Only the direction of propagation is changed, which is often described using the scattering vector, defined by the relation

$$\vec{q} = \frac{2\pi}{\lambda}(\vec{k} - \vec{k}_0) \tag{1}$$

(see also Fig. 7.1, and also the complete list of variables given at the end of this chapter). Consequently, what needs to be determined in an x-ray diffraction experiment is the probability of finding a scattered photon in a given solid angle element.

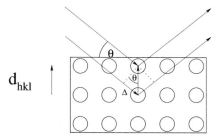

d_{hkl}

Figure 7.1. Derivation of the Bragg condition for x-ray diffraction.

The most elementary way to arrive at a statement of under what conditions intensity can be found in a given solid angle element can be derived from Fig. 7.1: in order to obtain a strong signal under a Bragg angle Θ_B measured relative to a given lattice plane, contributions of scattering events which occur in neighboring lattice planes must be in phase, i.e., the difference in the optical path length must be given by

$$n\lambda = 2d_{hkl}\sin\theta_B \tag{2}$$

Table 7.1. Different approaches for the measurement of x-ray diffraction data.

	White x-rays	**Monochromatic x-rays**
Single crystal	One position yields all reflexes (Laue method)	Rotate crystal to obtain all reflexes (rotating crystal method)
Powder	Each crystallite provides all reflexes (hard, if at all, to evaluate)	Each crystallite provides one reflex (Debye method)

which is exactly the Bragg condition. It should be stressed that this condition is necessary but not sufficient: it selects only the discrete angles under which diffraction may be observed and does not provide the angles under which it will be observed, because it is related only to the structural scattering factor discussed in the Appendix (Section A.3). It should also be noted that one can conclude from this formula that the smaller the wavelength of the photon, and therefore the higher its energy, the better the resolving power of the method, because more lattice planes/combinations of Miller indices hkl can be probed. Of course, in order to obtain a complete description of the structural arrangement, more than one reflection is needed. There are several experimental approaches to achieving this. The first distinction is whether one offers a broad distribution of wavelengths [e.g., the white light obtained from a synchrotron radiation (SR) source] or uses a monochromatic source (e.g., monochromatic SR beam or a characteristic line of an x-ray tube). The second distinction is whether a sufficiently large single crystal can be produced or not. The possible

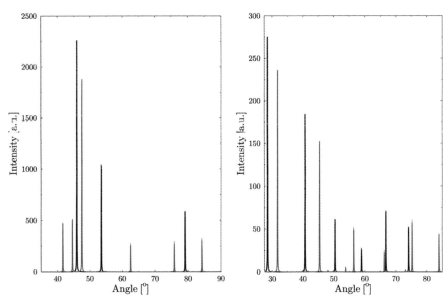

Figure 7.2. (a) Simulated Cu K_α XRD spectra of hcp (red) and fcc (black) Co. (b) Simulated Cu K_α XRD spectra of NaCl (red) and KCl (black).

approaches for obtaining a complete set of XRD data which emerge from the combination of these two criteria are summarized in Tab. 7.1.

Evidently, nanoparticles are not very good single crystals; therefore measurements in Laue geometry are not possible and one needs to perform powder diffraction experiments. Detailed information on common setups and instrumentation for the methods mentioned above can be found in the literature, e.g., Ref. [1] for laboratory source-based experiments, Ref. [2] for SR-based experiments on single crystals, and Ref. [3] for SR-based powder diffraction experiments.

What can one learn from a powder x-ray diffraction spectrum, as displayed in Fig. 7.2? In the description presented so far, we have just correlated a geometric arrangement of scattering centers and an angle under which the scattered photon is found one correlates [A] the geometric arrangement of scattering centers] and [B] the angle under which the scattered photon is found] to each other. Even though this information is not all that can be gained in an XRD experiment, it is of considerable use. The reason for this is that the shape and size of the unit cell of a given crystal can be deduced from the angular positions of the diffraction lines, as evident for the comparison of simulated hcp and "face centered cubic" Co spectra shown in Fig. 7.2(a) and discussed in more detail e.g., in Ref. [1]. Based on the combination of Bragg's law and the properties of the seven systems into which all crystals can be classified (cubic, tetragonal, orthorhombic, trigonal/rhombohedral, hexagonal, monoclinic, and triclinic), one obtains equations which allow indexing of the observed XRD peaks. For example, in a cubic system with lattice constant a, d_{hkl} is given by

$$d_{hkl} = \frac{a}{\sqrt{h^2+k^2+l^2}} \tag{3}$$

and thus one obtains

$$\frac{\sin^2\theta}{h^2+k^2+l^2} = \frac{\sin^2\theta}{s} = \frac{\lambda^2}{4a^2} = const \tag{4}$$

The next step consists in the determination of the number of atoms per unit cell. This is easily obtained, as the volume of the unit cell is readily calculated based on the parameters determined so far, and usually the density of a given material is obtained easily and its chemical composition can be determined. Therefore, the weight of a unit cell can be determined and must correspond to an integer number of atoms of the respective types.

What remains to be done is the assignment of the atoms to the respective positions, which can be done using the information contained in the intensity of the respective scattering peaks (after correcting for additional effects which can influence this intensity, such as multiplicity, Lorentz, absorption, and temperature factors). A vivid example of this is provided in Fig. 7.2(b), which shows the simulated Cu K_α XRD spectra of NaCl and KCl, which belong to the same space group but have a different lattice distance. Due to the latter, the angular positions of the respective Bragg peaks are shifted, but it is easy to identify the structures which correspond to each other. However, there are notable differences in absolute and relative intensities: in general, the intensity is consistently higher for the KCl spectrum. This is due to the higher scattering power of K relative to Na, and from the observation of how much this difference changes the intensity of a given peak can be derived which types of atoms are located at the lattice sites which contribute to a given Bragg peak. In fact, this is the key problem of structural determination for a new class of material using XRD, because a stepwise solution to the problem does not exist. Whereas in some cases comparison of the relative intensities of the respective Bragg reflexes and/or the information on chemical composition can suggest some possible site occupations, more often than not a brute force approach has to be used, distributing the atoms on the possible crystallographic sites and calculating the diffraction pattern which corresponds to this elementary cell of given symmetry and size. Naturally, the number of feasible combinations scales with the size of the elementary and the number of atoms which it contains. This is also the reason why CPU power is a critical factor in protein crystallography (cf. Ref. [2]), where elementary cells tend to be huge and contain several hundreds to thousands of atoms.

Note also that in the entire above description we have implicitly assumed that one is really dealing with a signal originating from a single phase of matter. Imagine the task of indexing reflexes in a mixture of several compounds in the right way! Also, especially with respect to the application of the method to nanoparticles, (see examples in Section 7.1.5 and theoretical background in A.2), the resolving power of XRD is strongly correlated to the number of elementary cells in the crystallites which are investigated, because this exerts direct influence on the width of the diffraction lines. Broadening of the Bragg reflexions is typically observed for particles with diameters below 100 nm. This is due to the fact that strictly speaking the Bragg condition is just a limiting case, as seen in the more detailed analysis of elastic scattering processes discussed in A.2. Of course, it is possible to attempt to make use of the broadening , e.g., by applying the Scherrer formula

$$D = \frac{0.9\lambda}{B \cos \theta_B} \qquad (5)$$

in order to determine the particle size. However, upon comparison of as-determined particles sizes and the ones observed directly, e.g., from high-resolution transmission electron microscopy (HRTEM) pictures, XRD tends to yield considerably larger particle sizes (cf., e.g., Ref. [4]).

With respect to these complications, it is fortunate that today the application of XRD rarely involves the actual determination of a completely new crystal structure. Instead, in most cases the XRD data obtained on a given compound are compared to the huge crystal structure databases like the International Crystal Structure Database (ICSD) available today, which has opened up the possibility of fitting a given XRD spectrum by a method known as the Rietveld method [5, 6]. In this approach, a structural model is refined to reproduce the experimental data, using intensity correction parameters, lattice parameters, and possible zero shifts of the detection system, the lineshape, crystal structure parameters, and the background. To achieve this fit, after indexing the observed diffraction patterns and determination of possible space groups, lattice parameters are refined. Already during this process, the presence of several phases

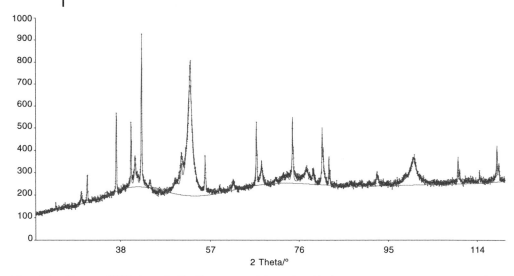

Figure 7.3. Measured XRD data of a $Co_{50}Fe_{50}$ nanoparticle produced by laser ablation. (Figure kindly provided by K. Moras, Technische Universität Clausthal, based on measurements by Dr. R. Kleeberg, Technische Universität Freiberg)

Table 7.2. Results of Rietveld analysis for a 10-nm $Fe_{50}Co_{50}$ nanoparticle. Quartz and wuestite are artifacts due to sample preparation.

	Relative contribution (in %)	Error (in % of absolute value)
Amorphous	34.10	3.90
Corundum	11.23	1.11
α-iron	40.07	2.43
Maghemite	3.78	0.72
Magnetite	3.91	0.75
Quartz	4.23	0.93
Wuestite	2.65	0.57

can be identified. The rough structure obtained from this processing step is then compared to isomorphous compounds or compounds which feature a similar structure contained in a database in order to obtain candidates for lattice site occupation. If the corresponding calculated diffraction pattern is similar to the observed one, refinement of the above parameters is performed. However, it should be noted that whereas obtaining a dissatisfying fit quality leads automatically to a change in the structural model, even a well-fitting result needs to be checked carefully for its chemical and physical relevance. Extreme values of the thermal displacement factor for a given site may indicate that it is occupied incorrectly; further parameters to be checked are bond lengths and bond angles, coordination numbers, and Madelung energies. An example of the result of a Rietveld refinement on a real XRD spectrum of a 10-nm nanoparticle with nominal composition $Fe_{50}Co_{50}$ is shown in Fig. 7.3 and summarized in Tab. 7.2. Bearing in mind the considerable broadening of the peaks and a notable, angle-dependent background contribution, it is evident that in such a system much of the clarity intrinsic to XRD on macroscopic systems is lost.

7.3
Small-Angle X-Ray Scattering: Learning About Particle Shape and Morphology

As in the XRD measurements discussed above, in a SAXS experiment elastic scattering processes into a given solid angle element are observed, but this time – as indicated by the name of the technique – the detector covers only small *scattering* angles (typically less than 1°). Looking at the Bragg condition discussed above, it is immediately clear that in a perfect, extended crystal this is a futile attempt – there is no scattering in that direction. However, this observation is based on destructive interference, which can only occur for a particle of dimension D if for a given x-ray another ray with a path difference $\lambda/2$ also exists, i.e., if

$$\lambda \approx D \sin \theta \tag{6}$$

This formula indicates that for a given x-ray wavelength the determination of the threshold angle under which scattering can just be observed yields information on the dimension of the particle. To illustrate this, let us assume that we are dealing with homogenous particles of uniform size and shape which are embedded in a homogenous matrix. In this case, it is possible to reduce the more complex description relying on local electron densities as described in appendix A.3, but also in more detail, e.g., in Refs. [7–9], to the form

$$I(\vec{q}) = CV^2 \left(\sum_i f_i (n_{i,P} - n_{i,M}) \right)^2 \left| \frac{1}{V} \int_V e^{i\vec{q}\vec{r}} d^3 r \right|^2 \tag{7}$$

where the last term represents a factor which is dependent on the particle shape and is the Fourier transform of a radial distribution function of the scattering centers in a given atom, the so-called "form factor" S_1. Some form factors for frequently encountered shapes are listed in Tab. 7.3. Note that in the above formula only the square of the scattering contrast is relevant, which means that the SAXS signal of a particle of chemical composition A in a matrix composed of B and particle B in matrix A are equivalent. However, this is only true if anomalous scattering effects [i.e., terms beyond the first term in Eq. (A6)] are negligible. The variation of scattering amplitudes, which is induced, e.g., in the vicinity of absorption edges of a given element due to these effects, can be used easily to separate the scattering contributions from matrix and particle, by subtraction of spectra which have been measured at two different photon energies, which cancels the contributions of the matrix, which stay constant, but not the varying ones of the particle. This approach is called ASAXS.

Table 7.3. SAXS parameter table.

Scatterer	Scattering cross section	Asymptotic form	Guinier radius parameters[a]
Sphere (radius R)	$NV^2(\Delta f)^2\left[\dfrac{J_1(qR)}{qR}\right]^2$	q^{-4}	$n_1=2,\ n_2=5,\ C=1$
Thin disk (radius R)	$\left(\dfrac{NV}{qR}\right)^2(\Delta f)^2\left[1-\dfrac{J_1(2qR)}{qR}\right]^2$	q^{-2}	$n_1=1,\ n_2=4,\ C=1$
Needle (length $2h$)	$NV^2(\Delta f)^2\left[\dfrac{\sinh(2qh)}{qh}-\dfrac{\sin^2(qh)}{(qh)^2}\right]$	q^{-1}	$n_1=2,\ n_2=5,\ C=1$
Spherical shell (radius R)	$(NV)^2(\Delta f)^2\left[\dfrac{\sin(qR)}{(qR)}\right]^2$	q^{-2}	—
Random fluctuation (correlation length l)	$(NV)^2(\Delta f)^2\left[\dfrac{1}{1+(ql)^2}\right]^2$	q^{-4}	—

[a]See Eqs. (11, 12).

It should be stressed that Eq. (7) is only valid for randomly oriented particles with a strictly defined shape, i.e., without any size distribution. If particles are arranged in a given way (e.g., if in a given volume around a particle center no other particle can be located, as in a pile of spheres), an interference function needs to be introduced, which in turn is the Fourier transform of a pair correlation function g(r) between the single particles. If in addition the particles follow a size distribution d(D), the full expression to be evaluated is:

$$I(\vec{q}) = \int d(D)CV^2\left(\sum_i f_i(n_{i,P}-n_{i,M})\right)^2 \left|\frac{1}{V}\int_V e^{i\vec{q}\vec{r}}d^3r\right|^2 \left|\left(1+\frac{1}{V}\int g(r)e^{i\vec{q}\vec{r}}d^3r\right)\right|^2 dD \tag{8}$$

It is easily seen that the exact description of the physical situation probed by the scattering experiment can become arbitrarily complicated (e.g., so far we have not yet considered the case that the orientation of anisotropic particles might follow yet another distribution, which will influence the measured intensity in a given spherical angle as well, and so on).

Bearing this in mind, it is of special relevance for the application of this method to find approximations/spectral features which allow the extraction of some of the particle's properties without having to reproduce the entire observed data using a suitable structural model. In fact, it turns out that the particle shape can in principle be determined from the asymptotic behavior of the observed scattering intensity. It is even possible to generalize these cases to Porods's law, which correlates asymptotic behavior of the scattering cross-section and total surface A of isotropic or randomly oriented anisotropic particles for values qr>5:

$$\frac{d\sigma}{d\Omega}(\vec{q}\vec{r}>5)q^4 = C(\Delta n_f)^2 2\pi A \tag{9}$$

Also, it is possible to develop the form factor S_1 in terms of powers of qr, which leads to the Guinier approximation

$$S_1(\vec{qr} \leq 1.2) = e^{-\frac{q^2 R_g^2}{3}} \tag{10}$$

In this approximation, a fit of the scattering behavior for small values of qr yields the Guinier radius R_g, which is defined by the relation

$$R_g^2 = \frac{\int r^2 \Delta n_f(\vec{r}) d^3 r}{\int \Delta n_f(\vec{r}) d^3 r} \tag{11}$$

and be considered as an analogon of the radius of gyration known from mechanics. A general expression for this radius as a function of particle radius R is

$$R_g = R \sqrt{\frac{n_1 + C^2}{n_2}} \tag{12}$$

where the parameters n_1, n_2, and C are given for selected particle shapes in Tab. 7.3.

Therefore, starting from the most simple assumption, i.e., particles A are present in matrix B, from the general shape of the scattering curve it is possible to gather information on particle shape and particle surface by fitting the high q and low q area, respectively, of the scattering curve. In combination with the integral scattering intensity Q_0, which is obtained by integration of Eq. (7) over the entire q-space and correlated to the total volume contribution constant C of phase A via the relation:

$$Q_0 = (2\pi)^3 \Delta n_f^2 C(1 - C)V \tag{13}$$

Using this set of relations, a good starting point for development of a structural model can be defined. As an example, consider the ideal calculated scattering cross-sections for a sphere and a disk, respectively, which are displayed in Fig. 7.4.

Evidently, the asymptotic behavior of the scattered intensity is characteristic for a given shape and limits applicable structural models. Summing up, the SAXS method is an extremely sensitive tool for extracting information on particle morphology. However, it has a drawback which is due to its strength: as the exact extracted intensity distribution is sensitive to a large number of factors, in general the structural refinement which is applied to it is based on a number of implicit assumptions on the nature of the particle. For example, an inherent assumption which is frequently made is that the particle size distribution should follow a log-normal distribution, which is then fit to the data. Therefore, for the successful extraction of particle composition and shape, often additional information or confirmation is needed in order to arrive at a unique structural solution. Further details on SAXS and instrumentation for suitable experimental setups are found in, e.g., Refs. [7–9].

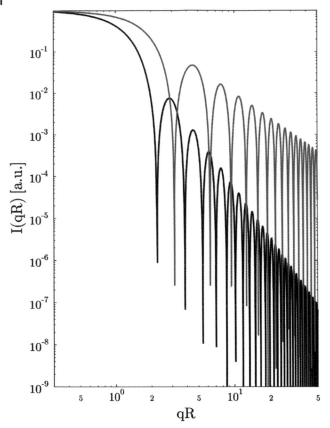

Figure 7.4. Scattering cross-section signal for sphere (black)
and spherical shell (red) of radius R. Note the asymptotic
behavior for large qR, from which direct statements on the
particle shape can be derived.

7.4
X-Ray Absorption: Exploring Chemical Composition and Local Structure

X-ray absorption spectroscopy (XAS) experiments measure the dependence of the
cross-section of the absorption process (i.e., the probability that it occurs) on the energy of the incoming photon. The insert in Fig. 7.5 displays the trend which is observed upon a variation of photon energy in big steps over an extended energy
range. It reflects the electronic structure of the corresponding element: steps in the
cross-section occur whenever the photon energy is sufficiently high to excite electrons from a deeper core level; at the highest energy from the 1s level, proceeding
with decreasing photon energy to 2s, $2p_{1/2}$ and $2p_{3/2,}$ and so on. However, there are
bound unoccupied states, and thus even at energies slightly lower than the ionization threshold it should be possible to excite the electrons into bound unoccupied

states. In fact, scanning an absorption edge in steps of the order of 1 eV, one observes a spectrum as shown in Fig. 7.5.

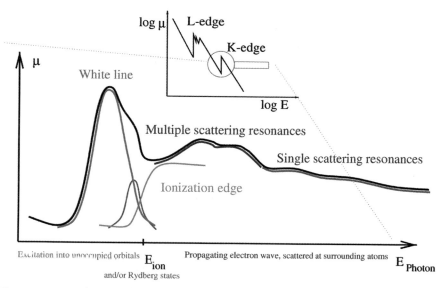

Figure 7.5. X-ray absorption fine structures and their (dominant) cause

The absorption edge, i.e., the onset of the increase in the absorption cross-section, lies in fact at lower energies than the ionization energy. Also, at energies higher than the ionization threshold oscillatory structures are observed. This observation can be explained in a simple model if one recalls that in fact the photoelectron propagating through matter can be treated as a spherical wave, whose wavenumber k is connected with the energy of the incoming photon E and the ionization energy E_0 via the relation

$$k = \sqrt{\frac{2m}{\hbar^2}(E - E_0)} \tag{14}$$

where m_e represents the electron mass and h the normalized Planck's constant. This photoelectron wave propagates through an environment in which it undergoes electron–electron scattering processes. Due to interference of outgoing and scattered wave, what one observes is an interference pattern, as schematically displayed in Fig. 7.6(a) for the case of constructive interference. On the other hand, varying the energy of the incoming photon varies the wavelength of the photoelectron; the interference pattern changes towards destructive interference, as shown in Fig. 7.6(b), and back to constructive interference. This explains the modulations observed in the absorption cross-section. Furthermore, the scattering probability is a function of the energy of the photoelectron, consequently in the energy region just above the absorption edge multiple scattering will be possible, and a large number of strong interference terms appear in the spectrum. These strong spectral features are called shape resonances. For historical reasons, the first two regions are known as x-ray absorption near edge structure (XANES) – also called near edge x-ray absorption fine structure (NEXAFS), the latter is known as the extended x-ray absorption fine structure (EXAFS).

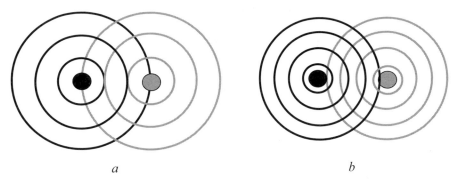

a   *b*

Figure 7.6. Constructive (a) and destructive (b) interference of outgoing and backscattered electron wave. Recall that the wavelength of the photoelectron depends on the energy of the incoming photon.

Even this rough understanding of the fine structure which appears in the absorption spectrum allows an estimation of the information one can obtain using this method: at the absorption edge, the unoccupied valence states are probed by the absorption process. But chemistry modifies the valence state/electronic structure of the elements, so one obtains information on the chemical environment of the absorber atom in the sample, which can be selected by choosing the excitation energy. At the same time, the interference pattern is characteristic for a given arrangement of the atoms which surround the absorbing atom and allows the extraction of information on its local coordination geometry without any need for long range order.

For the XANES structures, these effects are demonstrated at the Co K-edge in Fig. 7.7. In Fig. 7.7(a), a number of Co K-edge XANES spectra of compounds with varying formal oxidation state are displayed. Clearly, the onset of the structures shifts to higher energies with higher formal oxidation state. This is a rather systematic effect called "chemical shift." Also, intensities of absorption in the white line range vary notably, which is correlated to the density of (atom-projected) unoccupied states, which tends to be higher for higher formal oxidation states. In Fig. 7.7(b), calculated spectra of metallic Co phases are displayed. Clearly, these spectra do not show chemical shift, but they still show changes in their electronic structure. Perhaps even more interesting is the comparison of the shape resonances in the energy region between 7760 and 7840 eV (on the energy scale of the calculation), because it demonstrates clearly the localized point of view of the method: Co–Co distances in hcp and fcc Co are quite similar to each other, as demonstrated in Tab. 7.4. As a consequence, the shape resonances are also quite similar to each other, indicating that in fact the local environment and not long range order exert the dominating influence on the spectra.

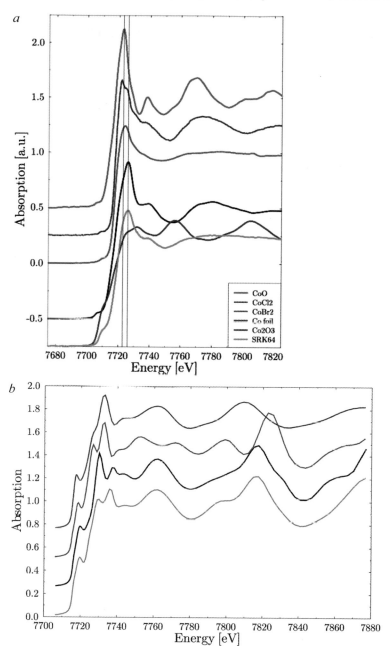

Figure 7.7. (a) Co K-edge XANES spectra of Co-compounds with varying formal oxidation state and electron affinity of the binding partner. Note the systematic increase in white line intensity for increasing electron affinity of the binding partner within the Co(II) compounds and the systematic shift of the onset and the maximum of absorption with increasing formal oxidation state. (b) Calculated Co K-edge XANES spectra of (bottom to top) hcp Co, fcc Co, bcc Co (hypothetical). and ε-Co. Note the high similarity of all but the bcc phase in the energy position of multiple scattering structures and the characteristic changes in the electronic band structure at the absorption edge.

Table 7.4. Local first and second shell coordination geometry of real and hypothetical metallic Co phases.

	hcp	fcc	bcc
First shell	6 @ 2.497 Å	12 @ 2.489 Å	8 @ 2.485 Å
	6 @ 2.507 Å		
Second shell	6 @ 3.538 Å	6 @ 3.520 Å	6 @ 2.870 Å

This is also the reason why a fingerprint approach to the interpretation of XANES spectra is so successful. Simple comparison with known reference substances allows, for example, the direct extraction of information on electronic structure, e.g., formal valency, and local coordination geometry in most cases. Due to their sensitivity to the local environment, XANES spectra are additive, i.e., a spectrum of a mixture of compounds A and B can be composed by weighted addition of the spectra of the pure reference compounds. This approach is called "quantitative analysis," discussed in more detail, e.g., in Ref. [10] and frequently used for speciation purposes.

Fitting a structural model to the interference pattern in an x-ray absorption spectrum is possible in the EXAFS region using a path-by-path approach that allows for the analytic extraction of structural parameters. The general idea of this approach is given in Fig. 7.7; a more detailed description of the process is given in Section A.4.

Further information on XAS and XAS instrumentation is found in several books, reviews on the method, and conference proceedings [11–14].

7.5
Applications

A huge selection of nanoparticle characterizations using the above x-ray techniques exclusively or in combination is available in the literature. Rather than discussing these examples in detail, I will discuss a selection of three cases which illustrate the respective strengths and problems of the techniques introduced above.

7.5.1
Co Nanoparticles with Varying Protection Shells

A lot of scientific interest has recently been focused on Co nanoparticles. The main reason for this lies in their favorable magnetic properties, which in turn open a broad field of possible applications. They can be used, for example, in ultra-high-density magnetic media, magnetoresistive devices, ferrofluids, and magnetic refrigeration systems. Also, with respect to the topic of this book, biomedical applications should be mentioned, such as contrast enhancement in magnetic resonance imaging, magnetic carriers for drug targeting, and catalysis [15–17].

Numerous techniques have been used in studies [18–21] on this class of nanoparticles, including XAS. In fact, Fig. 7.8 displays the Co K-edge XANES spectra of a

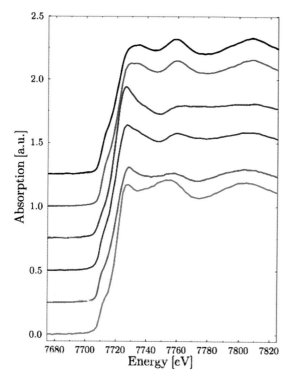

Figure 7.8. Co K-edge XANES spectra of Co nanoparticles (top to bottom): (i) synthesized by thermolysis of dicobaltoctocarbonyl in the presence of aluminumtrioctyl, Co:Al ratio 10:1; (ii) as (i) after exposure to air; (iii) as (i), but with aluminumtriethyl; (iv) as (i), but Co:Al ratio 5:1; (v) 8-nm nanoparticle, synthesized by laser codeposition of Co and C; (vi) as (i), but in the presence of an additional surfactant molecule, Korantin SH.

number of differently synthesized Co nanoparticles, including surfactant variations, while Fig. 7.9 shows a set of nanoparticles of different sizes which are all stabilized with the same surfactant molecules, CTAB (cetyltrimethylammonium bromide). On the one hand, these data show impressively the considerable sensitivity of the technique with respect to details of sample preparation. Clearly, there is no such thing as "the spectrum of an *x* nanometers in size Co nanoparticle," or "the spectrum of a Co nanoparticle stabilized by Y." Instead, both size- and surfactant-induced effects are clearly observed, as discussed for these and other types of nanoparticles in the recent literature [22–25].

On the other hand, the exact identification of the different chemical and structural phases involved is rather tedious. In the case of the structural metal phases, the main reason for this is the similarity of the local environment of the absorbing Co atoms. Looking at *ab initio* calculations of fcc and hcp Co phases displayed in Fig. 7.7(b), even the shape resonances, which are usually the most sensitive indicator of changes in coordination geometry, are found at rather similar energy positions, and, as discussed in Ref. [21], the observable differences in the EXAFS evalua-

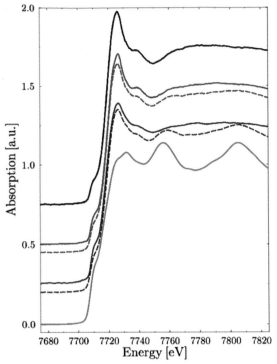

Figure 7.9. Co K-edge XANES spectra of CTAB-stabilized Co nanoparticles of (solid lines, top to bottom) 5.5-nm, 8-nm, and 11-nm diameter and hcp Co reference foil. Broken lines show the reproduction obtained by linear combination of the spectra of hcp Co and the 5.5-nm particle.

tion, too, are small and difficult to extract, especially if Co is also present in other environments, as is the case, e.g., in core–shell systems. The biggest changes which are observed occur in the electronic structure, directly at the absorption edge. However, this is exactly the region in which chemical interaction also plays an important role, and decoupling these two effects is problematic. This is even more the case bearing in mind that reference spectra for some of the metal phases, e.g., the ε-Co phase, which is stable only in nanoscopic systems, are not available. In contrast to that, whenever the particle size and homogeneity allow the extraction of reliable structural parameters, at least the nature of the core phase can be determined quite unambiguously using XRD, as indicated by the comparison of (simulated) XRD data of the different Co phases shown in Fig. 7.2(a).

However, even in a situation like this, a lot of information on the missing compounds can be derived based on the additivity of XANES spectra if, e.g., a series of samples varies only with respect to a single parameter, e.g., particle size. As an example, consider the changes in the series of CTAB (cetyltrimethylammonium bromide)-stabilized Co nanoparticles with particle sizes of 5.5, 8, and 11 nm only. More details on this sample system are found in Ref. [26]. Evidently, the intensity of the pre-edge structure of these spectra grows with the particle size, whereas the maximum absorption decreases. In general, resemblance to the hcp Co spectrum increases, as the observed changes in intensity occur at those energy positions where spectral features in the Co foil spectrum are located, but even the spectrum of the largest particles differs from pure hcp Co. This suggests that addition of the hcp Co spectrum and the smallest nanoparticle spectrum might reproduce the spectra of both the 11-nm and the 8-nm particles by linear superposition of the spectra of the smallest particle, whose diameter is 5.5 nm, and that of hcp Co as shown also in Fig. 7.9. From this fit, an additional hcp content of about 40% for the 11-nm particle and of about 18.5% for the 8-nm particle can be derived. The error range for these numbers can be estimated to be ±5%. As TEM pictures show homogenous particles and no indication of additional phases, this observation can be interpreted assuming a core–shell type structure of the particles. Assuming spherical particles and that the pre-edge intensity present in the spectrum of the smallest particle can be directly correlated to the amount of metallic Co present in this particle (40%), it is possible to extract the thickness of the respective cores and shells for the particles from the hcp Co contributions. The determined contribution of this phase is then given by the quotient of the cube of the core radius r_c and the cube of the total radius r_t. Performing this calculation, one obtains a core radius of 4.75 nm for the particles of 11 nm diameter, a core radius of 3.15 nm for the particles of 8 nm diameter, and a core radius of 2 nm for the 5.5-nm particles. All values indicate that the shell thickness is always 0.8 nm, which in turn enhances the plausibility of the assumptions. Further support for the formation of a core–shell system stems from an aging experiment discussed in Ref. [26].

Still, the nature of the shell needs to be identified. Likely candidates for the shell would be a Co oxide, such as CoO or Co_2O_3. (cf. Fig. 7.9). Due to the high similarity between the spectral features of the latter spectrum and the observed white line of the smallest nanoparticle, this may seem to be a good candidate, but the reproduction of the XANES spectra of the smallest particle by a linear superposition of the spectra of hcp Co and the corresponding oxide fails completely. In addition, the assumption of Co_2O_3 as being the compound forming the shell faces the problem that prolonged exposure to air converts the particles to CoO, and annealing removes the shell, which is inconsistent with assuming any type of oxide for the shell, because of the high stability of oxides. However, the comparison of the spectrum of the smallest particle to the ones of the different Co(II) reference spectra indicates clearly that, under the assumption that the shell material dominates the constituents of the smallest particle, this material must contain Co(III). Keeping in mind also that the shell is destroyed completely at very low temperatures of ~210 °C [26], more likely candidates for the shell are Co(III) complexes such as $Co[(NH_3)_6]Br_3$ and $Co[(NR_3)_6]Br_3$. The formation of these complexes and also the observed core–shell structure of the particles seem to be "typical" of CTAB as none of these effects have been reported for Co nanoparticles stabilized with other surfactants, whereas, interestingly, a similar result has been obtained in a recent XAS study on CTAB-stabilized CeO_2 nanoparticles, which were reported to possess a Ce^{3+} shell [27].

7.5.2
Pd$_x$Pt$_y$ Nanoparticles

While the samples discussed in this example are of minor importance for biomedical application, they are of tremendous significance for applications in catalysis, where binary and ternary Pt catalysts on the nanoscale play a crucial role and are the subject of numerous investigations [28–31]. A key question in this context is how the arrangement of atoms in such a nanoparticle will actually look: will there be an ordered alloy, a statistical distribution of neighbors, a core–shell system, or some internal segregation and grain formation? At the same time, this system is especially suited for discussion in this chapter, as it can be used nicely to characterize the problems often encountered when applying XRD techniques to binary metallic systems such as alloys and intermetallic compounds. As an example of these problems, have a look at the XRD spectra of three Pd$_x$Pt$_{100-x}$ nanoparticles with particle sizes of about 3.2±0.4 nm (as determined by TEM) shown in Fig. 7.10. Evidently, the differences between the obtained XRD spectra are rather small, and therefore a detailed determination of corresponding structures is hardly possible using this approach. A Rietveld analysis of this system will of course yield different results for different particles, but taking into account the lack of clarity of these structures, the statements derived from it may seem as questionable as the ability of XRD to distinguish between the different types of structures mentioned above. In fact, this is easily understood: the two metals under consideration are perfectly miscible in each concentration; both crystallize in fcc structure with lattice constants which vary by only 0.03 Å.

Still, the EXAFS data from the same materials lead to different interference signals and consequently (non-phase-corrected) radial distribution functions which show clear differences even under mere visual inspection, as shown in Fig. 7.11 for the two bimetallic types of nanoparticles. The results of the analysis are summarized in Tab. 7.5. As a general trend, a lattice contraction is observed, which is stronger in the bimetallic particles. This is a trend which is frequently encountered in the analysis of small nanoparticles in general and was especially reported for a number of similar bimetallic particles by several authors (Refs. [25, 28–31] and references therein). Another striking effect is the significant reduction of the determined coordination numbers. This, too, is frequently encountered in nanoparticle systems. Partly, this meets the expectations, because a considerable share of the constituent atoms of a nanoparticle is located at its surface and thus not fully coordinated; but the observed reduction cannot be fully explained by this argument. However, such estimates assume the presence of perfect particles and perfect particle surfaces, which might not be an adequate description of the situation considering, e.g., the theoretical results obtained in Ref. [32]. A similar approach to the explanation of drastically reduced coordination numbers obtained when analyzing Fe nanoparticles has been suggested, e.g., by Di Cicco et al. [33, 34]. Another problem one does encounter is that Debye–Waller factors may no longer be described correctly, which are connected in the standard analysis to the gaussian pair distribution function which is the basis for the inclusion of the effective pair distribution in the

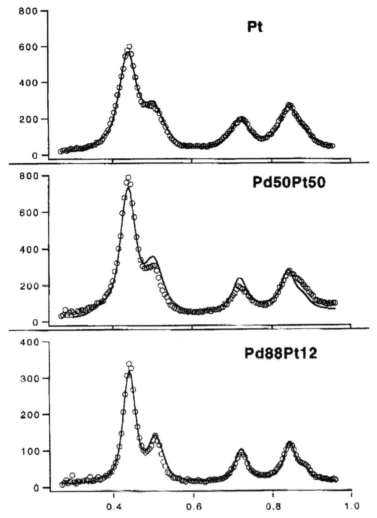

Figure 7.10. X-ray diffraction data of Pd_xPt_{100-x} nanoparticles. Note the extreme similarity between the data sets! (From Ref. [54].)

Debye–Waller factor. This assumption is clearly questionable in the case of nanoparticles. Rather, one might encounter a bond-length distribution which is shaped like an asymmetric double-well potential and thus far from the gaussian ideal. As a matter of fact, Babanov et al. [35] have performed a more general EXAFS analysis on Co nanoparticles and reached the conclusion that in a boundary layer of the particles a four-fold coordination might be observed. Apart from an asymmetric static radial distribution function discussed in Ref. [35], anharmonic Co–Co vibrations could lead to a dynamic asymmetry in this function, as discussed in detail in Ref. [36].

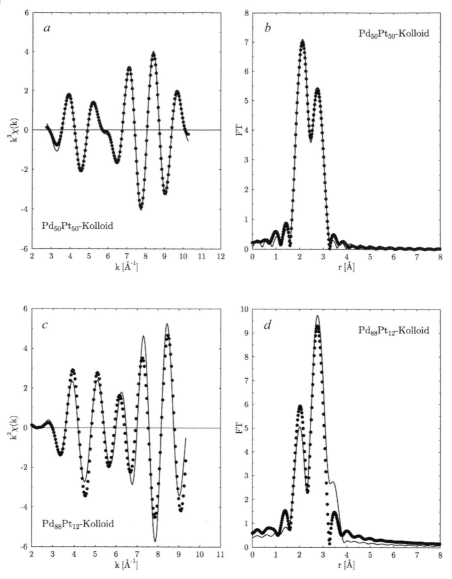

Figure 7.11. EXAFS analysis of Pd_xPt_{100-x} nanoparticles. (a, c) $\chi(k)$ function and fit; (b, d) modified Fourier transform and fit. (From Ref. [54].)

Even though the absolute coordination of the absorbing atoms is frequently underestimated, the relative coordination numbers can yield important information on the arrangement of the various types of atoms. As evident from Tab. 7.5, the relative coordination numbers of a Pt absorber meet the purely statistical prediction reasonably well for the Pd-rich particles, whereas the contribution of Pt neighbors is higher than such a model would predict. This result can be explained by a nonstatistical distribution, i.e., if partial segregation of Pt and Ru occurs. This makes both a statistical distribution of Pt and Ru and a core–shell structure unlikely and thus yields key information on particle morphology.

Table 7.5. Results of the EXAFS analysis of Pd_xPt_{100-x} nanoparticles.

Sample	Scatterer	$R[Å]$	N	$N_{Pt}:N_{Pd}$ (theory)	$\sigma_i^2 [Å^2]$	$\Delta E_0 [eV]$
Pt_{100}	Pt	2.75±0.02	5.9±0.3		0.007±0.001	9.0±1
$Pd_{50}Pt_{50}$	Pt	2.72±0.02	3.4±0.3	2.27 (1)	0.003±0.001	3.0±1
	Pd	2.70±0.02	1.5±0.3		0.002±0.001	3.0±1
$Pd_{88}Pt_{12}$	Pt	2.70±0.02	1.2±0.3	0.18 (0.14)	0.001±0.001	−8.0±1
	Pd	2.73±0.02	6.8±0.3		0.010±0.001	6.0±1

7.5.3
Formation of Pt Nanoparticles

In my opinion, one of the most interesting classes of processes to be investigated systematically in the near future in order to get closer to nature's records in the realms of wet-chemical nanosystem synthesis is the detailed characterization of nanoparticle formation beyond the kinetic aspects [37, 38], because detailed understanding of these processes, and developing ideas to control and steer them, will be needed to use the full power of wet-chemical nanoparticle synthesis, as carried out by nature itself. This personal interest is also the reason why I use this example for the application of (A)SAXS, even though, for example, its power to give information on particle shapes and in application to particle networks and porous particles has been demonstrated more clearly in other studies, e.g., Refs. [39–43].

The general mechanism for the formation of (metal) nanoclusters as described by Turkevich and Kim [44], consists of three steps: nucleation, growth, and agglomeration. Clearly, using small-angle scattering, a detailed characterization of the growth process should be possible. Such a characterization has been performed, e.g., for the synthesis of Pt nanoparticles using $Pt(acac)_2$ and $Al(alkyl)_3$ as educts, which is described in detail in Refs. [45, 46]. Using time-resolved ASAXS in the vicinity of the Pt L_{III} edge, the nucleation process between 0.8 and 1000 hours reaction time was studied. By changing the photon energy in this region, changes in the scattering amplitude of the Pt atoms due to anomalous scattering can be used to separate the unknown scattering contribution from the organic molecules in the solution and contributions of scattering on Pt atoms. The obtained difference scattering cross-section (E_2–E_1) provides unbiased information on the distribution of the Pt particle only.

Figure 7.12 displays the results after reaction times of 3.6 and 65.4 hours at room temperature. From the curve fit, it emerges that one is dealing with Pt particles with mean radii $\langle R \rangle = 5.8$ Å and a rather narrow monomodal lognormal particle size distribution. Using an icosahedral model, this particle size corresponds to 53 atoms

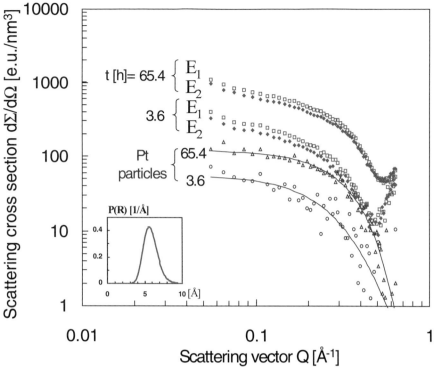

Figure 7.12. ASAXS data measured during the synthesis of Pt nanoparticles after reaction times of 3.6 and 65.4 hours at x-ray energies of $E_1 = 11.46$ keV and $E_2 = 11.54$ keV and fitted difference cross-sections of Pt nanoparticles with mean radii $\langle R \rangle = 5.8$ Å assuming monomodal lognormal particle size distribution. For further details see Refs. [45, 46].

and corresponds well to the second in the energetically favored "magic numbers" of atoms. As illustrated in Fig. 7.13, during the experiment the fraction of Pt atoms found in these stable particles grows, but the mean size of the particles and the width of the distribution remain the same. The amount of Pt converted into particles, $x = (m_{particle}/m_{total} - 0.206) / (1 - 0.206)$, follows the exponential time dependence $x = 1 - \exp(-t/t0)$ shown as the solid line in Fig. 7.13. The rate of nucleation into particles $dx/dt \sim [1-x(t)]$ is linearly proportional to the number of precursor molecules in the solution, $[1-x(t)]$. The rate-controlling step for the nucleation is the decomposition of a thermally unstable binuclear precursor molecule, whose formation was derived by XAS and NMR spectroscopy, as discussed in [45, 46]; it is not a subsequent diffusion-controlled agglomeration process of the single zero-valent Pt atoms into the particles. As the formation of this intermediate complex involves surfactant molecules, this result implies that potentially control of the nucleation process and properties of the obtained nanoparticle may be influenced in a controlled way by slight changes in the chemical approach, once an understanding of the reactions has been obtained. In fact, such delicate dependence has been noted, e.g., for Co nanoparticles discussed in Ref. [47].

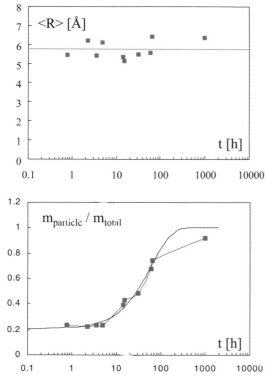

Figure 7.13. Particle radius $\langle R \rangle$ (a) and mass fraction of Pt transformed into particles (b) during the synthesis of Pt nanoparticles discussed in Refs. [45, 46].

It is interesting to compare these results to a recently performed study by Mencau et al. [48], which followed a similar concept to monitor the formation of CdS and ZnS particles *in situ* based on a combination of time-resolved SAXS and EXAFS. A two-step process in particle formation was observed, which can be interpreted as the exact analogon of the nucleation and growth step suggested by Turkovich: after about 5 minutes of reaction time, 5-nm-diameter particles appear. In the following two hours, they grow to their final equilibrium size, about 20 nm. Before the appearance of the 5-nm particles, no indication for the formation of smaller nuclei is found. In both cases, the time resolution achieved in this study is not sufficient to provide insight into the exact mechanism responsible for the addition of the subsequent atomic layers to the particle core and the nucleation process in spite of the fact that the entire synthesis takes quite a long time. In particular, it appears amazing that there is no indication for significant amounts of smaller particles, which seems to suggest that the lowest stable particle size is 5 nm, if one does in fact interpret this step as the nucleation. Understanding in detail what exactly is going on in this phase of particle formation should lead to significantly enhanced control of wet-chemical nanoparticle synthesis. It should be stressed that the application of *in situ* XAS techniques allowed in both cases some, but incomplete, insight into this time window, as discussed in detail in the articles cited above.

7.6
Summary and Conclusions

The above sections should have conveyed my readers that x-ray methods – especially if performed at a synchrotron radiation source – are a prime toolset with which to investigate nanostructured materials. This is especially true if they are applied in combination, because they have overlapping strengths and weaknesses. Whenever XRD is applicable (i.e., whenever the particles are sufficiently large and sufficiently ordered), it is the prime tool by which to determine the phase of the particles or at least the particle cores. Still, it is quite blind to the presence of additional amorphous phases and thin shells, and size determinations using this technique seem to tend to exaggerate the particle size. Determination of particle shapes is not possible.

In contrast to this, (A)SAXS is an excellent tool for the characterization of the size, shape, and morphology of the particles. Furthermore, a careful evaluation of the scattering contrast can also be used (see, e.g., the discussion in Ref. [7]) to determine roughly the chemical composition of the particle. Due to the extremely local nature of the information gained by the application of XAS, this technique can be considered a prime tool for the determination of the chemical composition of particles, even if they are extremely small or present in an amorphous phase. Also, structural information can be derived analytically from the EXAFS signal, even though this information is significantly less precise and more complicated to extract than in the case of XRD. The additional information in the XANES region is difficult to extract in general, especially if no macroscopic reference phases are available, although recent progress in the calculation of theoretical XANES spectra has to some extent created the capability to fill this gap. Still, there is a considerable way to go until this fundamental problem can be considered to have been eliminated. Theoretical calculations also seem to indicate that some rough information on particle shape may potentially also be obtained from XAS [49], but this is very difficult to extract from real data. Nevertheless, my – biased – belief is that the future of nanoparticle analysis lies in x-ray methods and their future development.

Acknowledgements

I am indebted to Prof. Dr. H. Bönnemann, Dr. G. Köhl, and Dr. K. Moras for input for this chapter.

Appendix: Formal Description of the Interaction of X-Rays with Matter

A.1
General Approach

The full hamiltonian of this physical system is composed of three components: the ground-state hamiltonian for matter

$$H_0 = \sum_j \{ \frac{1}{2m} (\vec{p}_j^2 - eU_N(\vec{r}_j)) \} + \frac{e^2}{2} \sum_{ij} \frac{1}{|\vec{r}_i - \vec{r}_j|} \tag{A1}$$

a hamiltonian describing the radiation field

$$H_{rad} = \sum_{\vec{k},\lambda} \hbar \omega_{\vec{k}} (a^+ (\vec{k}, \lambda) + \frac{1}{2}) \tag{A2}$$

and one describing the interaction between the matter and radiation field

$$H = -\frac{e}{mc} \sum_j \vec{A}(\vec{r}_j) \vec{p}_j + \frac{e^2}{2mc^2} \sum_j \vec{A}^2 (\vec{r}_j)$$

$$-\frac{e\hbar}{mc} \sum_j \vec{\sigma}_j [\vec{\nabla} \times \vec{A}(\vec{r}_j)] - \frac{e\hbar}{2(mc)^2} \frac{e^2}{c^2} \sum_j \vec{\sigma}_j [\vec{A}(\vec{r}_j) \times \vec{A}(\vec{r}_j)] \tag{A3}$$

(a complete list of variables and their meaning is provided at the end of this chapter.) As H' is small compared to H_0, it is possible to apply perturbation theory to determine the action of H' on the system. For different phenomena, different parts of H' are relevant: the terms in the interaction hamiltonian which are linear with respect to $A(r)$ correspond to absorption and emission, whereas terms which are quadratic in $A(r)$ correspond to two-photon processes like scattering. Note that this implies not only contributions of the second and fourth terms in Eq. (A3), but also second-order processes involving the first and third interaction terms, respectively. Applying perturbation theory to this interaction hamiltonian for the respective relevant terms (which are selected by the respective experiment), bearing in mind that the vector potential $A(r)$ can be written in second quantization as

$$\vec{A}(\vec{r}) = \sum_{\vec{k},\lambda} \sqrt{\left(\frac{2\pi\hbar c^2}{V\omega_{\vec{k}}} \right)} \left\{ \vec{\varepsilon}(\vec{k},\lambda) a(\vec{k},\lambda) e^{i\vec{k}\vec{r}} + \vec{\varepsilon}^* (\vec{k},\lambda) u^+ (\vec{k},\lambda) e^{-i\vec{k}\vec{r}} \right\} \tag{A4}$$

one obtains for absorption phenomena:

$$\mu = \left(\frac{2\pi e}{m} \right)^2 \frac{1}{\omega_0 V} \sum_{n_1} \left| \left\langle n_1 \left| \sum_j e^{i\vec{k}_0 \vec{r}_j} \vec{\varepsilon}_0 \vec{p}_j \right| n_0 \right\rangle \right|^2 \delta(E_{n0} + \hbar\omega_0 - E_{n1}) \tag{A5}$$

In contrast, one obtains for scattering processes:

$$w_{scat}(n_0 \vec{K}_0 \rightarrow n_1 \vec{K}_1) = \frac{2\pi}{\hbar} \left(\frac{c^2 \hbar 2\pi}{V\sqrt{\omega_0 \omega_1}} \right)^2 r_0^2 \delta(E_{n0} + \hbar\omega_0 - E_{n1} - \hbar\omega_1)*$$

$$* \left| \left\langle n_1 \left| \sum_j e^{i\vec{K}\vec{r}_j} \right| n_0 \right\rangle \vec{\varepsilon}_1 \vec{\varepsilon}_0 - \frac{i\hbar\omega_0}{mc^2} \left\langle n_1 \left| \sum_j e^{i\vec{K}\vec{r}_j} \vec{\sigma}_j \right| n_0 \right\rangle \vec{\varepsilon}_1 \times \vec{\varepsilon}_0 + \right.$$

$$+ \frac{1}{m} \sum_z \sum_{i,j} \left\{ \frac{\left\langle n_1 \left| (\vec{\varepsilon}_1 \vec{p}_i - i\hbar(\vec{k}_1 \times \vec{\varepsilon}_1) \vec{\sigma}_i) e^{i\vec{k}_1 \vec{r}_j} \right| z \right\rangle}{E_{n0} - E_z + \hbar\omega_0 + \frac{i}{2}\Gamma_z} * \left\langle z \left| (\vec{\varepsilon}_0 \vec{p}_j - i\hbar(\vec{k}_0 \times \vec{\varepsilon}_0) \vec{\sigma}_j) e^{i\vec{k}_0 \vec{r}_j} \right| n_0 \right\rangle + \right.$$

$$\left. + \left\langle n_1 \left| (\vec{\varepsilon}_0 \vec{p}_j - i\hbar(\vec{k}_0 \times \vec{\varepsilon}_0) \vec{\sigma}_j) e^{i\vec{k}_0 \vec{r}_j} \right| z \right\rangle * \frac{\left\langle z \left| (\vec{\varepsilon}_1 \vec{p}_i - i\hbar(\vec{k}_1 \times \vec{\varepsilon}_1) \vec{\sigma}_i) e^{i\vec{k}_1 \vec{r}_j} \right| n_0 \right\rangle}{E_{n0} - E_z + \hbar\omega_1} \right\} \right|^2 \tag{A6}$$

It should, however, be noted that if one considers scattering of photons whose energy is high compared to the typical binding energies of electrons, only the first term of the above formula contributes notably.

A.2
X-Ray Diffraction

In x-ray diffraction, elastic scattering processes are measured, which implies that wavenumbers k_1 and k_0 and thus wavelength and photon energy remain constant. At the same time, by definition of an elastic process, the initial and final states of the system at which scattering occurs stay identical. Working at photon energies that are high compared to the energy levels in the scattering system, the contribution of the first term in Eq. (A6) is dominant; thus one has to evaluate the matrix element

$$\left\langle n_0 \left| \sum_j e^{i\vec{K}\vec{r}_j} \right| n_0 \right\rangle = \int \Psi_{n0}^* \sum_j \delta(r - r_j) \Psi_{n0} \cdot e^{i\vec{K}\vec{r}} d^3 r = \int \rho_c(\vec{r}) e^{i\vec{K}\vec{r}} d^3 r \tag{A7}$$

which is proportional to the scattering amplitude $A(q)$. The observed scattering intensity is proportional to the square of the absolute value of this matrix element, and thus related to the charge distribution in the investigated material by a Fourier transform. In a periodic lattice, this is given by the superposition of the electron density in the unit cells centered at the different lattice sites R_l, i.e.,

$$\rho_c(\vec{r}) = \sum_l \rho(\vec{r} - \vec{R}_l) = \int \sum_l \delta(\vec{r}' - \vec{R}_l) \rho(\vec{r} - \vec{r}') d^3 r' = \int \sum_l n_L(\vec{r}') \rho(\vec{r} - \vec{r}') d^3 r' \tag{A8}$$

where n_L describes the density of the lattice points in the crystal, and it is possible to rewrite Eq. (A6) as

$$\left\langle n_0 \left| \sum_j e^{i\vec{K}\vec{r}_j} \right| n_0 \right\rangle = \int n_L(r') e^{i\vec{K}\vec{r}'} d^3 r' \int \rho(\vec{r}) e^{i\vec{K}\vec{r}} d^3 r = F_L(\vec{K}) F(\vec{K}) \tag{A9}$$

which leads to a separation of structure factors of lattice F_L and unit cell F, respectively. As the lattice is discrete and each lattice point in space can be expressed in terms of the base vectors, the expression for the contribution of the lattice structure factors to the scattering probability simplifies to

$$
\left| F_L(\vec{K}) \right|^2 = \left| \sum_{l_1, l_2, l_3}^{N_1-1, N_2-1, N_3-1} e^{i\vec{K}(l_1 \vec{a}_1 + l_2 \vec{a}_2 + l_3 \vec{a}_3)} \right|^2 =
$$

$$
\left| \prod_{j=1}^{3} \left(\sum_{l_j=0}^{N_j-1} e^{il_j \vec{K} \vec{a}_j} \right) \right|^2 = \prod_{j=1}^{3} \frac{\sin^2(N_j \vec{K} \vec{a}_j)/2}{\sin^2(\vec{K} \vec{a}_j)/2} \tag{A10}
$$

The function represented by this formula shows maxima at discrete values of momentum transfer only, and the experimentally observed diffraction peaks occur at these locations.

For large N_L, this factor even reduces to a Dirac delta function, which reflects the Bragg condition which was used in the above intuitive description of the x-ray diffraction principles, as

$$
\left| F_L(\vec{K}) \right|^2 = N_L^2 \sum_{\vec{G}} \delta(\vec{K} - \vec{G}) \tag{A11}
$$

The problem to be solved in x-ray diffraction is in a way inverse to the description developed so far, because from the observed diffraction pattern

$$
\frac{d\sigma}{d\Omega} \propto \sum_{\vec{G}} \delta(\vec{K} - \vec{G}) \left| F(\vec{K}) \right|^2 \tag{A12}
$$

one wants to determine the arrangement of atoms and thus the charge distribution in a given sample. Therefore, strictly speaking only the absolute value of the scattering amplitude is determined in a diffraction experiment, leaving the phase undetermined. Anomalous scattering experiments close to the resonance frequencies of the system, where all terms of Eq. (A6) contribute to the measured data, can be used to solve this problem.

A.3
Small-Angle Scattering

The major difference between the theoretical representation of the scattering process and the diffraction treatment discussed above is the way the electron density is described. Scattering under large angles implies that the distance between scattering centers which may interfere destructively is small, leading to the concept of local scattering, in which the lattice structure factor plays a dominant role. In contrast to that, for the small-angle scattering process much larger dimensions are relevant, and a description using an averaged electron density facilitates the analytical description of the process. Therefore, as above, the scattering amplitude in direction of a given scattering vector is obtained from

$$A(\vec{q}) = \int_V \rho(\vec{r}) e^{-i\vec{q}\vec{r}} d^3r \qquad (A13)$$

and the scattering intensity by

$$I(\vec{q}) = \int\int_V \rho(\vec{r}_1)\rho(\vec{r}_2) e^{-i\vec{q}(\vec{r}_1-\vec{r}_2)} d^3r_1 d^3r_2 \qquad (A14)$$

Introducing r as distance vector between r_1 and r_2 and the density of electron pairs with separation r, $\rho'(r)$, it is possible to re-interpret this entity which is measured by SAXS as the Fourier transform of a pair correlation function:

$$I(\vec{q}) = \int_V \rho'(\vec{r})^2 e^{-i\vec{q}\vec{r}_1} d^3r \qquad (A15)$$

and the aim of the investigation is to regain the pair correlation function, i.e., information as to what scattering power is found at what distance from absorbing atoms. The specific charm of this approach is that it is better adapted to the larger dimensions which need to be covered than the local scattering center approach discussed above, where for the representation of a given grain on the nanometer scale thousands to millions of local scattering centers need to be considered. However, while this approach is often more useful for practical data evaluation, the concept of local charge density distributions is sometimes less intuitive than the discussion in terms of local scattering centers, as used in Section 7.1.3.

A.4
X-Ray Absorption

In an x-ray absorption experiment, the annihilation of the photon is measured. Consequently, following Eq. (A5), the photoabsorption coefficient μ is proportional to the sum over all final states into which excitation can occur of the square of the transition matrix element between the respective initial state $<n_0|$ and final states $|n_1>$. The transition operator of the incoming photon can be well approximated by a dipole operator z, which leads directly to the selection rules $\Delta l = +/1$, $\Delta m = +/- 1,0$. Thus, e.g., K and L_1 excitation will probe p-states, $L_{2/3}$ excitation dominantly d and partly s-states, and so on. Therefore, XAS allows for the l-projected analysis of the electronic structure of a given system. In a good approximation the excitation into final states is atom-projected as well.

For final states which are (bound) valence states, the methods of molecular orbital theoretical chemistry can provide a wealth of information [12, 50] which makes the interpretation of spectra possible even if no references for fingerprinting are available, but this convenient description of the edge region grows more and more problematic with increasing photon energy. On the other hand, using a scattering theoretical approach, a complementary description is reached, and it is easy to transform the above formula into this kind of picture by inserting

$$-\frac{1}{\pi}\Im G(E) = \sum_{n1} |n_1 > \delta(E - E_{n0n1}) < n_1|$$ (A16)

which eliminates the need to calculate the final states. Next, one expresses the Green's function G(E) in terms of scattering theory, which leads to the operator equation

$$G=G^0+G^0VG=G^0+G^0TG^0$$ (A17)

However, in terms of the free propagator G^0 and the scattering matrix T the equation can only be solved iteratively, as G occurs on both sides of the equation. Alternatively, the Green's function is expressed in terms of a scattering T-matrix and the free propagator G^0 only. The T-matrix for the problem discussed here can be developed into a so-called path expansion (Ref. [51] and references therein), which may converge or not:

$$G = G^a + \sum_{i \neq 0} G^a t_i G^a + \sum_{i \neq j; i,j} G^a t_i G^0 t_j G^a + \sum_{i,j,k...} ...$$ (A18)

This means the T-matrix, reduced to local scattering events t occurring at atom i, i and j, and so on. To describe phenomena near the absorption threshold, the path development needs to be replaced by an implicit summation over all scattering paths, as achieved, e.g., by a full multiple scattering approach, because the scattering probability is high. At high energies relative to the absorption edge, the development can be restricted to the treatment of single scattering phenomena only, which yields the theoretical description of the processes by means of the EXAFS formula which allows the analytical extraction of structural parameters of a given material:

$$\chi(k) = \sum_j S_{0j}^2(k) \frac{N_j}{kr_j^2} F_j(k) e^{-2r_j/\lambda_j(k)} e^{-2\sigma_j^2 k^2} \sin(2kr_j + \Phi_{cj}(k))$$ (A19)

Each of the scattering paths j is connected with a set of specific variables, such as the multiplicity N_j with which it occurs, the distance to the scatterer r_j, the scattering phase and amplitude of the backscatterer Φ_j and F_j, and provides both an amplitude and a phase term.

The phase term (i.e., the sinus function) varies periodically with $2kr_j$ and contains a phase shift δ_c which the photoelectron suffers when leaving and re-entering the absorber atom and during its interaction with the scattering electron. This means there are only *two* variables which influence the phase function: r_j and E_0, as the latter cannot be directly observed, but can be used to finetune the k-scale. This allows for extraction of the distances with a precision of about ±1%.

The amplitude term is more complicated: Apart from the number of backscatterers in a given shell N_j, it is dependent of the scattering amplitude F_j of the involved neighbor. This allows the determination of the type of atom which contributes to a given path. Furthermore, there are two loss terms S_0^2 and $e^{-2r/\lambda(k)}$. The first one appears due to the fact that the EXAFS amplitude is reduced by multielectron excitation processes which are not contained in the one-electron picture developed above. The second, $e^{-2r/\lambda(k)}$, describes the effects of the finite lifetime of the final state, i.e., that inelastic processes can occur to the propagating electron wave and thus destroy the coherence needed to obtain interference of the outgoing and scattered photoelectron wave. In principle for an unknown material both parameters, especially the k-dependence of S_0^2, are more or less out of analytical reach. Usually, one tries determining S_0^2 for a well-known reference compound as a workaround; when doing so bear in mind that you should use an identical k-range. Finally, one has to bear in mind that even in an ideal crystalline system not all atoms are located at their ideal positions at finite temperatures. This leads to the well-known concept of introducing a Debye–Waller factor σ^2 to describe the average thermally induced distance distribution. However, in the case of the EXAFS formula this factor is actually composed of a static and a thermal contribution, where one implicitly postulates that the static contribution follows a gaussian distribution and no anharmonic effects are encountered. If one has access to temperature-dependent data, it is possible to isolate both contributions. Performing a cumulant analysis [52, 53], it is also possible to deal with asymmetric distance–distribution functions. Consequently, a lot of variables are involved in the amplitude determination, and consequently the actual coordination number to be determined is highly correlated with the other variables and thus often a rather soft parameter with quite some error.

Variables

$\vec{q}$	scattering vector
$\vec{k}_0$	wavevector of a particle (photon, electron) before interaction
$\vec{k}$	wavevector of a particle after interaction
λ	wavelength of a photon
n	a positive integer number
d_{hkl}	spacing between two crystal lattice planes with Miller indices h,k,l
θ_B	Bragg angle
D	diameter of a (nano)particle
B	width of a Bragg reflection
$I(\vec{q})$	observed (scattered) intensity corresponding to a given scattering vector
C	constant factor
V	volume (of a particle or elementary cell)
f	scattering amplitude
$\vec{r}$	vector defining a point in (three-dimensional real) space
$g(r)$	radial particle distribution function
$d(D)$	size distribution function
S_1	form factor $\left(= \left\| \frac{1}{V} \int_V e^{i\vec{q}\vec{r}} d^3r \right\|^2 \right)$
A	surface
$\frac{d\sigma}{d\Omega}$	differential (angular) cross-section

Δn_f	scattering contrast
R_g	Guinier radius
m	electron mass
$\hbar$	normalized Planck's constant
H_0, H_{rad}, H'	hamiltonians for ground state, radiation field, photon–matter interaction, respectively
e	elementary charge
$\vec{p}_j$	momentum operator of particle j
U_N	potential of the nuclei in a given system of atoms
a^+, a^-	creation and annihilation operator of the electromagnetic field
ω_k	radial frequency corresponding to a photon with wavevector $\vec{k}$
$\vec{A}(\vec{r})$	vector potential of the electromagnetic field
c	speed of light
σ_j	spin matrix
$\vec{\varepsilon}(\vec{k}, \lambda)$	polarization vector
n_0, n_1, z	initial/final/intermediate states
E_i	energy of state i
$w_{scat}(n_0\vec{k}_0 \rightarrow n_1\vec{k}_1)$	scattering probability for the process changing incoming state n_0, k_0 to outgoing state n_1, k_1
$\delta(...)$	Dirac's delta function
Γ_z	width of intermediate state z
Ψ_i	wave function of state i
$\rho(\vec{r})$	electron density
$\Im$	imaginary part of a complex function
G	Green's function
G^0	free electron propagator
T	global T-matrix
t_i	local T-matrix at scattering center i
$S_0{}^2$	amplitude reduction factor
σ^2	Debye–Waller factor
F_j	scattering amplitude
Φ_{cj}	scattering phase for absorber c and path j
λ_j	mean free electron path length

References

1 B. D. Cullity, *Elements of X-Ray Diffraction*, Addison-Wesley, Reading, **1978**.

2 H. D. Bartunik, Crystal structure analysis of biological macromolecules by synchrotron radiation diffraction, in *Handbook of Synchrotron Radiation*, Vol. 4, eds. S. Ebashi, M. E. Koch, E. Rubenstein, Elsevier, New York, **1991**.

3 D. F. Cox, Powder diffraction, in *Handbook of Synchrotron Radiation*, Vol. 3, eds. G. S. Brown, D. E. Moncton, North Holland, Amsterdam, **1991**.

4 H. Bönnemann, R. M. Richards, Nanoscopic metal particles – synthetic methods and potential applications, *Eur. J. Inorg. Chem.* **2001**, *10*, 2455–2480.

5 H. M. Rietveld, The Rietveld method – a historical perspective, *Aust. J. Phys.* **1988**, *41*, 113–116.

6 F. Izumi,The Rietveld method and its applications to synchrotron x-ray powder data, in *Application of Synchrotron Radiation to Materials Analysis*, eds. H. Saisho, Y. Gohshi, Elsevier, Amsterdam, **1996**.

7 K. Kajiwara, Y. Hiragi, Structure analysis by small-angle x-ray scattering, in *Application of Synchrotron Radiation to Materials Analysis*, eds. H. Saisho, Y. Gohshi, Elsevier, Amsterdam, **1996**.

8 O. Glatter, O. Kratky, *Small Angle X-Ray Scattering*, Academic Press, London, **1986**.

9 H. B. Stuhrmann, Small angle x-ray scattering of macromolecules in solution, in *Synchrotron Radiation Research*, eds. H. Winick, S. Doniach, Plenum Press, New York, **1980**.

10 H. Modrow, J. Hormes, F. Visel, R. Zimmer, Monitoring thermal oxidation of sulfur crosslinks in SBR-elastomers by quantitative analysis of sulfur K-edge XANES-spectra, *Rubber Chem. Technol.* **2001**, *74*, 281–294.

11 B. K. Teo, *EXAFS: Basic Principles and Data Analysis*, Springer Series Inorganic Chemistry Concepts, Vol. 9, Springer, Berlin, **1986**.

12 J. Stöhr, *NEXAFS Spectroscopy*, Springer Series in Surface Sciences, Vol. 25, Springer, Berlin, **1992**.

13 S. S. Hasnain, H. Kamitsubo, D. M. Mills (eds.), *Proceedings of the Eleventh International Conference on X-Ray Absorption Fine Structure: XAFS XI'. J. Synchrotron Radiat.* **2001**, *8*.

14 A. Prange, H. Modrow, X-ray absorption spectroscopy and its application in biological, agricultural and environmental research *Rev. Environ. Sci. Biotechnol.* **2003** *1*, 259–276.

15 C. B. Murray, S. Sun, W. Gauschler, H. Doyle, T. A. Betley, C. R. Kagan, Colloidal synthesis of nanocrystals and nanocrystal superlattices. *IBM J. Res. Dev.* **2001**, *47*, 45–56.

16 F. Fettar, F. S. Lee, F. Petroff, A. Vaures, P. Holody, L. F. Schelp, A. Fert, Temperature and voltage dependence of the resistance and magnetoresistance in discontinuous double tunnel junctions. *Phys. Rev. B* **2002**, *65*, 174415.

17 B. M. Berkovsky, V. F. Medvedev, M. S. Krakov, *Magnetic Fluids: Engineering Applications*, Oxford University Press, Oxford, **1993**.

18 S. Sun, C. B. Murray, Synthesis of monodisperse cobalt nanocrystals and their assembly into magnetic superlattices. *J. Appl. Phys.* **1999**, *85*, 4325–4330.

19 V. F. Puntes, K. M. Krishna, P. A. Alivisatos, Synthesis, self-assembly, and magnetic behavior of a two-dimensional superlattice of single-crystal epsilon-Co nanoparticles. *Appl. Phys. Lett.* **2001**, *78*, 2187–2189.

20 H. Modrow, S. Bucher, J. Hormes, R. Brinkmann, H. Bönnemann, Model for chainlength-dependent core-surfactant interaction in N(alkyl)$_4$Cl-stabilized colloidal metal particles obtained from x-ray absorption spectroscopy *J. Phys. Chem. B* **2003**, *107*, 3684–3689.

21 V. G. Palshin, R. Tittsworth, J. Hormes, E. I. Meletis, X. Nie, J. Jiang, H. Modrow, Size-dependence of the Co-phase in nanocrystalline thin films, in *From the Atomic to the Nano-Scale*, eds. C. T. Whelan, J. H. McGuire, Proceedings of the International Workshop, Old Dominion University Dec. 12–14, 2002, Old Dominion University, **2003**, pp. 99–125, ISBN 0–9742874–0-7.

22 P. Zhang, T. K. Sham, Tuning the electronic behavior of Au nanoparticles with capping molecules. *Appl. Phys. Lett.* **2002**, *81*, 736–738.

23 P. Zhang, T. K. Sham, X-ray studies of the structure and electronic behavior of alkanethiolate-capped gold nanoparticles: The interplay of size and surface effects *Phys. Rev. Lett.* **2003**, *90*, 245502.

24 L. X. Chen, T. Liu, M. C. Thurnauer, R. Csentcits, T. Rajh, Fe$_2$O$_3$ nanoparticle structures investigated by x-ray absorption near edge structure, modifications and model calculations. *J. Phys. Chem. B* **2002**, *106*, 8539–8546.

25 H. Modrow, Tuning nanoparticle properties – the x-ray absorption spectroscopic pouint of view. *Appl. Spectr. Rev.* **2004**, *39*, 183–290.

26 H. Modrow, N. Palina, C. S. S. R. Kumar, E. E. Doomes, M. Aghasyan, V. Palshin, R. C. Tittsworth, J. C. Jiang, J. Hormes, Characterization of size dependent structural and electronic properties of CTAB-stabilized cobalt nanoparticles by x-ray absorption spectroscopy. Accepted by *Phys. Scripta*, in print.

27 Z. Wu, J. Zhang, R. E. Benfield, Y. Ding, D. Grandjean, Z. Zhang, X. Ju, Structure and chemical transformation in cerium oxide nanoparticles coated by surfactant cetyltrimethylammonium bromide (CTAB): an x-ray absorption spectroscopic study. *J. Phys. Chem. B* **2002**, *106*, 4569–4577.

28 C. W. Hills, N. H. Mack, R. G. Nuzzo, The size dependent structural phase behaviors of supported bimetallic (Pt-Ru) nanoparticles. *J. Phys. Chem. B* **2003**, *107*, 2626–2636.

29 H. Bönnemann, W. Brijoux, J. Richter, R. Becker, J. Hormes, J. Rothe, The preparation of colloidal Pt/Rh alloys stabilized by NR_4^+ and PR_4^+ groups and their characterization by x-ray absorption spectroscopy. *Z. Naturforsch. B* **1995**, *50*, 333–338.

30 Y. Iwasawa (ed.), X-ray absorption fine structure for catalysts and surfaces, World Scientific Series on Synchrotron Radiation Techniques and Applications, Vol. 2, World Scientific, Singapore, **1996**.

31 A. I. Frenkel, C. W. Hills, R. G. Nuzzo, A view from the inside: complexity in the atomic scale ordering of supported metal nanoparticles. *J. Phys. Chem. B* **2001**, *105*, 12689–12703.

32 J. A. Larsson, M. Nolan, J. C. Greer, Interactions between thiol molecular linkers and the Au_{13} nanoparticle. *J. Phys. Chem. B* **2002**, *106*, 5931–5937.

33 A. Di Cicco, M. Berrettoni, S. Stiza, E. Bonetti, Microstructural defects in nanocrystalline iron probed by x-ray absorption spectroscopy. *Phys. Rev. B* **1994**, *65*, 12386–12397.

34 Di Cicco, A., Berrettoni, M., Stiza, S., Bonetti, E. EXAFS study of nanocrystalline iron. *Physica B* **1995**, *208/209*, 547–548.

35 Y. A. Babanov, I. V. Golovshchikova, F. Boscherini, T. Haubold, S. Mobilio, EXAFS study of nanocrystalline cobalt. *Nucl. Instrum. Methods A* **1995**, *359*, 231–233.

36 B. S. Clausen, J. N. Norskov, Asymmetric pair distribution functions in catalysts *Top. Catal.* **2000**, *10*, 221–230.

37 K. Malone, S. Weaver, D. Taylor, H. Cheng, K. P. Sarathy, G. Mills, Formation kinetics of small gold crystallites in photoresponsive polymer gels. *J. Phys. Chem. B* **2002**, *106*, 7422–7431.

38 A. Henglein, M. Giersig, Reduction of Pt(II) by H_2: effects of citrate and NaOH and reaction mechanism. *J. Phys. Chem. B* **2000**, *104*, 6767–6772.

39 T. Vad, H. G. Haubold, N. Waldöfner, H. Bönnemann, From Pt molecules to nanoparticles: in-situ (anomalous) small-angle x-ray scattering studies. *J. Appl. Cryst.* **2002**, *35*, 459–470.

40 P. Fratzl, Small-angle scattering in materials science – a short review of applications in alloys, ceramics and composite materials. *J. Appl. Cryst.* **2003**, *36*, 397–404.

41 K. Jokela, R. Serimaa, M. Torkkeli, V. Etelaniemi, K. Ekman, Structure of the grafted polyethylene-based palladium catalysts: WAXS and ASAXS study. *Chem. Mater.* **2002**, *14*, 5069–5074.

42 R. E. Benfield, D. Grandjean, J. C. Dore, H. Esfahanian, Z. H. Wu, M. Kroll, M. Geerkens, G. Schmid, Structure of assemblies of metal nanowires in mesoporous alumina membranes studied by EXAFS, XANES, x-ray diffraction and SAXS. *Faraday Discussions* **2004**, *125*, 327–342.

43 J. W. Andreasen, O. Rasmussen, R. Feidenhans'l, F. B. Rasmussen, R. Christensen, A. M. Molenbroek, G. Goerigk, An in situ cell for small-angle scattering experiments on nano-structured catalysts. *J. Appl. Cryst.* **2003**, *36*, 812–813.

44 J. Turkevich, G. Kim, Palladium: preparation and catalytic properties of particles of uniform size, *Science* **1970**, *169*, 873–875.

45 K. Angermund, M. Bühl, E. Dinjus, U. Endruschat, F. Gassner, H. G. Haubold, J. Hormes, G. Köhl, F. T. Mauschick, H. Modrow, R. Mörtel, R. Mynott, B. Tesche, T. Vad, N. Waldöfner, H. Bönnemann, Nanoscopic Pt colloids in the embryonic state. *Angew. Chem. Int. Ed.* **2002**, *41*, 4041–4044.

46 K. Angermund, M. Bühl, U. Endruschat, F. T. Mauschick, R. Mörtel, R. Mynott, B. Tesche, N. Waldöfner, H. Bönnemann, G. Köhl, H. Modrow, J. Hormes, E. Dinjus, F. Gassner, H. G. Haubold, T. Vad, M. Kaupp, In situ study on the wet chemical synthesis of nanoscopic Pt colloids by reductive stabilization. *J. Phys. Chem. B* **2003**, *107*, 7507–7515.

47 H. Bönnemann, W. Brijoux, R. Brinkmann, N. Matoussevitch, N. Waldöfner, N. Palina, H. Modrow, A size-selective synthesis of air stable colloidal magnetic cobalt nanoparticles. *Inorg, Chim. Acta* **2003**, *350*, 617–624.

48 F. Meneau, G. Sankar, N. Morgante, R. Winter, C. R. A. Catlow, G. N. Greaves, J. M. Thomas, Following the formation of nanometer-sized clusters by time-resolved SAXS and EXAFS techniques. *Faraday Discussions* **2003**, *122*, 203–210.

49 A. L. Ankudinov, J. J. Rehr, J. Low, S. Bare, Effect of hydrogen adsorption on x-ray absorption spectra of small Pt clusters *Phys. Rev. Lett*. **2001**, *86*, 1642–1645.

50 F. von Busch, J. Hormes, H. Modrow, N. B. Nestmann, Interaction of atomic core electrons with the molecular valence shell, in *Interactions in Molecules – Electronic and Steric Effects*, ed. S.D. Peyerimhoff, Wiley-VCH, Weinheim, **2003**.

51 J. J. Rehr, R. C. Albers, Theoretical approaches to x-ray absorption fine structure. *Rev. Mod. Phys*. **2000**, *72*, 621–654.

52 G. Bunker, Application of the ratio method of EXAFS analysis to disordered systems. *NIM* **1983**, *207*, 437.

53 H. Bertagnolli, T. S. Ertel. Röntgen-absorptionsspektroskopie an amorphen Festkörpern, Flüssigkeiten, katalytischen und biochemischen Systemen. *Angew. Chem.* **1994**, *106*, 15.

54 G. Köhl, Diploma Thesis, Bonn University, BONN-IB-96-35 (1996).

8

Single-Molecule Detection and Manipulation in Nanotechnology and Biology

Christopher L. Kuyper, Gavin D. M. Jeffries, Robert M. Lorenz, and Daniel T. Chiu

8.1
Introduction

The set of tools for visualizing the world at the molecular and nanoscopic level has been both broadened and refined in the past decades. Investigating individual molecules is becoming more routine, and a wide array of methods have revealed exciting new information that otherwise is unobtainable from ensemble studies. Examples of these tools include the family of scanning probe microscopies (SPM), such as scanning tunneling microscopy (STM) and atomic force microscopy (AFM), which have offered us exquisite details and visualizations of single atoms and molecules on surfaces [1–5]. The use of optical techniques to detect and study single molecules has also gained prominence in recent years owing to their ease of implementation and their ability to probe biological systems in solution under physiologically relevant conditions. Platforms such as confocal optics, epifluorescence detection, and total internal-reflection fluorescence (TIRF) microscopy have become increasingly common due to relatively straightforward design, impressive detection sensitivities, and easy integration into commercially available microscope systems [6–9]. Optically based single-molecule experiments are not only limited to detection and visualization; optical traps serve as tools for measuring forces and movements of individual biomolecules in solution [10–14]. Due to the broad range of optical techniques that have been developed for the detection and manipulation of single molecules, which range from near-field microscopy to the use of nonlinear optical methods [15–19], we will not be able to discuss all these examples in this chapter. Instead, we begin with an introduction of the common techniques used in optically based single-molecule detection (e.g., confocal, epifluorescence, TIRF) and manipulation (e.g., optical trapping) that are pertinent to biology and nanotechnology. We then continue with a discussion of a few select areas to which these techniques have been applied, both to illustrate the past accomplishments and the future potential of optically based single-molecule methodologies.

Nanofabrication Towards Biomedical Applications. C. S. S. R. Kumar, J. Hormes, C. Leuschner (Eds.)
Copyright © 2005 WILEY-VCH Verlag GmbH & Co. KGaA, Weinheim
ISBN 3-527-31115-7

8.2
Optical Detection of Single Molecules

8.2.1
Detecting Single Molecules with Confocal Fluorescence Microscopy

In confocal microscopy [Fig. 8.1(A)], coherent laser radiation is collimated and reflected off a dichroic mirror to fill the back aperture of a high numerical aperture (NA) objective (NA > 1.2). Exiting from the objective, laser light is focused to a diffraction-limited spot that has a beam waist of ~200–350 nm in diameter as defined by the Rayleigh criterion, which depends on the wavelength and NA of the objective. Along the same axis of the incoming light, fluorescence emitted from the focal point in the sample plane is recollected by the objective and passed through the dichroic mirror and a bandpass filter, which ensures that no extraneous Rayleigh scattering, or laser radiation, impinges on the detector. Prior to the bandpass filter, the fluorescent signal is focused by a tube lens into a pinhole located at the image plane of the microscope. To achieve maximum spatial resolution and signal collection efficiency, choice of the correct pinhole diameter is critical [20, 21]. In single-molecule experiments, optimally designed configurations with 100× high-NA objectives typically use pinholes 20–100 μm in diameter and produce an axial resolution of ~0.5–1 μm. With a waist of 200–350 nm and axial length of 1 μm, the ellipsoidal detection volume is on the order of ~10^{-15} L. As a result of extremely small detection volumes, background sources such as Raman scattering of water are reduced to improve the level of single-molecule signal. To record the signal, avalanche photodiodes are commonly used owing to remarkable quantum efficiency (QE, >70%) and low dark noise (< 25 counts s^{-1}), which along with low background give impressive signal-to-noise ratios (SNRs) as compared to other optical methods (e.g., epifluorescence). While confocal designs typically do point detection, the use of a scanning mirror or a high-resolution piezoelectric stage can produce images with single-molecule sensitivity. Although these images display high SNRs, full image capture takes minutes, which is slow compared to wide-field detection methods using a CCD camera (see Sections 8.2.2 and 8.2.3).

The distinct advantages of confocal microscopy have led to a wide array of studies related to single-molecule spectroscopy (SMS). Single-molecule sensitivity is an important advantage in fluorescence correlation spectroscopy, which detects and correlates individual bursts from molecules diffusing into and out of the detection volume to give diffusional information about any fluorescent species [22–29]. Furthermore, Zare and coworkers reported real-time observations of single molecules diffusing through the probe volume with 2 μs resolution [30, 31]. Although a key advantage of confocal detection lies in its ability to study freely diffusing molecules, it also has permitted easy study of single molecules immobilized in solids, polymer matrices, and gels, and on surfaces. With remarkable time resolutions and detection sensitivities, confocal experiments can study biological processes that occur on the millisecond time scale; specifically, single-enzyme kinetics [32, 33], conformational changes in enzymes and DNA [34–37], ribozyme function and dynamics [38, 39],

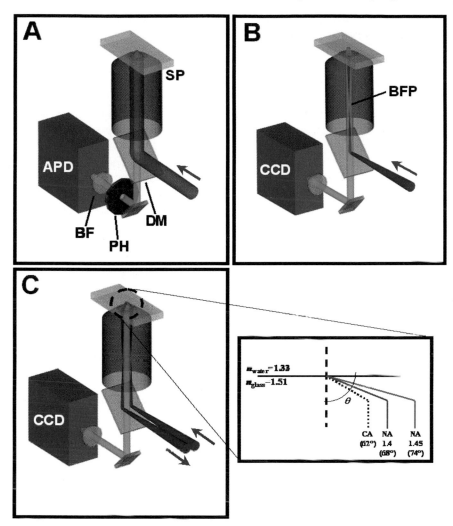

Figure 8.1. Three commonly used optical methods for detecting single molecules: (A) confocal point detection, (B) epifluorescence imaging, and (C) total internal-reflection fluorescence (TIRF) microscopy. In all three geometries, directed laser light is reflected off a dichroic mirror (DM) into the back aperture of the high numerical aperture (NA) objective and is then imaged onto the sample plane (SP). Single-molecule fluorescence is collected through the same objective and further filtered by a bandpass filter (BF). In confocal detection, the spectrally filtered light is further passed through a pinhole (PH) placed at the primary image plane. Both epifluorescence and TIRF designs require laser light to be focused at the back focal plane of the objective (BFP). For detection, confocal setups typically use avalanche photodiodes (APD), while epifluorescence and TIRF microscopy use high sensitivity CCD cameras; (D) shows the critical angle needed to achieve objective-type TIRF with an NA of 1.4 or 1.45.

protein complexes [40, 41], detection of DNA in nanopore technology [42], as well as orientational dynamics and optical properties of single dye molecules [43–47]. Not only can properties of single molecules be investigated, but the surrounding nanoenvironments can also be probed. For instance, heterogeneous rotational dynamics and lifetimes of single molecules in polymer films can be used to investigate molecular environments present within different regions of the film [48, 49]. Sensitive to changes in hydrophobicity, a new solvatochromic dye covalently bound to a polypeptide, for example, was used to monitor protein–protein interactions at the single-molecule level [50]. Inherent versatility, fast time resolution, and high SNR will continue to bolster the use of confocal optics in SMS; however, in contrast to epifluorescence and TIRF designs, which can study multiple point sources at once, confocal detection can only detect point source objects one at a time.

8.2.2
Visualizing Single Molecules with Epifluorescence Detection

Epifluorescence configurations are notably the most popular designs in optical microscopy. Lasers are commonly used for illumination in single-molecule studies, although recent reports describe single-molecule detection using mercury lamps and LEDs [51, 52]. Similar to confocal designs, epifluorescence illumination and fluorescence collection is carried out using the same high-NA objective and results in an area of illumination typically 50–100 µm in diameter for an objective with $100 \times$ magnification. With proper collimation optics, lamps and LEDs produce even illumination over a wide area because of the noncoherent nature of such light sources. To achieve even, wide-area illumination with lasers, however, some additional optics are required. For instance, a spinning holographic diffuser or ground glass plate is placed in the beam line prior to the objective to disrupt the coherence of the laser light and to eliminate specular patterns at the image plane [13, 53]. Unfortunately, these designs can suffer unnecessary loss of incoming laser light to reflection and scattering. A more efficient design [Fig. 8.1(B)] focuses laser light at the back focal plane of the high-NA objective, which results in emergence of a collimated laser beam at the object plane and provides wide-field illumination of the sample [54]. For imaging, charge-coupled device (CCD) cameras are most common and can reach time resolutions of tens to hundreds of milliseconds for single-molecule experiments, limited by the available signal and sensor readout time. While, per pixel, time resolution with CCD imaging is two to three orders of magnitude slower than with confocal point detection, multiple molecules can be studied simultaneously. In epifluorescence, background from out-of-focus light and Raman scattering from a large detection volume, however, does limit the detection of single molecule signal. One strategy to further reduce background is to use dyes that fluoresce in the red [55]. Fortunately, detection sensitivity steadily improves as newly developed back-thinned CCD cameras produce images with low noise and can achieve QEs of ~90%. Advances towards higher-QE chips with lower readout noise and faster readout speeds will continue to improve epifluorescence detection.

Epifluorescence detection can provide information about immobilized single molecules as well as individual species that move over distances of microns. Some interesting biological applications include: (i) The study of the brownian diffusion and the dynamics of single DNA molecules [56–62], (ii) the direct imaging of individual molecular motors (e.g., myosin and kinesin) and bioparticles moving along the surface of glass coverslip [63–66], (iii) the observation of discrete rotations of an active F1 ATPase molecule [67], and (iv) the monitoring of the rates of enzymatic digestion of DNA molecules by λ-exonuclease [68]. Moreover, with adequate signal, low noise, and stable optics, positions of fluorescent species can be accurately determined down to ~10 nm through Gaussian fitting of the fluorescent profile [68]. For single molecules, point-source fluorescence is below the diffraction limit (~200–350 nm) and can be highly magnified (100–500 ×) to result in an imaged spot that is spread over several micron-sized pixels (~7–25 μm diameter) with each pixel corresponding to sample distances of tens of nanometers. Remarkably, diffusional information from imaging single-molecule movements over time can be recorded in sol-gel films, porous materials, and in lipid bilayer membranes [53, 69–71]. Because of low SNR, single-molecule trajectories are typically accurate only to within 100–200 nm. While epifluorescence designs offer great advantage for studying molecules many microns above the coverslip, detection efficiencies for studies on surfaces are quite inferior to results produced using TIRF microscopy.

8.2.3
Total Internal-Reflection Fluorescence (TIRF) Microscopy

In comparison with point-detection techniques like confocal microscopy and wide-field visualization of single molecules in free solution using epifluorescence, TIRF microscopy boasts impressive reduction of background and noise to produce wide-field images of single molecules on surfaces. Total internal reflection occurs at an interface between high and low refractive index materials when incident light is directed at an angle greater than the critical angle defined by the two media. At the interface, light penetrates into the low refractive medium to produce an evanescent field that falls off exponentially, thus defining an illumination thickness of ~100–150 nm, depending on the wavelength used and the refractive indices of the two media [72]. Conveniently, excitation of any fluorescent species will only occur within the thin evanescent field. With a high power density within the evanescent field, which enhances single-molecule signal, and a dramatic reduction of background due to a decreased probe volume, improved detection efficiencies with TIRF are notable in comparison to epifluorescence.

Two commonly used TIRF configurations are (i) objective-type and (ii) prism-based. Objective-type TIRF not only requires the laser beam to be focused at the back focal plane of the objective like epifluorescence (see Section 8.2.2), but to ensure TIR the beam must also be directed towards the outer edge of the high-NA objective such that light is then incident at an angle greater than the critical angle [Fig. 8.1(C) with inset]. The critical angle in experiments using a glass coverslip ($n =$ 1.51) and aqueous solution ($n = 1.33$) is ~62°, thus requiring use of objectives with an NA of 1.4 (Θ~68°) or 1.45 (Θ~74°) to achieve TIR. Practically, the extra 6° provided by an NA 1.45 objective makes implementing TIRF a relatively straightforward task. Prism-based configurations, as the name implies, use a prism placed on top of the sample to direct an incoming beam at the necessary angle to achieve TIR. Unfortunately, these designs are not as versatile as objective-type TIRF because the location of the prism restricts the types of systems that can be studied. Although in a side-by-side comparison, Ambrose and coworkers reported better signal-to-background ratios (SBRs) for prism-based TIRF, objective-type TIRF produced significantly more photons from the single molecules [73]. In addition, objective-type TIRF designs improved both SNR and SBR by up to a factor of four when compared with epifluorescence [54, 74].

Even though single-molecule TIRF experiments are limited to studying molecules immobilized or close to surfaces, the enhancements of sensitivity prove quite useful for many applications. By degrading the focus of emission from single fluorophores and by changing the polarization of the excitation beam, orientation of emission dipoles can be determined from each molecule in all three dimensions; molecules with the dipole oriented in the z-axis showed a doughnut-like intensity profile [75, 76]. Furthermore, because of the thin sampling plane, TIRF has been used for isolated detection of individual binding events between molecules on the surface and free molecules in solution. For example, rates of absorption and conformational changes of individual λ-DNA molecules binding to clean, fused-silica surfaces can be affected by pH and buffer composition [77]. In addition, TIRF microscopy can observe binding of biotinylated DNA to streptavidin-coated coverslips and covalent attachment of DNA to linkers attached to the surface, and has permitted real-time visualization of protein–protein interactions [78, 79]. Other examples include monitoring of the movements of kinesin on microtubules [80, 81], and myosin along the surface of a coverslip [82, 83]. Even a single, fluorescently labeled RNA polymerase can be visualized in real-time while moving along a single strand of DNA [84]. Analyzing the rates of movement for the myosin, kinesin, and RNA polymerase uncovered new kinetic information about each system. Typically in TIRF, experiments look at signals from the solid/liquid interface, but lateral diffusion of dyes along the interface of two immiscible liquids can also be investigated [85].

8.2.4
Single-Molecule Surface-Enhanced Resonance Raman Spectroscopy

First reported in the 1990s by two groups, Nie and coworkers and Kneipp et al. [86, 87], single-molecule surface-enhanced resonance Raman spectroscopy (SM-SERRS) has emerged as a promising new method for studying single molecules. These early works reported impressive enhancement factors on the order of 10^{14}–10^{15} for single dye molecules (e.g., rhodamine 6G and crystal violet) adsorbed onto nanoclusters of silver particles. Surface-enhanced Raman analysis of single molecules offers several unique advantages: (i) in contrast to fluorescence spectroscopy, which excites molecular electronic states, Raman spectroscopy probes vibrational modes to elucidate structural information about a single molecule; (ii) SM-SERRS detection boasts spectra with impressive signals that can be two to three orders of magnitude brighter than fluorescence detection; and (iii) molecules studied using SM-SERRS can be observed for longer periods prior to photodestruction, due to use of an excitation wavelength that is to the red of the absorption wavelength of the dye and quenching of the excited state by metallic nanoclusters. Since the initial experiments, more papers have emerged, including studies on the significant enhancement factors in SM-SERRS [88], detection of single dye molecules in Langmuir-Blodgett monolayers [89–91], SM-SERRS raster-scanned images of single molecules [89, 92], monitoring of surface dynamics [93], investigations of single-molecule magnets [94, 95], and real-time observation of single protein molecules [96–99]. For the interested reader, we recommend the comprehensive reviews by Zander, Kneipp et al. [100, 101].

Notably, Hofkens and coworkers reported the use of SM-SERRS to study enhanced green fluorescent protein (EGFP), which consists of a unique arrangement of three amino acid residues that produce a highly fluorescent complex [99]. EGFP and the family of GFPs are known to convert between a protonated and de-protonated state, and for on/off blinking, the protonated form is believed to cause dark periods in fluorescence observation of the protein [102]. Real-time SM-SERRS spectra of EGFP could be obtained with a time resolution of 5 s and agreed (± 10 cm^{-1}) with bulk EGFP data. Interestingly, the group observed specific frequency jumps over time that were believed to correspond with protonated and deprotonated forms of the molecule. SM-SERRS represents a union of spectroscopy and nano-technology that provides detailed structural information on single molecules which would be unattainable using other methods. While SM-SERRS is limited in that spectra cannot be collected without the adsorption of molecules to suitable metallic nanoparticles, this method will undoubtedly continue to bring exciting new discoveries to the field of SMS.

8.3
Single-Molecule Manipulations Using Optical Traps

8.3.1
Force Studies Using Single-Beam Gradient Traps

Data collection with techniques such as confocal, epifluorescence, or TIRF microscopy relies on detection of single-dye-molecule fluorescence induced by light illumination. Lasers, however, are not only limited to exciting fluorescence; focused coherent radiation can be used to probe single molecules through mechanical manipulation with a laser optical trap. First reported by Ashkin and coworkers, a single-beam gradient trap consists of a Gaussian (TEM$_{00}$ mode) laser beam that is focused tightly onto the sample through a high-NA objective (e.g., NA 1.3) [103]. A particle in the vicinity of the focal spot will experience an attractive force that is directly proportional to the strength of the electric field and the polarizability of the molecule. As shown in Fig. 8.2(A), a particle will be held tightly within the three-dimensional

(3D) intensity gradient by the transverse force (F_T) that arises from the Gaussian intensity profile of the laser and the longitudinal force (F_L) from the tight focusing by the high-NA objective. In opposition to this attractive gradient force, however, is the scattering force, which is mainly in the direction of beam propagation and experimentally makes trapping in the z-direction more difficult than in the x–y plane.

Through careful consideration of force balance along with clever integration of optical traps into novel detection schemes, force and displacement measurements

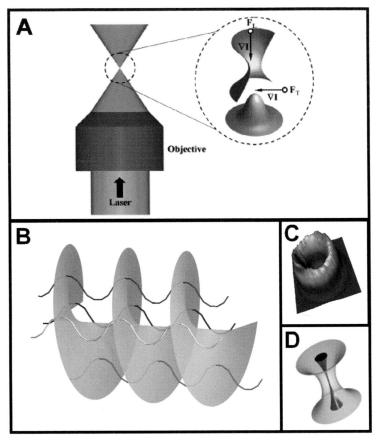

Figure 8.2. (A) Expanded drawing depicting the three-dimensional laser focus of a single-beam gradient trap. A Gaussian laser beam is directed through a high-NA objective to produce a tightly focused intensity gradient (∇I). A particle in the vicinity of the intensity gradient experiences a transverse force (F_T) from the TEM_{00} mode of the laser and a longitudinal force (F_L) from the tight focusing of the objective. (B) The Laguerre-Gaussian doughnut mode (LG_0^1) exhibits destructive interference along the beam axis due to the rotating phase of the beam. The colored ribbons represent individual light waves that comprise the rotating phase of the wavefront shown in gray. (C) The resulting intensity profile produced from the mode results in a doughnut-like gradient that maintains a dark core (D) when focused by a high-NA objective.

have been collected on a variety of individual biological molecules, particularly molecular motors. The pliant nature of the optical trap permits measurement of exquisitely small forces ($\sim 10^{-12}$ N) produced from biomolecules, which transduce chemical energy into a mechanical energy used to propel against the applied force of the trap. To measure forces generated by single molecules, a micron-sized polymer bead (usually polystyrene) is typically chemically or biochemically attached to the biomolecule and acts as a handle for subsequent optical manipulation. Nanometer-scale bead displacements, which arise from biomolecular forces, are sensed by a feedback mechanism that increases the laser power to a level needed to restore the position of the slightly displaced bead. Calibration of the displacements and power fluctuations provides the means to monitor detailed information about the energetics and motion of molecular motors at the single-molecule level. These exciting advances are discussed in a number of reviews [10–13, 104–106].

8.3.2
Optical Vortex Trapping

Since the initial observations of Ashkin and coworkers [103], advances in the area of optical trapping have progressed considerably, both in the area of single-molecule force measurements and in the development of more efficient and diverse types of traps to manipulate a broader spectrum of particles in the microscopic and nanoscopic scale, with emphasis towards biological applications. The single-beam gradient trap that uses a TEM_{00} laser mode is based on a balance of the gradient force that traps the particle and the opposing scattering force, which decreases the trapping efficiency of the gradient trap. To minimize this scattering force, Ashkin discussed the use of higher-order laser modes [107]. His investigation included the Laguerre–Gaussian doughnut mode (LG_0^1), which has a dark core [108]. The unusual look of this mode is a result of destructive interference in the beam along the optical axis, generated by the rotating phase of the beam, as illustrated in Fig. 8.2B [109]. Figure 8.2(C) shows the resulting intensity gradient.

In recent years, the development of these tailored beams has become a broad and active area of research [110]. Investigations into the generation of these beams (initially of the LG_0^1 doughnut mode) for optical trapping were undertaken by He et al. [109] using a computer-generated hologram (CGH) and later taken further by Gahagan et al. [111] and Arlt et al. [112]. The utility of CGHs was initially illustrated in 1992 by Bazhenov et al. [113], showing that holographic elements could induce a screw dislocation in the light wavefront. These CGH patterns acted as a diffractive optical element (DOE) in the beam path to produce the desired mode, which could then be selected and used. This same dislocation can be created by helically shifting the wavefront [114]. When focused down by a high-NA objective, the beam maintains its dark core (Fig. 8.2(D)).

In conventional laser tweezers, the types of particles that could be trapped must have a higher refractive index than the surrounding medium. Although the trapping of low-refractive-index particles was shown to be possible using a TEM_{00} beam, these schemes require the rapid scanning of the laser focus around the particle or utilize the interference pattern between two phase-shifted Gaussian beams [115, 116]. With the LG_0^1 mode, both low-index and high-index particles can be held stably at the laser focus [117, 118]. Theoretical calculations of trapping efficiencies and potentials have been reported [119], which indicate that trapping efficiencies are on average greater using a LG_0^1 beam than a TEM_{00} beam [120, 121]. The lateral trapping forces are reported not to change, and the increase in efficiency is largely due to the lack of scattering force that leads to more efficient trapping in the longitudinal direction.

Utilizing laser modes other than the usual TEM_{00} is advantageous in fields where the particle is nontransparent or very sensitive to photodamage. Reducing photodamage will allow a greater selection of biological materials to be optically manipulated, as well as decrease heating in the sample. Another observation of vortex trapping is the ability to generate movement of the trapped particle by the rotating phase of the beam [122, 123]. This transfer of angular momentum to absorptive particles permits the controlled rotation of the trapped particle, a useful tool in cellular studies [124, 125].

8.3.3
Optical Arrays

Some exciting work in the area of optical trapping has been in the development of optical tweezer arrays. Multiple traps can be constructed by adopting a time-sharing scheme, in which computer-controlled galvano- or piezoelectric mirrors scan rapidly the laser beam so as to park the laser focus in multiple locations at the object plane [115]. This scheme relies on the finite diffusion time it takes the particle to leave its originally trapped position and is not ideal for the trapping of small particles or a large number of particles. To overcome this limitation, holographic optical tweezer arrays (HOT arrays) were developed and can be generated in either two or three dimensions. Using the same holographic techniques utilized in creating the optical vortex trap, multiple optical traps can be generated from a single laser beam. These traps can be stationary or dynamic and can be conventional optical tweezers, vortex traps, or traps based on other higher-order laser modes. A number of methods exist to produce these HOT arrays, including the use of liquid crystal displays (LCDs) [126], spatial light modulators (SLMs) [127], and diffractive optics (physical and holographic) [128]. The use of LCDs or SLMs involves splitting a single beam into several separate beams, where each beam can have identical or different properties in their optical wavefronts [129]. The individual beams, when focused down by a high-NA objective, generate trapping forces in three dimensions [127]. One interesting application of HOT arrays is in the formation of nanoscopic materials, both physical (i.e., permanent) and virtual (i.e., transient). The ability to align particles into an ordered and predefined array, which can be controlled dynamically in time, should lead to interesting applications in areas of biomanipulation, organization, and construction, as well as in microfabrication and nanotechnology [130].

8.4
Applications in Single-Molecule Spectroscopy

8.4.1
Conformational Dynamics of Single DNA Molecules in Solution

Owing to the intercalation of multiple fluorophores, single DNA molecules with lengths ranging from ~5 to ~60 µm can be easily visualized in solution using fluorescence microscopy. Although useful information can be obtained from simply observing DNA motions in solution, exciting discoveries about DNA conformational dynamics have resulted from the use of single-beam optical traps to manipulate fluorescent DNA molecules either directly or through attached bead handles [13]. With bead handles, polymer motion, relaxation dynamics, and novel manipulations can be studied in real time to further our understanding of polymer behavior in so-

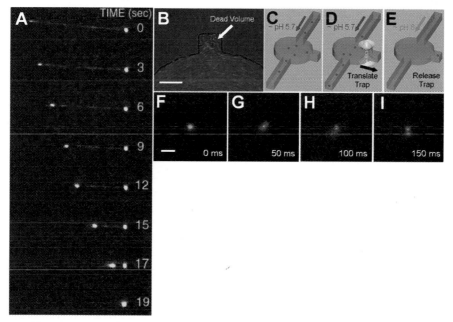

Figure 8.3. (A) Sequence of images showing the condensation of a single concatemer of λ-DNA in a solution of protamine; (B) fluorescence image displaying slow flow of 1-µm fluorescent beads within the side notch of a circular microchamber (scale bar: 50 µm). (C–E) The procedure used to initiate decondensation of single DNA molecules consisted of (i) flowing condensed DNAs in a pH 5.7 solution into the circular microchamber (C), (ii) individually trapping and translating the DNA into the side notch (D), and (iii) initiating decondensation with the optical trap after a pH 8 solution replaced the pH 5.7 solution by diffusion. (F–I) Decondensation of a single λ-DNA molecule occurred over 150 ms. After $t = 0$, each frame was acquired with 50-ms time resolution. Slight blurring occurred because internal motions of the DNA were faster than the image acquisition rate of the camera. Reprinted with permission from Brewer *et al.*, Science (Washington, D. C.) 286, 120, Copyright 1999 AAAS.

lution [59, 60, 131]. In addition to understanding physical properties of the polymer, optical trapping and fluorescence visualization have been valuable towards unraveling condensation and decondensation kinetics of single DNA molecules. Packaging of DNA in cells plays a critical role in space conservation and organization, and is associated with a range of important biological functions such as gene activation and transcription. In nature, for example, λ phage virus DNA with a contour length of ~17 μm is packed tightly into the phage head measuring only ~50 nm in diameter. While condensed DNA can be observed *in vivo*, studies *in vitro* provide more detailed information about the structural properties of DNA condensates. DNA molecules with micrometer-scale contour lengths have been observed *in vitro* to pack tightly into highly ordered toroidal, or doughnut-like, structures in the presence of several different polycations and proteins. DNA condensation is believed to be induced primarily by an electrostatic neutralization of the negatively charged DNA backbone that allows tight packing as result of a sufficient decrease in repulsive energies [132].

To probe the processes underlying DNA condensation, static visualization techniques [e.g., electron microscopy (EM) and AFM] have been used to obtain high-resolution images of DNA condensates [132–134]. These approaches provide static images of condensed DNA that contain high information content. Little information, however, is offered on the kinetics of the condensation process, which has been implicated in determining the final structure of the DNA condensate [135]. Using dynamic light scattering, the bulk measurements on the kinetics of DNA condensation and decondensation have been collected, yet these studies were unable to visualize individual events at the single-molecule level [136]. Recently, several SMS studies have observed condensation and decondensation of single DNA molecules in solution [57, 61, 62, 137]. Balhorn and coworkers combined microfluidics and optical trapping to isolate single concatemers of λ-DNA (attached to beads) in a flow of protamine, a protein known to condense DNA in sperm [61] [Fig. 8.3(A)]. Using fluorescence for visualization, dynamic changes in length of the DNA revealed information on the kinetics of condensation and decondensation of the molecule. Two studies conducted by Yoshikawa and coworkers monitored condensation and decondensation of (i) individual T4 DNA (~166 kilobase pairs) in solution using polyethylene glycol and Mg^{2+} [57] and (ii) single optically trapped DNA translated between condensing and decondensing environments [137]. Recently, Chiu and coworkers reported that the commonly used DNA fluorescent intercalator dye YOYO-1 can act as a condensing agent under moderately acidic pH conditions [62]. Individual YOYO-intercalated λ-DNA molecules (~48.5 kbp), for example, were collapsed into toroidal structures ranging from 100 to 150 nm in diameter at pH 5.7. Using microfluidics, the solution environment could be quickly changed around an optically trapped DNA condensate [Fig. 8.3(B–E)]. Shuttering of the trap initiated conformational transitions of single, condensed YOYO-intercalated DNA molecules to an extended, random coil state that occurred over a time period of ~150 ms [Fig. 8.3(F–I)]. Interestingly, the studies performed by Balhorn and Yoshikawa observed completion of DNA condensation and decondensation on the time scale of seconds. Slower completion times can possibly be attributed to slow mass transfer in the fluidic designs, which required packing (or unpacking) to occur during solution exchange. For Chiu and coworkers, the solution was exchanged prior to initiating the decondensation, thus possibly allowing access to faster uncoiling dynamics. Easy visualization with SMS will allow further investigation into these biologically interesting and important processes.

8.4.2
Probing the Kinetics of Single Enzyme Molecules

To achieve single-molecule detection, researchers must tailor experimental approaches to overcome limiting factors such as noise, background, detection efficiencies, and low signals from molecules under investigation. In single-molecule enzymology, each of the limiting factors dictates the experimental direction towards elucidating catalytic activity of individual enzyme molecules. A relatively straightforward design used to overcome detection hurdles relies on using the enzyme to amplify the signal since each enzyme can catalyze the production of many fluorescent product molecules from nonfluorescent substrate molecules. Using this strategy, the first single-enzyme study was reported in 1961 by Rotman, where he incubated single β-galactosidase enzymes in aqueous microdroplets containing nonfluorescent substrates for hours until enough fluorescent product had accumulated and could be detected [138]. Unfortunately, with an incubation time of hours, better time resolution was needed to gain detailed information about catalytic rates of single enzymes. More recently, Xue and Yeung have studied single-enzyme behavior with a time resolution of minutes by injecting low concentrations of enzyme solution into micron-sized capillaries followed by incubation of the injected single lactate dehydrogenase (LDH-1) molecules in millimolar (mM) concentrations of lactate and nonfluorescent NAD^+ [139]. Interestingly, the data showed catalytic rates of individual enzymes were constant over periods of ~2 hours, yet among each enzyme rates varied by up to a factor of four, which was hypothesized to be due to different long-lived conformations of each enzyme. Similarly, Dovichi and coworkers investigated the activity of single calf intestine alkaline phosphatase in capillaries and observed heterogeneity in catalytic rates, which in this case was believed to be a result of differences in chemical modification (e.g., posttranslational glycosylation) of the protein rather than different conformations of identical enzymes [140]. A follow-up analysis reported that highly pure enzymes have identical rates, suggesting heterogeneity is a product of chemical differences among single enzymes [141]. While these amplification techniques can provide information about static heterogeneity of enzyme catalysis, real-time detection of single-enzymatic turnover events is unattainable with characteristically low time resolutions and insufficient detection efficiencies.

Using confocal optics, single-enzyme turnovers have been monitored in real time. Notably, Xie and coworkers detected individual turnover events from single cholesterol oxidase molecules trapped in pores of an agarose gel [Fig. 8.4(A)] containing oxygen and micromolar to millimolar concentrations of cholesterol [32]. Flavin adenine dinucleotide (FAD), a fluorescent group covalently attached to the enzyme, is reduced by cholesterol to a nonfluorescent form, $FADH_2$, and then subsequently oxidized by oxygen to render the molecule fluorescent again. Each cycle of this on–off fluorescence represents a single turnover event. As configured, the experiment permitted direct study of each catalytic half-reaction. Direct monitoring of these on–off cycles [Fig. 8.4(B)] was achieved with a time resolution of ~10 ms and provided detailed information about static disorder of single enzymes as well as a new component, dynamic disorder. Detailed analysis of the on-times over durations of seconds to minutes revealed that the rate of an enzyme fluctuates over time – a property termed "dynamic disorder" and believed to be a property dependent upon conformational changes of the enzyme. In addition to monitoring fluorescence from the enzyme, turnover events can be observed from production of single fluorophores from quenched nonfluorescent substrates. Rigler and coworkers used confocal microscopy to study catalysis of dihydrorhodamine 6G by single horseradish peroxidase molecules immobilized onto a coverslip [33]. Here, fluorescent signals rendered by the enzyme-product complex were detected with ~20-ms time resolution and used to calculate rates of single-enzyme catalysis. In addition to experimental data, theoretical discussions are also providing interesting insight into these exciting new observations [142–145].

SMS has opened the doors to new observables in single-enzyme catalysis. Interestingly, investigations into the effect of the environment surrounding the enzyme

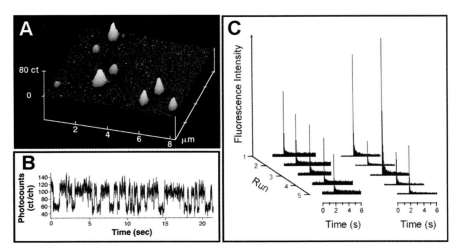

Figure 8.4. (A) Scanning confocal image of individual, fluorescent cholesterol oxidase molecules in an agarose gel; (B) fluorescence trace of on–off cycles produced from an active cholesterol oxidase molecule Reprinted with permission from Lu *et al.*, Science (Washington, D. C.) 282, 1877, Copyright 1998 AAAS, (C) catalytic activity of alkaline phosphatase contained within single lipid vesicles. Every 60 s, the accumulated fluorescent products (fluorescein) formed from the nonfluorescent substrate (fluorescein diphosphate) were probed at 488 nm and then bleached to reset the reaction clock. The left panel of (C) shows catalysis in a vesicle with a radius of 1.5 μm, and the right panel catalysis in a vesicle with a radius of 500 nm. Reprinted from Chem. Phys., 247, Chiu *et al.*, Manipulating the biochemical nanoenvironment around single molecules contained within vesicles, 133, Copyright (1999), with permission from Elsevier.

have received less attention. Possibly, changing surface properties or the volume surrounding an enzyme could modulate catalytic activity. In the bulk, enzymes trapped in chemically modified porous materials can exhibit significant rate enhancements as compared to enzyme in free solution [146, 147]. Furthermore, single-molecule experiments also suggest that surface properties and surface-to-volume ratios can affect activity. For instance, Tan and Yeung found that single LDH-1 molecules in a porous polycarbonate membrane showed discrete intervals of varied catalytic activity; however, when studied in fused silica supports, where adsorption of LDH-1 is not favored, rates of catalysis were constant over the same time period [148]. Another platform, designed by Zare and coworkers, investigated environmental effects by encapsulating enzymes within single vesicles meant to mimic a biological environment [149, 150]. With an optical trap immobilizing the vesicle, confocal detection monitored the catalysis of fluorescein diphosphate by alkaline phosphatase. Each fluorescence trace in Fig. 8.4(C) corresponded to an accumulation of product followed by probing and bleaching of the dye product formed by the enzyme. In a vesicle with 1.5 μm radius the catalytic rate appeared relatively homogeneous, yet for a vesicle with a 500 nm radius the incubated signal was quite variable and believed to be related to the surface-to-volume ratio of the environment surrounding the enzyme. With the exquisite detail provided by single-molecule detection, exciting potential lies in studying details of enzymatic catalysis affected by a variety of environmental factors.

8.4.3
Single-Molecule DNA Detection, Sorting, and Sequencing

The sensitivity offered by single-molecule detection has provided interesting possibilities in analytical chemistry, including DNA analysis, proteomics, sensors, and in the screening of rare molecules within a complex mixture (e.g., from combinatorial synthesis). Of these varied analytical applications, studies involving DNA molecules are the most widespread and developed. DNA molecules take center stage in many areas, such as medicine, forensics, environmental studies, and basic genetics, yet many hurdles still remain. Development of polymerase chain reaction (PCR) technology has overcome early problems with collecting enough DNA for experiments. Still, the technique requires a specific working protocol or set of PCR conditions for each DNA sample of interest and it is challenging to produce a high-quality DNA sample, so as not to amplify impurities in addition to the desired DNA sequence. With the amplified DNA, gel electrophoresis, the standard analytical tool for sizing DNA, is used as a means of sequencing, fingerprinting, creating restriction maps, and genotyping the sample. Recent innovations have improved or even eliminated these techniques in studies of single-molecule DNA and have been greatly facilitated by the appearance of new highly fluorescent dyes such as YOYO-1 (500× increase of quantum yield upon binding to DNA) [151]. This family of dyes form stable complexes with DNA via intercalation and generate little background when unbound, thus permitting the detection of single molecules of DNA in dilute samples under a wide range of experimental conditions [151]. For example, using the enhanced sensitivity enabled by this dye as well as a known staining ratio of dye molecules to base pairs, it is possible to correlate the fluorescent signal detected directly with the size of the DNA fragments (tens to hundreds of kilobase pairs) present at femtomolar concentrations. With tailored microfluidic designs and by using a planar sheet of excitation laser [152], it is possible to increase volumetric sample throughput (2000 fragments per second has been demonstrated) or to sort the DNA molecules after detection with controlled changes in electroosmotic flow [153]. This method of sizing DNA fragments from the fluorescence signal of individual DNA also saves time by bypassing electrophoretic separation.

Continuous improvements are made to these techniques for single DNA analysis. Examples are: a simple modification to the geometry of the microchannel by adding a taper before the detection area to focus and create a thin sample stream that optimally flows through the tightly focused laser probe volume enhances detection efficiency by three-fold [154]; and the use of various millisecond imaging techniques to measure electrophoretic mobilities without complete separation of the DNA molecules to increase throughput [155]. By combining microfabrication and single DNA imaging, Craighead and coworkers have created entropic traps by introducing alternating regions of micrometer- and nanometer-sized constrictions along the path of DNA migration driven by an applied electric field, and made the interesting observation that trapped DNA molecules escaped with a characteristic lifetime and that longer DNA molecules escaped the entropic traps faster than the shorter molecules [156]. Similar to fragment sizing in its technological requirements, single-molecule DNA restriction mapping has also been achieved [157], which yields important sequence information from the number and character of the cleavage sights and the resulting DNA fragments.

In addition to analytical assays such as DNA fragment sizing, with the complete sequencing of numerous genomes, most notably the human genome, the field of comparative genomics has evolved, and with it the need for rapid, accurate, and sensitive sequencing technology. Single-molecule DNA-sequencing techniques are still in their infancy, but proof-of-concept experiments demonstrate the possibility of an eventual viable method that could potentially sequence up to 2000 bp s^{-1} (greatly surpassing the fastest sequencing technique today) [158–162]. Several single-molecule sequencing schemes exist. To illustrate the general approach, an example of one scheme consists of: (i) fluorescent labeling or incorporation of fluorescently tagged nucleotides into the DNA, with different excitation and emission wavelengths for each of the four bases (or for at least two bases); (ii) handling of the tagged DNA molecules, which is implemented through a biotin-streptavidin bond to a microsphere, and can be controlled through suction with capillaries or using optical trapping; (iii) sequential degradation or cleavage of the DNA one base at a time with a 3′–5′ exonuclease; and (iv) efficient detection of each individual fluorescently labeled nucleotide as it becomes detached from the DNA after cleavage. This method will benefit from better labeling efficiency and an improved understanding of the suitable conditions required for single-base cleavage. Another example involves the simultaneous sequencing of numerous single molecules through the incorporation of dye-tagged nucleotides by DNA polymerase anchored to a coverslip with fluorescence microscopy detection [163]. Additionally, a scheme known as nanopore DNA sequencing utilizes the transmembrane protein α-hemolysin and electrically based detection to distinguish bases according to the distinct change in current flow [164]. Regardless of the exact detail of these single-molecule sequencing strategies, all of them face daunting challenges. Nevertheless, it seems this ambitious goal will be achieved with impressive advancements in single-molecule technologies.

8.4.4
Single-Molecule Imaging in Living Cells

The ultimate nanoscale molecular machine, perhaps, is the biological cell. The next step in understanding cellular function and behavior is to examine each component of this machine in detail as they perform their respective tasks. Proteins carry out the bulk of the work in the cell and operate within molecular networks whose functions are controlled by gene expression, energy transduction, and membrane transport. To probe these processes, single-molecule imaging of cells is often done using TIRF (see Section 8.2.3) and sometimes with epifluorescence. Remarkably, cell imaging not only determines where proteins are localized and distributed in the cell, but it can also identify associated structures and visualize how fast the proteins move, bind, and unbind within signaling pathways. Furthermore, visualizing single molecules in cells does not require the synchronization of reaction species like bulk experiments and can reveal information about rare intermediates and "memory effects" that are normally lost through ensemble averaging.

In cellular imaging, the most versatile approach is to induce expression of green fluorescent protein (GFP) in the host cell through genetic manipulation [165]. One caveat that is pertinent to single-molecule studies is the overexpression of the GFP-labeled proteins, which complicates the identification and tracking of individual molecules. Unfortunately, GFP suffers from a higher rate of photobleaching than some of the robust organic dyes, and the size of GFP (~27 kDa) can affect the diffusion rate of the tagged protein and perturb its function due to steric hindrance.

The greatest challenge in imaging single molecules in cells lies in the presence of high background noise. Presence of molecules such as flavins and NADH [55], for example, can generate high levels of autofluorescence. A number of schemes may be used to minimize background noise, including the use of excitation in the red (which also minimizes phototoxicity to the cell), time gating, and TIRF. Careful handling and culturing of cells in proper media also helps in reducing background fluorescence [166]. Some examples (Fig. 8.5) of how single-molecule imaging have been applied include (i) tracking of molecules on the cell surface [55, 166–169] as well as within the cytoplasm [55, 167, 169, 170] and the nucleus [169, 170]; (ii) the study of signal transduction [171]; (iii) visualization of viral infection of cells [172]; and (iv) measurement of disassociation kinetics between a ligand and a receptor [171]. Future advancements will rely upon both new techniques in microscopy and the development of new optical probes, such as fluorescent proteins with enhanced photostability and species that can be excited in the red [173, 174]. The capability to image molecular process at the single-molecule level will offer new levels of information about cellular processes and will increase our understanding of biological systems.

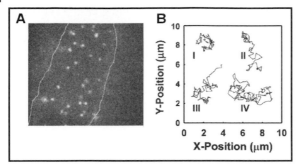

Figure 8.5. (A) Epifluorescence micrograph depicting single molecules on the surface of a cell; (B) experimentally measured single-molecule diffusional paths. Reprinted from TrAC, 22, W. E. Moerner, Optical measurements of single molecules in cells, 544, Copyright (2003, with permission from Elsevier.

8.5
Single-Molecule Detection with Bright Fluorescent Species

8.5.1
Optical Probes

In the nanoscopic single-molecular world, direct observations can be made using nonoptical techniques such as the scanning probe microscopies (SPM), patch clamp, or through the use of redox cycling. Most of these approaches, however, are unsuitable for real-time monitoring of single molecules in solution. For such studies, optical-based single-molecule detection has proved to be remarkably versatile [7, 86]. Although advancements in the hardware used to monitor single molecules have been impressive, measurements are ultimately limited by the reporter molecules, which in most cases are organic dyes. For single-molecule experiments, the dye ideally should possess the following optical characteristics: (i) excellent photostability, (ii) high quantum efficiency and absorption cross-section, (iii) fast cycling between ground and excited states to produce a high rate of photon emission, (iv) low probability of intersystem crossing from the singlet excited state to the triplet dark state, (v) a narrow emission peak that is well separated from the absorption peak, and (vi) an excitation peak that matches well with commonly used laser wavelengths, preferably towards the red where background noise and autofluorescence from biological samples is minimal [175]. In addition to these optical properties, the dye should also contain easily modified functional groups so it can be tailored with the desired chemical functionality necessary for the end application [176].

Currently, no dye satisfies fully this set of demanding criteria, but dramatic improvements to conventional dyes (e.g., fluorescein) have been made. For improved optical properties, examples include the Alexa family of dyes from molecular probes and the family of carbocyanine dyes (CyDyes) from Amersham biosciences. For chemical functionality there is now a wide selection, which includes dyes that change fluorescence characteristics selectively when bound to biologically important ions (e.g., calcium), have diverse chemical functional groups and coupling chemistries, and are compatible with long-term cell culture to report particular cellular functions or states [177]. One particularly powerful reporter is the family of GFPs, owing to the ease by which these fluorescent reporters can be genetically manipulated [25, 176, 178]. In this case, the marriage of genetics with a good fluorescent reporter and high-sensitivity optical imaging has provided unprecedented insight into cellular function.

8.5.2
Quantum Dots

Advances in inorganic crystal generation have enabled nanoscopic particles to be generated for use as molecular probes. Semiconductor nanocrystals, more commonly called quantum dots (QDs), have unique optical characteristics making them behave neither as small molecules nor as bulk solids. Composed usually of a CdSe core with a CdS or ZnS shell, these nanoparticles typically range in size from 2 to 10 nm [179]. QDs have an advantage over organic dyes in that they can be tailored to obtain the desired florescent properties: high quantum yield, high photostability, high extinction coefficient, and narrow emission spectra, but with a broad excitation band (multiple reporters can be excited at a single wavelength), and long fluorescence lifetime [179–182]. These properties make emissions intense and ideal for single-molecule studies; however, synthetic shells added to increase solubility and biocompatibility can make the particles bulky in comparison to other probes. The major emerging area of QDs is in single-molecule bioconjugate work for *in vitro* and *in vivo* studies of biological systems, which can be performed in some cases without interrupting cellular processes [183–185]. Unfortunately, unlike GFPs, QDs can be attached to proteins only after they are expressed.

8.6
Nanoscale Chemistry with Vesicles and Microdroplets

While the area of nanoscale science has experienced tremendous growth in both the fabrication of nanostructures and the imaging of such nanoscale objects, few if any experiments have demonstrated the ability to control chemical reactions in the nanoscale, confined within femtoliter (10^{-15} L) volumes of solution. This capability is especially pertinent with advancements in single-molecule detection. Although we can detect single molecules, most of these studies rely on the use of bulk solutions that must contain sufficiently high concentrations of the molecules of interest so a single molecule can be isolated easily [186, 187]. To manipulate chemically and selectively only one or a small number of molecules at a time, a strategy is needed to localize, confine, and chemically transform in solution the selected molecules of interest.

Two approaches have been pursued to control chemical transformations of molecules within ultrasmall volumes (femtoliters or less). Reported by Zare and coworkers, the first approach relies on the use of lipid vesicles, in which the molecules of interest are first encapsulated within the vesicle [149]. Individual vesicles can then be mechanically manipulated (e.g., with optical trap or micropipettes). To initiate a chemical transformation at this small length scale, two vesicles each containing different types of reactive molecules are brought into contact and fused via application of a short (μs) and intense (kV cm^{-1}) electric field through a pair of carbon-fiber microelectrodes [150, 188, 189]. In addition to the use of an electric field, the fusion of liposomes can be accomplished on an individual basis using a focused laser beam and on the bulk scale using chemicals [190]. A second method to achieve controlled nanoscale chemical transformations is to use aqueous micro- and nanodroplets that are dispersed in an immiscible medium [191]. These droplets can be generated on chip in a microfluidic system with excellent control, and the molecules of interest can be encapsulated within the droplets during their formation. The direct mechanical manipulation of such droplets, however, is nontrivial and is best achieved through the use of vortex trap [110, 117]. Like vesicles, these individual droplets can be fused together either spontaneously or with a small applied force (Fig. 8.6) so their respective contents will mix and a reaction can be initiated [191]. The advantage to using lipid vesicles lies in their biomimetic nature and, as demonstrated by Orwar and coworkers, the remarkable range of topology and shape that these liposomes exhibit [192–194]. The usefulness of the droplet platform is based on the ease and control with which droplets can be produced in a microfluidic platform, as well as on the wide range of interesting interfacial phenomena that may be studied and exploited in using droplets. One particular useful example is the possibility to concentrate molecules within individual aqueous droplets to very high levels [195], which offers new possibilities in understanding spatially confined single-molecule reactions and the effects of macromolecular crowding.

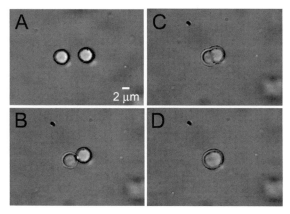

Figure 8.6. Sequence of images showing the directed fusion of two aqueous microdroplets in oil. The droplets measured ~4 μm in diameter, corresponding to a volume of 2×10^{-13} L. Reprinted from TrAC, 22, D. T. Chiu, Micro- and nano-scale chemical analysis of individual sub-cellular compartments, 528, Copyright (2003), with permission from Elsevier.

8.7
Perspectives

Single-molecule studies and technologies are poised to offer both fundamental understandings of the molecular and nanoscopic world and to provide new tools that probe the inner workings of individual living cells. Over the past years, we have witnessed the increased application of established single-molecule methods in biology, such as AFM and high-sensitivity fluorescence microscopy, as well as the birth of new techniques that were made possible by advances in optics and instrumentation. The rapid pace with which this area has progressed will surely continue in the coming years to offer us yet more striking views of the nanoscale machinery and molecular engines that make biology work.

Acknowledgments

C.L.K. thanks the National Science Foundation for a graduate research fellowship. This work was supported by the National Institutes of Health (R01 GM65293) and the Keck Foundation.

References

1 W. Ho, Single-molecule chemistry, *J. Chem. Phys.* **2002**, *117*, 11033–11061.

2 A. Ikai, STM and AFM of bio/organic molecules and structures, *Surf. Sci. Rep.* **1996**, *26*, 263–332.

3 M. A. Poggi, L. A. Bottomley, P. T. Lillehei, Scanning probe microscopy, *Anal. Chem.* **2002**, *74*, 2851–2862.

4 P. T. Lillehei, L. A. Bottomley, Scanning probe microscopy, *Anal. Chem.* **2000**, *72*, 189R–196R.

5 L. A. Bottomley, Scanning probe microscopy, *Anal. Chem.* **1998**, *70*, 425R–475R.

6 S. Nie, R. N. Zare, Optical detection of single molecules, *Ann. Rev. Biophys. Biomol. Struct.* **1997**, *26*, 567–596.

7 X. S. Xie, J. K. Trautman, Optical studies of single molecules at room temperature, *Annu. Rev. Phys. Chem.* **1998**, *49*, 441–480.

8 S. Weiss, Fluorescence spectroscopy of single biomolecules, *Science (Washington, D. C.)* **1999**, *283*, 1676–1683.

9 W. E. Moerner, A dozen years of single-molecule spectroscopy in physics, chemistry, and biophysics, *J. Phys. Chem. B* **2002**, *106*, 910–927.

10 C. Bustamante, J. C. Macosko, G. J. L. Wuite, Grabbing the cat by the tail: manipulating molecules one by one, *Nat. Rev. Mol. Cell. Biol.* **2000**, *1*, 130.

11 K. Svoboda, S. M. Block, Biological applications of optical forces, *Ann. Rev. Biophys. Biomol. Struct.* **1994**, *23*, 247–285.

12 A. D. Mehta, M. Rief, J. A. Spudich, D. A. Smith, R. M. Simmons, Single-molecule biomechanics with optical methods, *Science (Washington, D. C.)* **1999**, *283*, 1689–1695.

13 C. L. Kuyper, D. T. Chiu, Optical trapping: a versatile technique for biomanipulation, *Appl. Spectrosc.* **2002**, *56*, 300A–312A.

14 M. Ludwig, M. Rief, L. Schmidt, H. Li, F. Oesterhelt, M. Gautel, H. E. Gaub, AFM, a tool for single-molecule experiments, *Appl. Phys. A* **1999**, *68*, 173–176.

15 J. K. Trautman, J. J. Macklin, L. E. Brus, E. Betzig, Near-field spectroscopy of single molecules at room temperature, *Nature (London)* **1994**, *369*, 40–42.

16 S. Xie, R. C. Dunn, Probing single molecule dynamics, *Science (Washington, D. C.)* **1994**, *265*, 361–364.

17 W. Denk, J. H. Strickler, W. W. Webb, 2-photon laser scanning fluorescence microscopy, *Science (Washington, D. C.)* **1990**, *248*, 73–76.

18 T. Plakhotnik, D. Walser, M. Pirotta, A. Renn, U. P. Wild, Nonlinear spectroscopy on a single quantum system: two-photon absorption of a single molecule, *Science (Washington, D. C.)* **1996**, *271*, 1703.

19 E. J. Sanchez, L. Novotny, G. R. Holtom, S. Xie, Room-temperature fluorescence imaging and spectroscopy of single molecules by two-photon excitation, *J. Phys. Chem. A* **1997**, *101*, 7019–7023.

20 T. Wilson, *Confocal Microscopy*, Academic Press, San Diego, **1990**.

21 J. B. Pawley, *Handbook of Biological Confocal Microscopy*, Plenum Press, New York, **1995**.

22 M. Eigen, R. Rigler, Sorting single molecules: application to diagnostics and evolutionary biotechnology, *Proc. Natl. Acad. Sci. U. S. A.* **1994**, *91*, 5740–5747.

23 P. Schwille, U. Haupts, S. Maiti, W. W. Webb, Molecular dynamics in living cells observed by fluorescence correlation spectroscopy with one- and two-photon excitation, *Biophys. J.* **1999**, *77*, 2251–2265.

24 M. J. Wirth, D. Swinton, Single-molecule study of an adsorbed oligonucleotide undergoing both lateral diffusion and strong adsorption, *J. Phys. Chem. B* **2001**, *105*, 1472–1477.

25 A. Zumbusch, Single-molecule spectroscopy of the green fluorescent protein, *Single Mol.* **2001**, *2*, 287–288.

26 M. Boehmer, J. Enderlein, Fluorescence spectroscopy of single molecules under ambient conditions: methodology and technology, *Chem. Phys. Chem.* **2003**, *4*, 792–808.

27 M. Diez, M. Boersch, B. Zimmermann, P. Turina, S. D. Dunn, P. Graeber, Binding of the β-subunit in the ATP synthase from *Escherichia coli*, *Biochemistry* **2004**, *43*, 1054–1064.

28 S. A. Sanchez, J. E. Brunet, D. M. Jameson, R. Lagos, O. Monasterio, Tubulin equilibrium unfolding followed by time-resolved fluorescence and fluorescence correlation spectroscopy, *Protein Sci.* **2004**, *13*, 81–88.

29 A. Schenk, S. Ivanchenko, C. Roecker, J. Wiedenmann, G. U. Nienhaus, Photodynamics of red fluorescent proteins studied by fluorescence correlation spectroscopy, *Biophys. J.* **2004**, *86*, 384–394.

30 S. Nie, D. T. Chiu, R. N. Zare, Probing individual molecules with confocal fluorescence microscopy, *Science (Washington, D. C.)* **1994**, *266*, 1018–1021.

31 S. Nie, D. T. Chiu, R. N. Zare, Real-time detection of single molecules in solution by confocal fluorescence microscopy, *Anal. Chem.* **1995**, *67*, 2849–2857.

32 H. P. Lu, L. Xun, X. S. Xie, Single-molecule enzymatic dynamics, *Science (Washington, D. C.)* **1998**, *282*, 1877–1882.

33 L. Edman, Z. Foeldes-Papp, S. Wennmalm, R. Rigler, The fluctuating enzyme: a single molecule approach, *Chem. Phys.* **1999**, *247*, 11–22.

34 S. Wennmalm, L. Edman, R. Rigler, Conformational fluctuations in single DNA molecules, *Proc. Natl. Acad. Sci. U. S. A.* **1997**, *94*, 10641–10646.

35 D. M. Warshaw, E. Hayes, D. Gaffney, A.-M. Lauzon, J. Wu, G. Kennedy, K. Trybus, S. Lowey, C. Berger, Myosin conformational states determined by single fluorophore polarization, *Proc. Natl. Acad. Sci. U. S. A.* **1998**, *95*, 8034–8039.

36 T. Ha, A. Y. Ting, J. Liang, W. B. Caldwell, A. A. Deniz, D. S. Chemla, P. G. Schultz, S. Weiss, Single-molecule fluorescence spectroscopy of enzyme conformational dynamics and cleavage mechanism, *Proc. Natl. Acad. Sci. U. S. A.* **1999**, *96*, 893–898.

37 H. Yang, G. Luo, P. Karnchanaphanurach, T.-M. Louie, I. Rech, S. Cova, L. Xun, X. S. Xie, Protein conformational dynamics probed by single-molecule electron transfer, *Science (Washington, D. C.)* **2003**, *302*, 262–266.

38 X. Zhuang, H. Kim, M. J. B. Pereira, H. P. Babcock, N. G. Walter, S. Chu, Correlating structural dynamics and function in single ribozyme molecules, *Science (Washington, D. C.)* **2002**, *296*, 1473–1476.

39 E. Tan, T. J. Wilson, M. K. Nahas, R. M. Clegg, D. M. J. Lilley, T. Ha, A four-way junction accelerates hairpin ribozyme folding via a discrete intermediate, *Proc. Natl. Acad. Sci. U. S. A.* **2003**, *100*, 9308–9313.

40 K. Weninger, M. E. Bowen, S. Chu, A. T. Brunger, Single-molecule studies of snare complex assembly reveal parallel and antiparallel configurations, *Proc. Natl. Acad. Sci. U. S. A.* **2003**, *100*, 14800–14805.

41 C. Hofmann, T. J. Aartsma, H. Michel, J. Koehler, Direct observation of tiers in the energy landscape of a chromoprotein: a single-molecule study, *Proc. Natl. Acad. Sci. U. S. A.* **2003**, *100*, 15534–15538.

42 E. L. Chandler, A. L. Smith, L. M. Burden, J. J. Kasianowicz, D. L. Burden, Membrane surface dynamics of DNA-threaded nanopores revealed by simultaneous single-molecule optical and ensemble electrical recording, *Langmuir* **2004**, *20*, 898–905.

43 T. Ha, T. A. Laurence, D. S. Chemla, S. Weiss, Polarization spectroscopy of single fluorescent molecules, *J. Phys. Chem. B* **1999**, *103*, 6839–6850.

44 C. G. Hubner, A. Renn, I. Renge, U. P. Wild, Direct observation of the triplet lifetime quenching of single dye molecules by molecular oxygen, *J. Chem. Phys.* **2001**, *115*, 9619–9622.

45 B. Bowen, N. Woodbury, Single-molecule fluorescence lifetime and anisotropy measurements of the red fluorescent protein, DsRed, in solution, *Photochem. Photobiol.* **2003**, *77*, 362–369.

46 J. Hernando, M. van der Schaaf, E. M. H. P. van Dijk, M. Sauer, M. F. Garcia-Parajo, N. F. van Hulst, Excitonic behavior of rhodamine dimers: a single-molecule study, *J. Phys. Chem. A* **2003**, *107*, 43–52.

47 D. S. Ko, Photobleaching time distribution of a single tetramethylrhodamine molecule in agarose gel, *J. Chem. Phys.* **2004**, *120*, 2530–2531.

48 L. A. Deschenes, D. A. Vanden Bout, Single molecule studies of heterogeneous dynamics in polymer melts near the glass transition, *Science (Washington, D. C.)* **2001**, *292*, 255–258.

49 R. A. L. Vallee, M. Cotlet, J. Hofkens, F. C. De Schryver, K. Muellen, Spatially heterogeneous dynamics in polymer glasses at room temperature probed by single molecule lifetime fluctuations, *Macromolecules* **2003**, *36*, 7752–7758.

50 X. Tan, P. Nalbant, A. Toutchkine, D. Hu, E. R. Vorpagel, K. M. Hahn, H. P. Lu, Single-molecule study of protein-protein interaction dynamics in a cell signaling system, *J. Phys. Chem. B* **2004**, *108*, 737–744.

51 M. Unger, E. Kartalov, C.-S. Chiu, H. A. Lester, S. R. Quake, Single-molecule fluorescence observed with mercury lamp illumination, *BioTechniques* **1999**, *27*, 1008–1014.

52 J. S. Kuo, C. L. Kuyper, P. B. Allen, D. T. Chiu, High power LED as excitation source for fluorescence applications, *Electrophoresis* **2004** in press.

53 K. S. McCain, D. C. Hanley, J. M. Harris, Single-molecule fluorescence trajectories for investigating molecular transport in thin silica sol-gel films, *Anal. Chem.* **2003**, *75*, 4351–4359.

54 M. Tokunaga, K. Kitamura, K. Saito, A. H. Iwane, T. Yanagida, Single molecule imaging of fluorophores and enzymic reactions achieved by objective-type total internal reflection fluorescence microscopy, *Biochem. Biophys. Res. Commun.* **1997**, *235*, 47–53.

55 W. E. Moerner, Optical measurements of single molecules in cells, *Trends Anal. Chem.* **2003**, *22*, 544–548.

56 T. T. Perkins, D. E. Smith, S. Chu, Direct observation of tube like motion of a single polymer chain, *Science (Washington, D. C.)* **1994**, *264*, 819–822.

57 K. Yoshikawa, Y. Matsuzawa, Nucleation and growth in single DNA molecules, *J. Am. Chem. Soc.* **1996**, *118*, 929–930.

58 D. T. Chiu, R. N. Zare, Biased diffusion, optical trapping, and manipulation of single molecules in solution, *J. Am. Chem. Soc.* **1996**, *118*, 6512–6513.

59 S. R. Quake, H. Babcock, S. Chu, The dynamics of partially extended single molecules of DNA, *Nature (London)* **1997**, *388*, 151–154.

60 Y. Arai, R. Yasuda, K.-I. Akashi, Y. Harada, H. Miyata, K. Kinosita, H. Itoh, Tying a molecular knot with optical tweezers, *Nature (London)* **1999**, *399*, 446–448.

61 L. R. Brewer, M. Corzett, R. Balhorn, Protamine-induced condensation and decondensation of the same DNA molecule, *Science (Washington, D. C.)* **1999**, *286*, 120–123.

62 C. L. Kuyper, G. P. Brewood, D. T. Chiu, Initiating conformation transitions of individual YOYO-intercalated DNA molecules with optical trapping, *Nano Lett.* **2003**, *3*, 1387–1389.

63 I. Sase, H. Miyata, S. i. Ishiwata, K. Kinosita, Jr., Axial rotation of sliding actin filaments revealed by single-fluorophore imaging, *Proc. Natl. Acad. Sci. U. S. A.* **1997**, *94*, 5646–5650.

64 T. Sakamoto, I. Amitani, E. Yokota, T. Ando, Direct observation of processive movement by individual myosin v molecules, *Biochem. Biophys. Res. Commun.* **2000**, *272*, 586–590.

65 A. B. Asenjo, N. Krohn, H. Sosa, Configuration of the two kinesin motor domains during ATP hydrolysis, *Nat. Struct. Bio.* **2003**, *10*, 836–842.

66 H. Hess, C. M. Matzke, R. K. Doot, J. Clemmens, G. D. Bachand, B. C. Bunker, V. Vogel, Molecular shuttles operating undercover: a new photolithographic approach for the fabrication of structured surfaces supporting directed motility, *Nano Lett.* **2003**, *3*, 1651–1655.

67 K. Adachi, R. Yasuda, H. Noji, H. Itoh, Y. Harada, M. Yoshida, K. Kinosita, Jr., Stepping rotation of F1-ATPase visualized through angle-resolved single-fluorophore imaging, *Proc. Natl. Acad. Sci. U. S. A.* **2000**, *97*, 7243–7247.

68 A. M. van Oijen, P. C. Blainey, D. J. Crampton, C. C. Richardson, T. Ellenberger, X. S. Xie, Single-molecule kinetics of λ-exonuclease reveal base dependence and dynamic disorder, *Science (Washington, D. C.)* **2003**, *301*, 1235–1238.

69 T. Schmidt, G. J. Schutz, W. Baumgartner, H. J. Gruber, H. Schindler, Imaging of single molecule diffusion, *Proc. Natl. Acad. Sci. U. S. A.* **1996**, *93*, 2926–2929.

70 P. C. Ke, C. A. Naumann, Hindered diffusion in polymer-tethered phospholipid monolayers at the air-water interface: a single-molecule fluorescence imaging study, *Langmuir* **2001**, *17*, 5076–5081.

71 C. Seebacher, C. Hellriegel, F.-W. Deeg, C. Braeuchle, S. Altmaier, P. Behrens, K. Muellen, Observation of translational diffusion of single terrylenediimide molecules in a mesostructured molecular sieve, *J. Phys. Chem. B* **2002**, *106*, 5591–5595.

72 D. Axelrod, Total internal-reflection fluorescence microscopy, *Metods. Cell Biol.* **1989**, *30*, 245–270.

73 W. P. Ambrose, P. M. Goodwin, J. H. Jett, A. Van Orden, J. H. Werner, R. A. Keller, Single-molecule fluorescence spectroscopy at ambient temperature, *Chem. Rev.* **1999**, *99*, 2929–2956.

74 M. F. Paige, E. J. Bjerneld, W. E. Moerner, A comparison of through-the-objective total internal reflection microscopy and epi-fluorescence microscopy for single-molecule fluorescence imaging, *Single Mol.* **2001**, *2*, 191–201.

75 R. M. Dickson, D. J. Norris, W. E. Moerner, Simultaneous imaging of individual molecules aligned both parallel and perpendicular to the optic axis, *Phys. Rev. Lett.* **1998**, *81*, 5322–5325.

76 A. Bartko, R. M. Dickson, Imaging three-dimensional single molecule orientations, *J. Phys. Chem. B* **1999**, *103*, 11237–11241.

77 S. H. Kang, M. R. Shortreed, E. S. Yeung, Real-time dynamics of single-DNA molecules undergoing adsorption and desorption at liquid-solid interfaces, *Anal. Chem.* **2001**, *73*, 1091–1099.

78 M. A. Osborne, C. L. Barnes, S. Balasubramanian, D. Klenerman, Probing DNA surface attachment and local environment using single-molecule spectroscopy, *J. Phys. Chem. B* **2001**, *105*, 3120–3126.

79 R. Yamasaki, M. Hoshino, T. Wazawa, Y. Ishii, T. Yanagida, Y. Kawata, T. Higurashi, K. Sakai, J. Nagai, Y. Goto, Single molecular observation of the interaction of GroEL with substrate proteins, *J. Mol. Biol.* **1999**, *292*, 965–972.

80 R. D. Vale, T. Funatsu, D. W. Pierce, L. Romberg, Y. Harada, T. Yanagida, Direct observation of single kinesin molecules moving along microtubules, *Nature (London)* **1996**, *380*, 451–453.

81 A. Seitz, H. Kojima, K. Oiwa, E.-M. Mandelkow, Y.-H. Song, E. Mandelkow, Single-molecule investigation of the interference between kinesin, tau and map2c, *EMBO J.* **2002**, *21*, 4896–4905.

82 T. Funatsu, Y. Harada, M. Tokunaga, K. Saito, T. Yanagida, Imaging of single fluorescent molecules and individual ATP turnovers by single myosin molecules in aqueous solution, *Nature (London)* **1995**, *374*, 555–559.

83 K. Oiwa, J. F. Eccleston, M. Anson, M. Kikumoto, C. T. Davis, G. P. Reid, M. A. Ferenczi, J. E. T. Corrie, A. Yamada, H. Nakayama, D. R. Trentham, Comparative single-molecule and ensemble myosin enzymology: sulfoindocyanine ATP and ADP derivatives, *Biophys. J.* **2000**, *78*, 3048–3071.

84 Y. Harada, T. Funatsu, K. Murakami, Y. Nonoyama, A. Ishihama, T. Yanagida, Single-molecule imaging of RNA polymerase-DNA interactions in real-time, *Biophys. J.* **1999**, *76*, 709–715.

85 F. Hashimoto, S. Tsukahara, H. Watarai, Lateral diffusion dynamics for single molecules of fluorescent cyanine dye at the free and surfactant-modified dodecane-water interface, *Langmuir* **2003**, *19*, 4197–4204.

86 S. Nie, S. R. Emory, Probing single molecules and single nanoparticles by surface-enhanced Raman scattering, *Science (Washington, D. C.)* **1997**, *275*, 1102–1106.

87 K. Kneipp, Y. Wang, H. Kneipp, L. T. Perelman, I. Itzkan, R. R. Dasari, M. S. Feld, Single-molecule detection using surface-enhanced Raman scattering (SERS), *Phys. Rev. Lett.* **1997**, *78*, 1667–1670.

88 J. Jiang, K. Bosnick, M. Maillard, L. Brus, Single-molecule Raman spectroscopy at the junctions of large Ag nanocrystals, *J. Phys. Chem. B* **2003**, *107*, 9964–9972.

89 C. J. L. Constantino, T. Lemma, P. A. Antunes, R. Aroca, Single-molecule detection using surface-enhanced resonance Raman scattering and Langmuir-Blodgett monolayers, *Anal. Chem.* **2001**, *73*, 3674–3678.

90 T. Lemma, R. F. Aroca, Single-molecule surface-enhanced resonance Raman scattering on colloidal silver and Langmuir-Blodgett monolayers coated with silver overlayers, *J. Raman Spectrosc.* **2002**, *33*, 197–201.

91 P. Goulet, N. Pieczonka, R. Aroca, Single-molecule SERRS of mixed perylene Langmuir-Blodgett monolayers on novel metal island substrates, *Can. J. Anal. Sci. Spectrosc.* **2003**, *48*, 146–152.

92 Z. Wang, S. Pan, T. D. Krauss, H. Du, L. J. Rothberg, The structural basis for giant enhancement enabling single-molecule Raman scattering, *Proc. Natl. Acad. Sci. U. S. A.* **2003**, *100*, 8638–8643.

93 A. Weiss, G. Haran, Time-dependent single-molecule Raman scattering as a probe of surface dynamics, *J. Phys. Chem. B* **2001**, *105*, 12348–12354.

94 J. M. North, L. J. van de Burgt, N. S. Dalal, A Raman study of the single molecule magnet Mn_{12}-acetate and analogs, *Solid State Commun.* **2002**, *123*, 75–79.

95 J. M. North, N. S. Dalal, Raman and infrared modes of the single molecule magnet Fe_8Br_8 and analogs, *J. Appl. Phys.* **2003**, *93*, 7092–7094.

96 H. Xu, E. J. Bjerneld, M. Kaell, L. Borjesson, Spectroscopy of single hemoglobin molecules by surface enhanced Raman scattering, *Phys. Rev. Lett.* **1999**, *83*, 4357–4360.

97 A. R. Bizzarri, S. Cannistraro, Surface-enhanced resonance Raman spectroscopy signals from single myoglobin molecules, *Appl. Spectrosc.* **2002**, *56*, 1531–1537.

98 E. J. Bjerneld, Z. Foeldes-Papp, M. Kaell, R. Rigler, Single-molecule surface-enhanced Raman and fluorescence correlation spectroscopy of horseradish peroxidase, *J. Phys. Chem. B* **2002**, *106*, 1213–1218.

99 S. Habuchi, M. Cotlet, R. Gronheid, G. Dirix, J. Michiels, J. Vanderleyden, F. C. De Schryver, J. Hofkens, Single-molecule surface enhanced resonance Raman spectroscopy of the enhanced green fluorescent protein, *J. Am. Chem. Soc.* **2003**, *125*, 8446–8447.

100 C. Zander, Single-molecule detection in solution: a new tool for analytical chemistry, *Fresenius J. Anal. Chem.* **2000**, *366*, 745–751.

101 K. Kneipp, H. Kneipp, I. Itzkan, R. R. Dasari, M. S. Feld, Surface-enhanced Raman scattering and biophysics, *J. Phys. Condens. Matter* **2002**, *14*, R597–R624.

102 R. M. Dickson, A. B. Cubitt, R. Y. Tsien, W. E. Moerner, On/off blinking and switching behavior of single molecules of green fluorescent protein, *Nature (London)* **1997**, *388*, 355–358.

103 A. Ashkin, J. M. Dziedzic, J. E. Bjorkholm, S. Chu, Observation of a single-beam gradient force optical trap for dielectric particles, *Opt. Lett.* **1986**, *11*, 288–290.

104 A. Ashkin, History of optical trapping and manipulation of small-neutral particle, atoms, and molecules, *IEEE J. Select. Topics Quantum Electr.* **2000**, *6*, 841–856.

154 B. B. Haab, R. A. Mathies, Single-molecule detection of DNA separations in microfabricated capillary electrophoresis chips employing focused molecular streams, *Anal. Chem.* **1999**, *71*, 5137–5145.

155 M. R. Shortreed, H. Li, W.-H. Huang, E. S. Yeung, High-throughput single-molecule DNA screening based on electrophoresis, *Anal. Chem.* **2000**, *72*, 2879–2885.

156 J. Han, S. W. Turner, H. G. Craighead, Entropic trapping and escape of long DNA molecules at submicron size constriction, *Phys. Rev. Lett.* **1999**, *83*, 1688–1691.

157 B. Schafer, H. Gemeinhardt, V. Uhl, K. O. Greulich, Single-molecule DNA restriction analysis in the light microscope, *Single Mol.* **2000**, *1*, 33–40.

158 J. H. Werner, H. Cai, J. H. Jett, L. Reha-Krantz, R. A. Keller, P. M. Goodwin, Progress towards single-molecule DNA sequencing: a one color demonstration, *J. Biotechnol.* **2003**, *102*, 1–14.

159 M. Sauer, B. Angerer, W. Ankenbauer, Z. Foldes-Papp, F. Gobel, K. T. Han, R. Rigler, A. Schulz, J. Wolfrum, C. Zander, Single-molecule DNA sequencing in submicrometer channels: state of the art and future prospects, *J. Biotechnol.* **2001**, *86*, 181–201.

160 M. Sauer, B. Angerer, K. T. Han, C. Zander, Detection and identification of single dye-labeled mononucleotide molecules released from an optical fiber in a microcapillary: First steps towards a new single-molecule DNA sequencing technique, *PCCP Phys. Chem. Chem. Phys.* **1999**, *1*, 2471–2477.

161 M. A. Augustin, W. Ankenbauer, B. Angerer, Progress towards single-molecule sequencing: enzymatic synthesis of nucleotide-specifically labeled DNA, *J. Biotechnol.* **2001**, *86*, 289–301.

162 Z. Foldes-Papp, B. Angerer, W. Ankenbauer, R. Rigler, Fluorescent high-density labeling of DNA: error-free substitution for a normal nucleotide, *J. Biotechnol.* **2001**, *86*, 237–253.

163 I. Braslavsky, B. Hebert, E. Kartalov, S. R. Quake, Sequence information can be obtained from single DNA molecules, *Proc. Natl. Acad. Sci. U. S. A.* **2003**, *100*, 3960–3964.

164 M. Akeson, D. Branton, J. J. Kasianowicz, E. Brandin, D. W. Deamer, Microsecond time-scale discrimination among polycytidylic acid, polyadenylic acid, and polyuridylic acid as homopolymers or as segments within single RNA molecules, *Biophys. J.* **1999**, *77*, 3227–3233.

165 R. Brock, G. Vamosi, G. Vereb, T. M. Jovin, Rapid characterization of green fluorescent protein fusion proteins on the molecular and cellular level by fluorescence correlation microscopy, *Proc. Natl. Acad. Sci. U. S. A.* **1999**, *96*, 10123–10128.

166 R. Iino, I. Koyama, A. Kusumi, Single-molecule imaging of green fluorescent proteins in living cells: E-cadherin forms oligomers on the free cell surface, *Biophys. J.* **2001**, *80*, 2667–2677.

167 G. I. Mashanov, D. Tacon, A. E. Knight, M. Peckham, J. E. Molloy, Visualizing single molecules inside living cells using total internal reflection fluorescence microscopy, *Methods* **2003**, *29*, 142–152.

168 M. Vrljic, S. Y. Nishimura, S. Brasselet, W. E. Moerner, H. M. McConnell, Translational diffusion of individual class II MHC membrane proteins in cells, *Biophys. J.* **2002**, *83*, 2681–2692.

169 T. A. Byassee, W. C. W. Chan, S. M. Nie, Probing single molecules in single living cells, *Anal. Chem.* **2000**, *72*, 5606–5611.

170 U. Kubitscheck, Single protein molecules visualized and tracked in the interior of eukaryotic cells, *Single Mol.* **2002**, *3*, 267–274.

171 Y. Sako, T. Uyemura, Total internal reflection fluorescence microscopy for single-molecule imaging in living cells, *Cell Struct. Func.* **2002**, *27*, 357–365.

172 G. Seisenberger, M. U. Ried, T. Endress, H. Buning, M. Hallek, C. Brauchle, Real-time single-molecule imaging of the infection pathway of an adeno-associated virus, *Science (Washington, D. C.)* **2001**, *294*, 1929–1932.

173 R. E. Campbell, O. Tour, A. E. Palmer, P. A. Steinbach, G. S. Baird, D. A. Zacharias, R. Y. Tsien, A monomeric red fluorescent protein, *Proc. Natl. Acad. Sci. U. S. A.* **2002**, *99*, 7877–7882.

174 N. G. Gurskaya, A. F. Fradkov, A. Terskikh, M. V. Matz, Y. A. Labas, V. I. Martynov, Y. G. Yanushevich, K. A. Lukyanov, S. A. Lukyanov, GFP-like chromoproteins as a source of far-red fluorescent proteins, *FEBS Lett.* **2001**, *507*, 16–20.

175 A. N. Kapanidis, S. Weiss, Fluorescent probes and bioconjugation chemistries for single-molecule fluorescence analysis of biomolecules, *J. Chem. Phys.* **2002**, *117*, 10953–10964.

176 J. Zhang, R. E. Campbell, A. Y. Ting, R. Y. Tsien, Creating new fluorescent probes for cell biology, *Nat. Rev. Mol. Cell. Biol.* **2002**, *3*, 906–918.

177 K. A. Willets, O. Ostroverkhova, M. He, R. J. Twieg, W. E. Moerner, Novel fluorophores for single-molecule imaging, *J. Am. Chem. Soc.* **2003**, *125*, 1174–1175.

178 R. Tsien, Rosy dawn for fluorescent proteins, *Nat. Biotech.* **1999**, *17*, 956–957.

179 M. Bruchez, Jr., M. Moronne, P. Gin, S. Weiss, A. P. Alivisatos, Semiconductor nanocrystals as fluorescent biological labels, *Science (Washington, D. C.)* **1998**, *281*, 2013–2016.

180 X. Michalet, F. Pinaud, T. D. Lacoste, M. Dahan, M. P. Bruchez, A. P. Alivisatos, S. Weiss, Properties of fluorescent semiconductor nanocrystals and their application to biological labeling, *Single Mol.* **2001**, *2*, 261–276.

181 Y. Ebenstein, T. Mokari, U. Banin, Fluorescence quantum yield of CdSe/ZnS nanocrystals investigated by correlated atomic-force and single-particle fluorescence microscopy, *Appl. Phys. Lett.* **2002**, *80*, 4033–4035.

182 B. O. Dabbousi, J. Rodriguez-Viejo, F. V. Mikulec, J. R. Heine, H. Mattoussi, R. Ober, K. F. Jensen, M. G. Bawendi, (CdSe).ZnS core-shell quantum dots: synthesis and optical and structural characterization of a size series of highly luminescent materials, *J. Phys. Chem. B* **1997**, *101*, 9463–9475.

183 W. C. W. Chan, S. Nile, Quantum dot bioconjugates for ultrasensitive nonisotopic detection, *Science (Washington, D. C.)* **1998**, *281*, 2016–2018.

184 M. Han, X. Gao, J. Z. Su, S. Nie, Quantum-dot-tagged microbeads for multiplexed optical coding of biomolecules, *Nat. Biotech.* **2001**, *19*, 631–635.

185 X. Gao, S. Nie, Molecular profiling of single cells and tissue specimens with quantum dots, *Trends Biotech.* **2003**, *21*, 371–373.

186 M. D. Barnes, K. C. Ng, W. B. Whitten, J. M. Ramsey, Detection of single rhodamine 6G molecules in levitated microdroplets, *Anal. Chem.* **1993**, *65*, 2360–2365.

187 M. D. Barnes, N. Lermer, C. Y. Kung, W. B. Whitten, J. M. Ramsey, S. C. Hill, Real-time observation of single-molecule fluorescence in microdroplet streams, *Opt. Lett.* **1997**, *22*, 1265–1267.

188 C. F. Wilson, G. J. Simpson, D. T. Chiu, A. Stroemberg, O. Orwar, N. Rodriguez, R. N. Zare, Nanoengineered structures for holding and manipulating liposomes and cells, *Anal. Chem.* **2001**, *73*, 787–791.

189 A. Stromberg, F. Ryttsen, D. T. Chiu, M. Davidson, P. S. Eriksson, C. F. Wilson, O. Orwar, R. N. Zare, Manipulating the genetic identity and biochemical surface properties of individual cells with electric-field-induced fusion, *Proc. Natl. Acad. Sci. U. S. A.* **2000**, *97*, 7–11.

190 S. Kulin, R. Kishore, K. Helmerson, L. Locascio, Optical manipulation and fusion of liposomes as microreactors, *Langmuir* **2003**, *19*, 8206–8210.

191 D. T. Chiu, Micro- and nano-scale chemical analysis of individual sub-cellular compartments, *Trends Anal. Chem.* **2003**, *22*, 528–536.

192 A. Karlsson, M. Karlsson, R. Karlsson, K. Sott, A. Lundqvist, M. Tokarz, O. Orwar, Nanofluidic networks based on surfactant membrane technology, *Anal. Chem.* **2003**, *75*, 2529–2537.

193 M. Karlsson, K. Sott, M. Davidson, A.-S. Cans, P. Linderholm, D. Chiu, O. Orwar, Formation of geometrically complex lipid nanotube-vesicle networks of higher-order topologies, *Proc. Natl. Acad. Sci. U. S. A.* **2002**, *99*, 11573–11578.

194 A. Karlsson, R. Karlsson, M. Karlsson, A.-S. Cans, A. Stromberg, F. Ryttsen, O. Orwar, Molecular engineering: networks of nanotubes and containers, *Nature (London)* **2001**, *409*, 150–152.

195 M. He, C. Sun, D. T. Chiu, Concentrating solutes and nanoparticles within individual aqueous microdroplets, *Anal. Chem.* **2004**, *76*, 1222–1227.

9

Nanotechnologies for Cellular and Molecular Imaging by MRI

Patrick M. Winter, Shelton D. Caruthers, Samuel A. Wickline, and Gregory M. Lanza

9.1
Introduction

Developments in cellular and molecular biology are extending the horizons of medical imaging from gross anatomic description towards delineation of cellular and biochemical signaling processes. The emerging fields of cellular and molecular imaging aim to diagnose disease noninvasively on the basis of *in vivo* detection and characterization of complex pathological processes, such as induction of inflammation or angiogenesis. Techniques have been developed recently to achieve molecular and cellular imaging with most imaging modalities, including nuclear [1, 2], optical [2, 3], ultrasound [4], and MRI [5, 6]. This chapter focuses on two techniques developed for detection of atherosclerosis by MRI: cellular imaging of macrophages associated with inflammatory lesions and molecular imaging of angiogenesis that is induced in developing vascular plaques. A selection of the contrast agent formulation and imaging methods will be discussed, as well as the optimization of these techniques for successful cellular and molecular imaging *in vivo*.

Because individual cells and biochemical molecules are too small to be imaged directly with noninvasive techniques, specific and sensitive site-targeted contrast agents are needed to visualize the epitopes of interest. Historically, nuclear imaging has dominated the fields of cellular and molecular imaging due to the extremely high sensitivity and the relative simplicity of conjugating radioactive tags onto biochemical molecules. For instance, fluorodeoxyglucose (FDG) activity can be imaged with PET scanners to characterize such diverse disease states as tumor metabolism [7] and mental disorders [8]. Radiolabeled somatostatin analogs have also been developed to allow receptor imaging for detection of neuroendocrine tumors [9]. In addition, cellular apoptosis can be detected with technetium-labeled annexin-V, which binds to phosphatidyl serine expressed on the surface of apoptotic cells [10]. Nuclear imaging agents have also been designed to detect gene transfection by imaging the resultant protein products [11].

While nuclear imaging predominated in the early development of cellular and molecular imaging, other modalities, such as optical, magnetic resonance imaging (MRI), and ultrasound, have been pursued to take advantage of the increased resolution, signal-to-noise ratio, and contrast associated with these techniques. The inher-

Nanofabrication Towards Biomedical Applications. C. S. S. R. Kumar, J. Hormes, C. Leuschner (Eds.)
Copyright © 2005 WILEY-VCH Verlag GmbH & Co. KGaA, Weinheim
ISBN 3-527-31115-7

ently lower sensitivity of these methods, however, often requires the use of complex targeted imaging agents in order to incorporate sufficient signal generation motifs. For instance, targeted perfluorocarbon nanoparticles, the first reported molecular imaging agent for ultrasound applications, can significantly increase the reflectivity of thrombi [12–14]. Many other acoustic contrast agents have been developed for similar targeting applications, such as endothelial integrins, tissue factor and thrombi [15–21].

MRI is emerging as a particularly advantageous modality for molecular and cellular imaging given its high spatial resolution and the opportunity to extract both anatomical and physiological information simultaneously [22, 23]. The majority of targeted MRI agents consist of nanoparticles in order to carry sufficient amounts of paramagnetic or superparamagnetic material for *in vivo* visualization [24]. In cardiovascular disease, novel targeted contrast agents are being developed to detect unstable lesions through identification of fibrin deposited within plaque microfissures [25–28], adhesion or thrombogenic molecules expressed on endothelium of vulnerable plaques [15, 16, 29–31], matrix metalloproteinases in the cores of progressing lesions [32], or angiogenic activity (i.e., expanding vasa vasorum) supporting plaque development [33].

9.2
Cardiovascular Disease

Atherosclerosis, like many chronic human diseases including cancer and diabetes, develops slowly over many years. Unlike most other diseases, however, atherosclerosis is often diagnosed only after an acute, fatal event. Of the 720,000 cardiac deaths per year in America, approximately 60% are "sudden deaths," occurring without any advanced warning of pathology [34]. Atherosclerosis starts as "fatty streak" lesions *in utero* [35], and can produce plaques prone to rupture by the early teens [36]. Rupture of unstable atherosclerotic plaques can lead to thrombosis, vascular occlusion, and subsequent myocardial infarction or stroke [37–39].

Atherosclerotic plaques grow in discrete stages consisting of repeated episodes of rupture, thrombosis, and healing (Fig. 9.1) [40], leading inevitably to a final event causing complete vascular obstruction [41]. A "vulnerable plaque" is defined as a lesion exhibiting physical and biochemical properties that predispose it to rupture and thrombosis [40]. Typically, these plaques consist of a large lipid core covered by a thin fibrous cap harboring relatively few smooth muscle cells, a population of activated macrophages, and abundant angiogenesis (Fig. 9.2) [40]. The large lipid core is known to destabilize lesions by directing mechanical stress to the fragile shoulder-regions of the plaque [42]. Exposure of the lipid core, even through a small localized rupture, can induce the clotting cascade initiating with tissue factor [43]. The lack of smooth muscle cells also weakens the cap [44], facilitating plaque rupture. The accumulation of macrophages as well as other inflammatory cells, which usually secrete high levels of metalloproteinases (MMP), can undermine the fibrous cap, potentially exposing the thrombotic lipid core [45, 46]. Up-regulation of angiogenesis can lead to erosion of the extracellular matrix and replacement with physically fragile neovascular beds, weakening the fibrous cap and promoting plaque rupture [47–49].

In current clinical practice, the diagnosis and characterization of most atherosclerotic plaques is achieved with invasive x-ray catheterization. Highly stenotic lesions, typically with 50% or greater narrowing of the luminal diameter, are identified for immediate therapeutic intervention, while less stenotic lesions are generally deemed clinically insignificant. Ironically, these plaques are often the very lesions prone to rupture leading to heart attack and stroke [50, 51]. Stress tests employing nuclear or ultrasound imaging are also extensively used for detection of flow-limiting vascular obstructions during a range of metabolic challenges. Similar to invasive x-ray angiography, however, these techniques are insensitive to plaques with low-grade stenosis, which are those prone to rupture. Because rupturing atherosclerotic plaques are frequently manifest in arteries with only modest (40–60%) stenosis [52], they remain diagnostically elusive with routine clinical imaging techniques. If recognized and localized, a window of opportunity extending from days to months exists to intervene and/or stabilize plaques medically before more serious clinical sequelae ensue [41]. The principal difficulty revolves around the fact that atherosclerosis produces numerous plaques throughout the vascular system and it is not feasible to treat each lesion individually. Of all the <50% stenotic lesions, only a small fraction may rupture and lead to clinical events (Fig. 9.3) [40]. A significant new opportunity exists for delineating which lesions are prone to rupture and applying therapeutic treatments to only the areas of pathology that pose an immediate danger.

Molecular imaging techniques are being pursued with the goals of both primary and secondary prevention of atherosclerosis. Primary prevention focuses on the identification of patients at the earliest stages of atherosclerotic development, leading to the application of lifestyle and/or pharmacological therapies to prevent further plaque development and perhaps even significant regression of existing unstable or vulnerable lesions. Secondary prevention aims to detect plaques that are prone to impending (e.g., days to weeks) rupture and thromboembolism in patients with advanced atherosclerotic disease, allowing aggressive intervention concentrated on only the most dangerous areas of pathology.

9.3
Cellular and Molecular Imaging

To achieve effective cellular and molecular imaging with MRI, targeted contrast agents must be designed to accomplish long circulating half-life, sensitive, and selective binding to the epitope of interest, prominent contrast-to-noise enhancement, acceptable toxicity, ease of clinical use, and applicability with standard commercially available imaging systems.

Discoveries in molecular biology have elucidated cellular and molecular markers for a wide variety of diseases. These signatures may serve as targets for contrast agents to provide sensitive and specific imaging of the earliest manifestations of

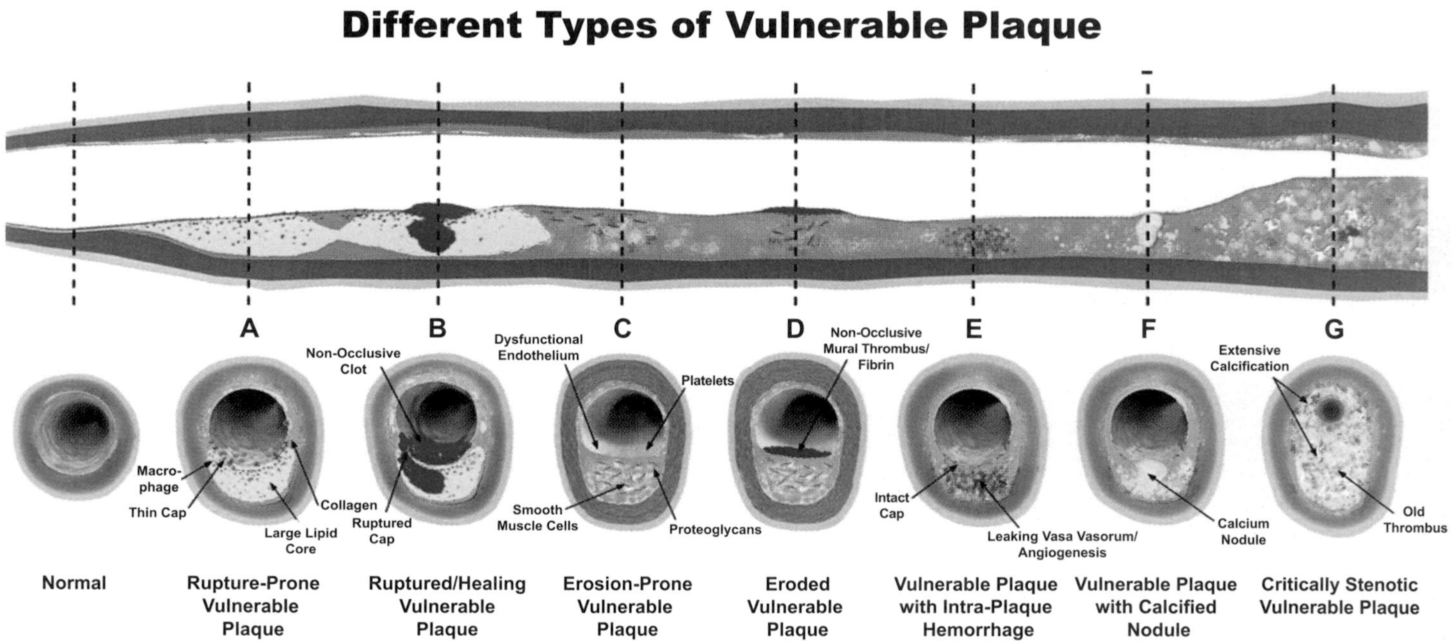

Different Types of Vulnerable Plaque

Figure 9.1. Atherosclerotic plaques develop in discrete steps, leading to a wide range of lesion types in a single patient. Although all these plaque types may lead to thromboembolism, only the most advanced plaques produce luminal stenosis. From left to right: (A) rupture-prone plaque with large lipid core and thin fibrous cap infiltrated by macrophages; (B) ruptured plaque with subocclusive thrombus and early organization; (C) erosion-prone plaque with proteoglycan matrix in a smooth muscle cell-rich plaque; (D) eroded plaque with subocclusive thrombus; (E) intraplaque hemorrhage secondary to leaking vasa vasorum; (F) calcific nodule protruding into the vessel lumen; (G) chronically stenotic plaque with severe calcification, old thrombus, and eccentric lumen. (Reprinted with permission from Ref. [40]).

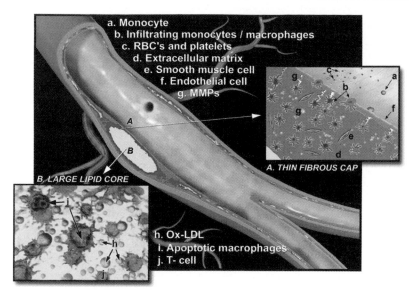

Figure 9.2. The most common type of nonstenotic vulnerable plaque. The thin fibrous cap, extensive macrophage infiltration, depletion of smooth muscle cells, and large lipid core all contribute to the plaque's propensity for rupture. (Reprinted with permission from Ref. [40].)

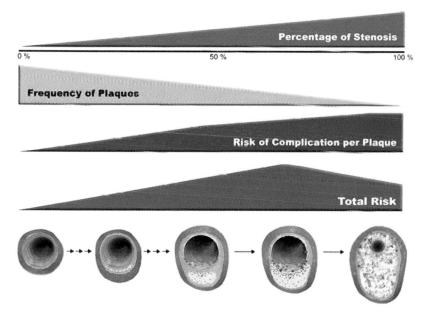

Figure 9.3. Correlation between frequency of plaques, degree of stenosis, and risk of complication per plaque as a function of plaque progression. Although the average absolute risk of severely stenotic plaques may be higher than the average absolute risk of mildly stenotic plaques, there are more plaques with mild stenoses than plaques with severe stenoses. (Reprinted with permission from Ref. [40].)

pathology. A variety of different types of ligands can be utilized for targeting, including antibodies, peptides, polysaccharides, aptamers, etc. The emerging fields of cellular and molecular imaging utilize nanotechnologies to develop targeted contrast agents for noninvasive detection of the molecular signatures of disease. Unlike blood pool agents, a targeted contrast agent produces signal enhancement from pathological tissue that would otherwise be difficult to distinguish from surrounding normal tissue. Cellular and molecular imaging techniques have been developed for most clinical imaging modalities, including nuclear, MRI, ultrasound, computed tomographic, and optical. MRI, however, may enjoy several advantages over the other modalities, such as high resolution, high anatomical contrast, high signal-to-noise ratio, widespread clinical availability, and lack of ionizing radiation.

These imaging techniques do not depend upon the native tissue contrast or resolution associated with the underlying imaging modality. Instead, they enhance the conspicuity of very small areas of disease by directly targeting the cellular or molecular processes responsible for the pathological transformation of tissues. Moreover, by targeting the underlying molecular processes, these imaging techniques can provide unique information to characterize the complex metabolic state of the disease. This information could provide crucial insight for various therapeutic aspects, including evaluating potential treatment options, and predicting and monitoring response to therapy.

Image contrast in MRI often depends upon the relaxation characteristics of the tissues [53]. There are three primary relaxation times: longitudinal relaxation (T_1), intrinsic transverse relaxation (T_2), and apparent transverse relaxation (T_2^*), each with an associated relaxation rate: $1/T_1$, $1/T_2$, and $1/T_2^*$ [54,55]. MRI contrast agents are designed to shorten the relaxation times of the tissue of interest and therefore increase the relaxation rates [56]. T_1 relaxation describes the regrowth of MRI signal during the pulse sequence, and so shortening T_1 allows the signal to recover more quickly. Therefore, contrast agents designed to affect T_1 tend to provide increased MRI signal. On the other hand, both T_2 and T_2^* relaxation describes the decay of MRI signal and shortening these relaxation times decreases the MRI signal. Typically, contrast agents increase the relaxation rate in a linear fashion. Therefore, plotting $1/T_1$ vs. contrast agent concentration yields a straight line. The slope of this line corresponds to the relaxivity of the contrast agent and is given in terms of s^{-1} mM^{-1}. The relaxivity is a measure of the potency of the contrast agent. A compound with higher relaxivity can provide higher contrast at a given concentration or identical contrast at a lower concentration. Relaxivity is typically measured relative to the paramagnetic or superparamagnetic ion concentration, such as gadolinium or iron. For molecular and cellular imaging applications, however, the relaxivity per nanoparticle, i.e., particle or molecular relaxivity, is more useful for comparing contrast agent effect per binding site.

9.4
Cellular Imaging with Iron Oxides

Ultrasmall superparamagnetic iron oxide (USPIO) nanoparticles are potent MRI contrast agents. The iron produces strong local disruptions in the magnetic field of MRI scanners, which lead to increased T_2^* relaxation. This increased relaxation causes decreased image intensity in areas with iron oxide accumulation, termed "susceptibility artifacts". Because of the extremely large change in MRI signal induced by superparamagnetic particles, they have been developed for a wide variety of contrast agent applications, including imaging vasculature, bowel, liver, spleen, lymphatics, bone marrow, and tumors, and stem cell therapies [57–59]. In particular, particles with a 15- to 25-nm diameter have a very long circulating half-life and are preferentially taken up by macrophages in the body when coated with dextran. These properties have allowed dextran-coated USPIO nanoparticles to be employed for *passive* targeted imaging of pathological inflammatory processes, such as unstable atherosclerotic plaques, by MRI (Fig. 9.4) [60]. USPIO-labeled macrophages have been shown to preferentially accumulate in unstable and ruptured plaques (75% demonstrating uptake), but not in stable lesions (only 7% showing USPIO uptake) [61].

Several aspects of the USPIO nanoparticle must be optimized for successful visualization of atherosclerotic plaques with MRI. The USPIO formulation must adhere to certain physical and chemical requirements. For instance, particles lacking the dextran coating are not effectively phagocytosed by inflammatory cells. Also, particles with a diameter similar to low-density lipoprotein (LDL), 15–25 nm, have a circulating half-life on the order of hours, providing sufficient time for cellular accu-

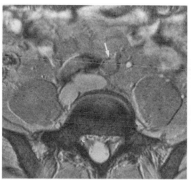
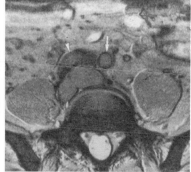

A   *B*

Figure 9.4. Axial 2D gradient-echo MRI at the level of the common iliac arteries, before (A) and 24 hours after (B) USPIO administration. Precontrast image shows small foci of low signal intensity within the wall of the left common iliac artery, possibly representing calcification (A, arrow). On the postcontrast image a pronounced signal loss of the entire cross-section of the artery is seen (B, arrow). The right common iliac artery is not displayed at an acceptable image quality, as it takes an oblique course through the image plane (B, arrowhead). (Reprinted with permission from Ref. [60].)

mulation and effective delivery to atherosclerotic plaques [62]. In addition, the dose of USPIOs has a dramatic influence on the appearance by MRI. Increasing the dose by a factor of four produces signal loss in ten times more slices [62].

The imaging protocols are also critically important for USPIO detection by MRI. The susceptibility artifacts created by accumulation of USPIOs are most sensitively imaged with T_2^*-weighted imaging sequences. T_1-weighted and proton density imaging sequences are far less sensitive to the susceptibility artifacts induced by USPIO uptake in tissues [61]. The T_2^*-weighted sequences typically utilize gradient echo techniques with long echo times. The long echo time accentuates signal loss due to the presence of USPIOs, but these sequences also tend to suffer from a low signal-to-noise ratio. Often, highly specialized coils, such as phased array coils or application-specific surface coils, are employed in order to maximize the available signal to noise [60,61].

Blood vessels are typically surrounded by fatty tissue, which can interfere with imaging structures within the arterial wall. On MRI, fat appears as a bright signal, which often displays a spatial misregistration artifact relative to other tissues in the body, called a "chemical shift" artifact. This artifact can often overlap signals from the vessel wall and obscure the atherosclerotic plaques under investigation. In order to avoid this problem, imaging sequences employing fat suppression or selective excitation techniques are used for imaging USPIO uptake [60–62]. With these sequences, fat appears dark because its signal is either suppressed or not excited. In a similar manner, bright blood imaging sequences, such as 2D fast low-angle shot (FLASH) sequences, are often used for detection of USPIOs [60, 61]. The bright blood signal allows clearer definition of the vessel lumen, and signal voids in the arterial wall caused by USPIO uptake are much easier to distinguish.

In addition to the imaging protocol itself, the choice of imaging time after USPIO injection is critically important. The long circulating half-life of dextran-coated USPIO nanoparticles is necessary to achieve adequate loading into inflammatory cells, but it can also interfere with obtaining high-quality images. Up to approx. 24 hours after USPIO administration, the blood concentration is high enough to create image artifacts [61, 62], which can obscure visualization of the vessel wall. On the other hand, too long a delay (~72 hours) after USPIO injection can result in no detectable susceptibility artifacts [61].

9.5
Molecular Imaging with Paramagnetic Nanoparticles

As an alternative approach, we proposed and evaluated a ligand-targeted, lipid-encapsulated, nongaseous perfluorocarbon nanoparticle emulsion for molecular imaging applications (Fig. 9.5) [12, 15, 26, 28]. The nanoparticles can be formulated to carry paramagnetic gadolinium ions and targeted to a number of important biochemical epitopes, such as fibrin, tissue factor and $\alpha_v\beta_3$-integrin. Fibrin deposition is one of the earliest signs of plaque rupture, allowing detection of the "culprit" lesion before a high-grade stenosis has been formed [63]. Tissue factor is another

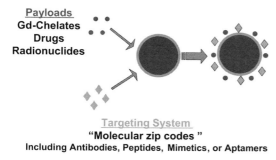

Figure 9.5. Generalized molecular imaging paradigm for ligand-targeted paramagnetic nanoparticles.

prothrombotic agent and is expressed on the surface of vascular smooth muscle cells, which contribute to restenosis following vascular injury or stent placement [64].

While fibrin and tissue factor can be utilized to delineate unstable cardiovascular diseases, the $\alpha_v\beta_3$-integrin is a general marker of angiogenesis and plays an important role in a wide variety of disease states [65]. The $\alpha_v\beta_3$-integrin is a well-characterized heterodimeric adhesion molecule that is widely expressed by endothelial cells, monocytes, fibroblasts, and vascular smooth muscle cells. In particular, $\alpha_v\beta_3$-integrin plays a critical part in smooth muscle cell migration and cellular adhesion [66,67], both of which are required for the formation of new blood vessels. The $\alpha_v\beta_3$-integrin is expressed on the luminal surface of activated endothelial cells but not on mature quiescent cells [68]. We have demonstrated the utility of $\alpha_v\beta_3$-integrin-targeted nanoparticles for the detection and characterization of angiogenesis associated with growth factor expression [69], tumor growth [70], and atherosclerosis [33].

Angiogenesis plays a critical role in plaque growth and rupture [71]. Normal large-caliber arteries in humans receive oxygen and nutrients from blood that is delivered by the adventitial vasa vasorum, not necessarily from the vessel lumen itself. The normal vasa vasorum carries blood in the same direction as arterial flow and extends perpendicular branches around the vessel wall to supply deep tissues. In regions of atherosclerotic lesions, angiogenic vessels proliferate from the vasa vasorum to meet the high metabolic demands of plaque growth [72]. Atherosclerosis promotes both inflammation and angiogenesis in the arterial wall, probably via a positive feed-back type of system. Inflammatory cells within the lesion stimulate angiogenesis through local molecular signaling, which in turn promotes neovascular growth, thereby providing an avenue for more inflammatory cells to enter the plaque. This process yields a strong correlation between the extent of plaque angiogenesis and local accumulation of inflammatory cells [73].

9.5.1
Optimization of Formulation Chemistry

Successful development of a targeted nanoparticle contrast agent requires optimization of numerous formulation procedures, including the number and chemical structure of the paramagnetic chelates and the targeting ligands. Even the choice of target should be optimized. The target molecule must be expressed at relatively high concentrations in diseased tissue and at very low levels in normal tissue. Also, the target must be accessible to the contrast agent construct. For instance, a 200-nm-diameter nanoparticle could encounter and bind to receptors on the luminal side of an endothelial cell, but perhaps not directly to DNA fragments inside the nucleus of a neuronal cell. The target ligand should have high affinity for the target of interest and minimal binding to similar receptors that may be expressed on normal tissues. In addition, the ligand must retain stability and potency throughout the process of nanoparticle formulation. For example, antibodies or their fragments could denature and lose their targeting abilities during extreme chemical processing steps, such as sterilization of the final nanoparticle product. Other targeting ligands, such as carbohydrates, aptamers, or peptidomimetics, may be less sensitive to these formulation procedures and retain their bioactivity.

Molecular imaging with liquid perfluorocarbon nanoparticles was originally pursued to target fibrin deposits on ruptured atherosclerotic plaques [12,25–28]. By incorporating an anti-fibrin antibody into the particles, a dense layer of nanoparticles can be selectively targeted to the surface of a fibrin clot (Fig. 9.6) [25]. This model provides a robust platform for exploring the consequences of changes to the formulation chemistry of the nanoparticle.

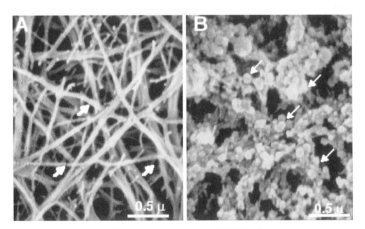

Figure 9.6. Scanning electron micrographs (×30,000) of control fibrin clot (A) and fibrin-targeted paramagnetic nanoparticles bound to clot surface (B). Arrows indicate (A) fibrin fibril; (B) fibrin-specific nanoparticle-bound fibrin epitopes. (Reprinted with permission from Ref. [25].)

Unlike the USPIO agents discussed previously, the image enhancement characteristics of commercially available paramagnetic contrast agents are too small to visualize the very sparse concentrations, picomolar to nanomolar, of epitopes relevant to molecular imaging. By incorporating vast numbers of paramagnetic complexes (>50,000) onto each particle, the signal enhancement possible for each binding site is magnified dramatically, by a factor of >10^6. The increased paramagnetic influence arises from two mechanisms: the relaxivity per particle increases linearly with respect to the number of gadolinium complexes, and the relaxivity of each gadolinium complex increases due to the slower tumbling of the molecule when attached to the much larger particle. By studying dilutions of nanoparticles in water, we have observed increased T_1 relaxivity with increasing gadolinium payloads (Fig. 9.7) [25]. We also verified that the increased relaxivity seen in solution corresponded to increased signal intensity on T_1-weighted images of particles bound to fibrin clots (Fig. 9.8) [25]. Thus, improvements in the formulation chemistry can be assessed in simple phantoms of nanoparticle dilutions and verified in models of physiological systems.

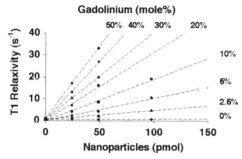

Figure 9.7. T_1 relaxation rates as a function of nanoparticle number expressed in picomoles for formulations ranging from 0 to 50 mol% Gd-DTPA-BOA in the outer 2% lipid monolayer. (Reprinted with permission from Ref. [25].)

The fibrin clot model also provided a means to evaluate the paramagnetic influence of different gadolinium chelate structures [27]. The T_1 relaxivity (r_1) of nanoparticles formulated with gadolinium diethylene-triamine-pentaacctic acid bis-oleate [74] (Gd-DTPA-BOA) or gadolinium diethylene-triamine-pentaacetic acid-phosphatidylethanolamine (Gd-DTPA-PE [75]) has been measured at selected magnetic field strengths: 0.47 T, 1.5 T, and 4.7 T. At all magnetic field strengths, r_1 of the Gd-DTPA-PE formulation was approximately two times greater than for the Gd-DTPA-BOA agent [27], indicating that small adjustments to the molecular configuration of a paramagnetic chelate can substantially improve the fundamental relaxation properties.

Variable temperature relaxometry measurements have shown that r_1 of the Gd-DTPA-BOA emulsion was largely independent of temperature [27]. In contrast, r_1 increased at higher temperatures for the Gd-DTPA-PE emulsion. These temperature-dependence data suggest that the water exchange rate with the paramagnetic ion is higher for the Gd-DTPA-PE chelate than for Gd-DTPA-BOA. At the higher

A

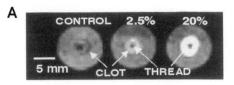

B

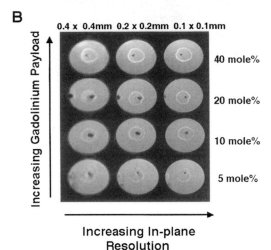

Figure 9.8. (A) Low-resolution MRI of fibrin clots targeted with nanoparticles presenting a homogeneous, T_1-weighted enhancement that improves with increasing gadolinium level (0, 2.5, and 20 mol%). (B) High-resolution scans of fibrin clots revealing a thin layer of nanoparticles along the surface. Peak signal increases with increasing gadolinium concentration (5, 10, 20, and 40 mol%) and decreasing voxel size (0.4, 0.2, and 0.1 mm). (Reprinted with permission from Ref. [25].)

temperature, the r_1 of Gd-DTPA-PE nanoparticles increased due to the faster water exchange and increased kinetic activity. The Gd-DTPA-BOA nanoparticles, however, may experience somewhat restricted water access and did not benefit from the increased kinetic activity of water at the higher temperature. This increased water exchange may result from the elevated position of the chelate relative to the phosphate headgroup level for Gd-DTPA-PE nanoparticles [75] and probably contributed to the marked increase of the relaxivity of the Gd-DTPA-PE nanoparticles.

Improvements in contrast agent formulation, such as increased relaxivity, must be confirmed with nanoparticles bound to a physiological target. In addition to assessing the paramagnetic effects of bound nanoparticles, these experiments are crucial for determining whether changes to gadolinium-chelate chemistry affect the final targeting ability of the formulation. T_1 relaxation (R_1) maps of fibrin clots treated with either Gd-DTPA-BOA or Gd-DTPA-PE nanoparticles were collected at 1.5 T (Fig. 9.9) [27]. These colorized maps show higher R_1 values at the surface of the Gd-DTPA-PE clot, demonstrating increased paramagnetic effect with this nanoparticle formulation even when bound to fibrin fibrils. Compared with the clot interior, the surface layer of Gd-DTPA-PE nanoparticles increased R_1 at the clot margin by 72%,

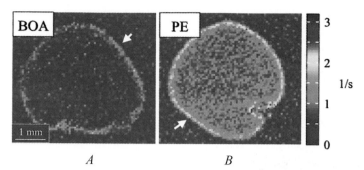

Figure 9.9. T_1 relaxation maps obtained at 1.5 T of human plasma clots targeted with Gd-DTPA-BOA (A) or Gd-DTPA-PE (B) nanoparticles. The Gd-DTPA-PE nanoparticles induce a much higher R_1 at the clot surface (white arrows) compared to Gd-DTPA-BOA nanoparticles. (Reprinted with permission from Ref. [27].)

compared to only a 48% increase with the Gd-DTPA-BOA agent. The density of Gd-DTPA-BOA and Gd-DTPA-PE nanoparticles over the clot surface was identical [27], indicating that changing the paramagnetic chelate to Gd-DTPA-PE did not adversely affect the contrast agent binding.

While we have demonstrated that increased relaxivity augments visualization of targeted epitopes, these techniques can only improve molecular imaging to a finite extent. Using MRI signal modeling programs, we can theoretically evaluate a range of contrast agent relaxivities at different magnetic field strengths (Fig. 9.10) [76]. At higher magnetic fields, the signal-to-noise ratio increases, leading to increased contrast-to-noise and a lower detectable limit of contrast agent binding. Therefore, we expect that the application of higher clinical field strengths, i.e., 3 T, will considerably improve the performance of current molecular imaging agents and help propel these techniques into clinical practice. Increasing the paramagnetic relaxivity of a molecular imaging agent, however, does not improve the lower detection limit in a linear fashion, but rather yields diminishing returns at high relaxivity values, as shown by us [76] and others [77]. Therefore, efforts to improve the performance of molecular imaging contrast agents through increasing the ionic relaxivity will provide only limited success.

The chemistry employed to bind the targeted ligand to the particle surface can also dramatically affect the efficacy of the final contrast agent. In some cases, the active binding site of the ligand may become occupied or obscured after attachment to the nanoparticle. Obviously, such agents would yield poor molecular imaging results. In addition, the incorporation of flexible polymer spacers, i.e. polyethylene glycol, between the targeting ligand and the nanoparticle surface may improve targeting efficiency. These flexible "tethers" permit a wider range of motion for the targeting ligand, potentially increasing the likelihood of encountering and binding to the target of interest.

In addition to the number of gadolinium ions per particle, the number of targeting ligands per particle must also be optimized. A nanoparticle with numerous

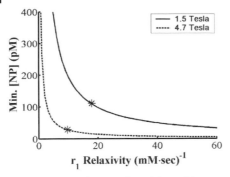

Figure 9.10. Mathematical modeling of the minimum contrast agent concentration needed for diagnostic imaging with increasing ionic r_1 relaxivity. The asterisks (*) indicate the ionic-based r_1 relaxivities of Gd-DTPA-BOA-conjugated nanoparticles at both 1.5 T and 4.7 T field strengths. (Reprinted with permission from Ref. [76].)

binding ligands, i.e., high ligand valency, tends to provide a more efficacious contrast agent. The combination of ligand valency and binding affinity, i.e., avidity, allows the contrast agent to bind rapidly and tenaciously to the intended biomarker. Incorporating too many targeting ligands on each nanoparticle, however, can in certain cases produce steric hindrance, inhibiting binding to the desired epitope. An excessively dense cover of ligands on the particle surface may also interfere with the relaxivity by impeding water interaction with the gadolinium complexes.

9.5.2
Optimization of MRI Techniques

In contradistinction to USPIO agents which are designed to increase the T_2^* relaxation of targeted tissue, paramagnetic compounds are typically used to increase T_1 relaxation. The T_2^* effect employed with USPIOs is usually visualized as decreased image intensity with T_2^*-weighted MRI. Paramagnetic agents, however, will produce increased tissue signal when interrogated with T_1-weighted MRI. Therefore, the contrast effects of USPIO and paramagnetic agents are very different, and the MRI pulse sequences and parameters are also distinct for the two methods.

We investigated the influence of MRI resolution on T_1-weighted signal enhancement with fibrin-targeted nanoparticles. We observed that increased resolution produced increased image enhancement due to a reduction in partial volume dilution [Fig. 9.8(B)] [25]. The resolution, however, cannot be increased indefinitely, because increasing MRI resolution requires either increased scan times or decreased signal-to-noise. Therefore, the signal-to-noise and contrast-to-noise must be balanced in order to achieve acceptable image quality and diagnostic contrast within an acceptable clinical scan time.

In addition to image resolution, the image contrast depends heavily on the repetition time (TR) used in the scanning sequence [78]. On a T_1-weighted image, the contrast between two tissues with different T_1 relaxation times can be maximized by

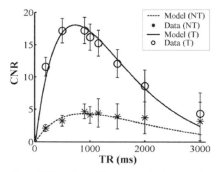

Figure 9.11. Theoretical and actual contrast-to-noise ratios (CNRs) obtained from cultured smooth muscle cells with different repetition times (TR). Cells treated with targeted nano-particles (T) display a much higher CNR compared to nontargeted particles (NT). The CNR reaches a maximum at some optimum value of TR. (Reprinted with permission from Ref. [76].)

imaging at the optimum value of TR (Fig. 9.11) [76]. Altering the TR by 300–400 ms away from the optimum value can reduce the contrast-to-noise by up to 25% [76].

As with the USPIO particles, adequate circulation must be allowed in order for ligand targeted nanoparticles to reach and bind to the molecular marker of interest. The pharmacokinetics and pharmacodynamics can be influenced by surface chemistry, *in vivo* stability, size, and the biological milieu. Approximately two to three hours are required to saturate a vascular accessible biochemical target, reflecting the time required for all blood to complete one passage through a remote vascular bed. The time to maximum signal can be decreased by increasing the amount of contrast agent injected, but the dose is typically limited to the minimum effective dose. We generally inject 0.5–1.0 ml kg^{-1} for animal experiments, which is much lower than the amount required to generate an appreciable blood pool signal [70]. This avoids the lingering questions of whether the contrast is targeted to or circulating through a region of interest. Such confounding issues can be particularly troublesome for agents with high signal intensity and/or prolonged half-life. The dramatic effects of particle composition and chemistry on elimination kinetics and the resulting imaging protocol can be seen by comparing the optimum imaging timepoints of three vastly different contrast agents. $\alpha_v\beta_3$-Integrin-targeted nanoparticles yield significant MRI signal enhancement two hours after injection [70] compared to 24 hours for a liposome agent [79] and 36 hours for an avidin-DTPA complex [80].

While bright blood imaging sequences are preferred for distinguishing the negative contrast effects of USPIO accumulation, black blood techniques are better suited for visualization of MRI signal enhancement caused by targeted paramagnetic nanoparticles (Fig. 9.12) [33]. For instance, one black blood MRI method involves collecting cross-sectional images of a vessel and placing saturation bands on either side of the current imaging plane. These saturation bands null all MRI signals on either side of the imaged slice, including signal from the blood. A short delay is inserted between application of the saturation bands and image data acquisition, allowing the saturated blood to flow into the imaged slice, which provides no signal.

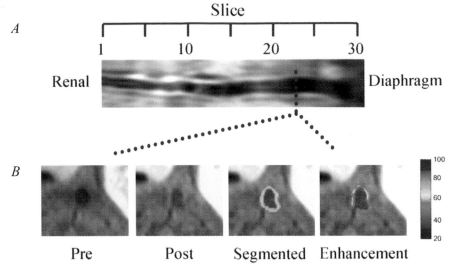

Figure 9.12. *In vivo* fat-suppressed, black blood imaging of rabbit abdominal aorta from renal arteries to diaphragm (A). Transverse images (B) before (Pre) and after (Post) treatment, after semiautomated segmentation (Segmented), and with color-coded signal enhancement in percent (Enhancement). Fat suppression and black blood imaging allow discrimination of arterial wall from vessel lumen and surrounding fatty tissue. (Reprinted with permission from Ref. [33].)

By placing saturation bands on both sides of the current slice, both arterial and venous blood appears black on the final image. As with USPIO imaging, the chemical shift artifact from fat can obscure visualization of the vessel wall. Therefore fat suppression techniques are often employed in order to null the fat signal and provide clear delineation of vessel wall anatomy (Fig. 9.12) [33].

9.5.3
In Vivo Molecular Imaging of Angiogenesis

Fundamentally, validation of a molecular imaging contrast agent relies upon corroboration between MRI signal enhancement and histological staining of the targeted epitope. Angiogenic expression of $\alpha_v\beta_3$-integrin was detected by colocalized histological staining of $\alpha_v\beta_3$-integrin and PECAM, a generalized vascular marker. This biochemical feature was extensively observed in histological sections from cholesterol-fed rabbits, but much more sparsely detected in animals on a control diet (Fig. 9.13) [33]. Likewise, MRI signal enhancement was widely observed with $\alpha_v\beta_3$-integrin-targeted nanoparticles in the aortic wall of cholesterol-fed rabbits, but not control diet animals (Fig. 9.14) [33].

The targeting ability of the final nanoparticle contrast agent must be validated through controlled experimental conditions. Comparing the MRI signal enhancement achieved with targeted vs. nontargeted particles provides an assay for separating

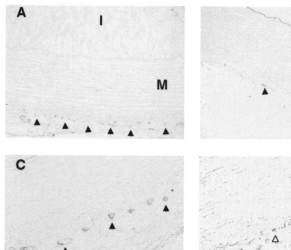

Figure 9.13. Immunohistochemistry of $\alpha_v\beta_3$-integrin in aorta sections from cholesterol-fed (A, C, D) rabbits and rabbits fed on a control diet (B). Costaining of $\alpha_v\beta_3$-integrin (C, solid arrows) and PECAM (D, open arrows) in aorta from cholesterol-fed animal delineates neovasculature. Prevalence of $\alpha_v\beta_3$-integrin staining in control rabbit (B, solid arrows) is far less prominent than in cholesterol-fed animal (A). M, media; Av, adventitia. (Reprinted with permission from Ref. [33].)

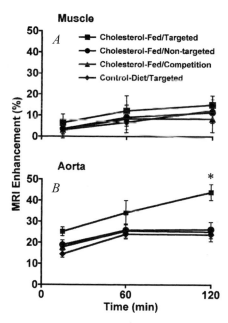

Figure 9.14. Quantitative analysis of MRI signal enhancement (percent) from aorta (A) and skeletal muscle (B) after treatment with $\alpha_v\beta_3$-targeted or nontargeted nanoparticles in cholesterol-fed or control diet groups. *$P<0.05$ for cholesterol-fed/targeted vs. all other groups. (Reprinted with permission from Ref. [33].)

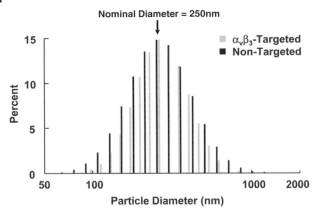

Figure 9.15. Particle size distribution of $\alpha_v\beta_3$-targeted (gray) and nontargeted nanoparticles (black) demonstrating identical physical characteristics of the two formulations. (Reprinted with permission from Ref. [70].)

particle accumulation through passive entrapment in the hyperpermeable angiogenic vasculature vs. active targeting to the biochemical epitope of interest (Fig. 9.14) [33]. Differences in particle size, however, may lead to differences in MRI signal enhancement between targeted and nontargeted nanoparticles. Therefore, the particle size distributions for the two formulations must be identical (Fig. 9.15) [33].

Further confirmation of active particle targeting can be achieved with *in vivo* competition procedures (Fig. 9.14) [33]. Targeted, high-avidity nanoparticles lacking the paramagnetic chelate are injected in order to occupy all available binding sites, but they do not provide MRI signal enhancement. Subsequent injection of targeted paramagnetic nanoparticles will produce image enhancement only through passive particle accumulation. Analysis of MRI signal enhancement in experimental control animals, e.g., control diet rabbits, as well as control tissues, e.g., skeletal muscle, provides even greater discrimination between the ability of the particle to actively target the epitope of interest and passive accumulation of the agent (Fig. 9.14) [33].

In a similar manner, we have also demonstrated molecular imaging of angiogenesis in nascent Vx-2 tumors implanted in the hindlimb of rabbits [70]. Corroboration between MRI signal enhancement and histological staining of angiogenic $\alpha_v\beta_3$-integrin expression reaffirms the sensitivity of $\alpha_v\beta_3$-targeted nanoparticles. Comparing the amount of MRI signal enhancement achieved with targeted nanoparticles vs. nontargeted or competition formulations provides more supporting evidence for the specificity of our molecular imaging contrast agent. In addition, we have shown that some masses which appeared to be viable tumor on T_2-weighted MRI were revealed to be only tumor remnants by histology and consisted mostly of inflammatory cells. These masses did not display MRI enhancement with $\alpha_v\beta_3$-targeted nanoparticles, further demonstrating the specificity of these methods for molecular imaging of active angiogenesis.

9.6
Conclusions

Cellular and molecular imaging represents an emerging clinical diagnostic paradigm that could alter the practice of medicine within the next decade. Today, characterization of disease by histological biopsy sections suffers from errors induced through limited sampling, which can fail to appropriately characterize the full extent and severity of disease. However, with noninvasive imaging, a 3D representation of the disease process can be developed, leading to improved disease delineation. These methods clearly provide a tremendous opportunity to detect and treat disease early. Targeted imaging agents can have many forms, use a variety of homing ligands, and be compatible with various imaging modalities. All of these characteristics can ultimately improve or degrade the overall efficacy of the final contrast agent. Therefore, each component must be carefully tested and optimized for clinical use. Clearly, the success of molecular imaging agents has the potential to change the future practice of medicine as these technologies reach the clinic.

The fields of cellular and molecular imaging and the associated nanotechnologies are quickly evolving and constantly changing. The rapid progression of genomic and proteomic sciences together with advances in basic molecular and cellular research should continue to uncover new and useful molecular targets. Interdisciplinary efforts are expected to lead ultimately to the development of new biomarkers of disease and surrogate end points for therapeutic studies.

References

1 S. H. Britz-Cunningham, S. J. Adelstein, Molecular targeting with radionuclides: state of the science. *J Nucl. Med.* **2003**, *44*, 1945–1961.

2 H. R. Herschman, Molecular imaging: looking at problems, seeing solutions. *Science* **2003**, *302*, 605–608.

3 R. Y. Tsien, Imagining imaging's future. *Nat. Rev. Mol. Cell Biol.* **2003**, Suppl. pp. 16–21.

4 Lanza GM, Wickline SA. Targeted ultrasonic contrast agents for molecular imaging and therapy. *Curr. Probl. Cardiol.* **2003**, *28*, 625–653.

5 S. A. Wickline, G. M. Lanza, Molecular imaging, targeted therapeutics, and nanoscience. *J. Cell Biochem. Suppl.* **2002**, *39*, 90–97.

6 S. A. Wickline, G. M. Lanza, Nanotechnology for molecular imaging and targeted therapy. *Circulation* **2003**, *107*, 1092–1095.

7 L. Kostakoglu, H. Agress, S. J. Goldsmith Jr., Clinical role of FDG PET in evaluation of cancer patients. *Radiographics* **2003**, *23*, 315–40; quiz 533.

8 K. Herholz, PET studies in dementia. *Ann. Nucl. Med.* **2003**, *17*, 79–89.

9 D. Kwekkeboom, E. P. Krenning, M. de Jong, Peptide receptor imaging and therapy. *J. Nucl. Med.* **2000**, *41*, 1704–1713.

10 F. G. Blankenberg, H. W. Strauss, Noninvasive strategies to image cardiovascular apoptosis. *Cardiol. Clin.* **2001**, *19*, 165–172, x.

11 F. M. Bengel, M. Anton, N. Avril, T. Brill, N. Nguyen, R. Haubner, F. Gleiter, B. Gansbacher, M. Schwaiger, Uptake of radiolabeled 2′-fluoro-2′-deoxy-5-iodo-1-beta-D-arabinofuranosyluracil in cardiac cells after adenoviral transfer of the herpesvirus thymidine kinase gene: the cellular basis for cardiac gene imaging. *Circulation* **2000**, *102*, 948–950.

44 M. M. Kockx, G. R. De Meyer, N. Buyssens, M. W. Knaapen, H. Bult, A. G. Herman, Cell composition, replication, and apoptosis in atherosclerotic plaques after 6 months of cholesterol withdrawal. *Circ. Res.* **1998**, *83*, 378–387.

45 A. C. Newby, A. B. Zaltsman, Fibrous cap formation or destruction – the critical importance of vascular smooth muscle cell proliferation, migration and matrix formation. *Cardiovasc. Res.* **1999**, *41*, 345–360.

46 P. K. Shah, Role of inflammation and metalloproteinases in plaque disruption and thrombosis. *Vasc. Med.* **1998**, *3*, 199–206.

47 P. Libby, Molecular bases of the acute coronary syndromes. *Circulation* **1995**, *91*, 2844–2850.

48 M. J. McCarthy, I. M. Loftus, M. M. Thompson, L. Jones, N. J. London, P. R. Bell, A. R. Naylor, N. P. Brindle, Angiogenesis and the atherosclerotic carotid plaque: an association between symptomatology and plaque morphology. *J. Vasc. Surg.* **1999**, *30*, 261–268.

49 O. J. de Boer, A. C. van der Wal, P. Teeling, A. E. Becker, Leucocyte recruitment in rupture prone regions of lipid-rich plaques: a prominent role for neovascularization? *Cardiovasc. Res.* **1999**, *41*, 443–449.

50 C. M. Ballantyne, Clinical trial endpoints: angiograms, events, and plaque instability. *Am. J. Cardiol.* **1998**, *82*, 5M–11M.

51 K. Yokoya, H. Takatsu, T. Suzuki, H. Hosokawa, S. Ojio, T. Matsubara, T. Tanaka, S. Watanabe, N. Morita, K. Nishigaki, G. Takemura, T. Noda, S. Minatoguchi, H. Fujiwara, Process of progression of coronary artery lesions from mild or moderate stenosis to moderate or severe stenosis: a study based on four serial coronary arteriograms per year. *Circulation* **1999**, *100*, 903–909.

52 J. A. Ambrose, M. A. Tannenbaum, D. Alexopoulos, C. E. Hjemdahl-Monsen, J. Leavy, M. Weiss, S. Borrico, R. Gorlin, V. Fuster, Angiographic progression of coronary artery disease and the development of myocardial infarction. *J. Am. Coll. Cardiol.* **1988**, *12*, 56–62.

53 P. Winter, N. Bansal, Magnetic resonance, general medical. In: Meyers RA, ed. *Encyclopedia of Analytical Chemistry: Applications, Theory, and Instrumentation.* Chichester; New York: Wiley; **2000**:201–236.

54 S. C. Bushong, *Magnetic Resonance Imaging: Physical and Biological Principles*, 2nd ed. St. Louis: Mosby; **1996**.

55 F. H. Epstein, J. R. Brookeman, Physics of MRI – basic principles. In: A. Lardo, Z. Fayad, N. A. F. Chronos, V. Fuster, eds. *Cardiovascular magnetic resonance: established and emerging applications.* New York: Martin Dunitz; **2003**:1–15.

56 R. M. Weisskoff, P. Caravan, MR contrast agent basics. In: A. Lardo, Z. Fayad, N. A. F. Chronos, V. Fuster, eds. *Cardiovascular magnetic resonance: established and emerging applications.* New York: Martin Dunitz; **2003**:17–38.

57 Y. X. Wang, S. M. Hussain, G. P. Krestin, Superparamagnetic iron oxide contrast agents: physicochemical characteristics and applications in MR imaging. *Eur. Radiol.* **2001**, *11*, 2319–2331.

58 H. J. Weinmann, W. Ebert, B. Misselwitz, H. Schmitt-Willich, Tissue-specific MR contrast agents. *Eur. J. Radiol.* **2003**, *46*, 33–44.

59 J. W. Bulte, I. D. Duncan, J. A. Frank, In vivo magnetic resonance tracking of magnetically labeled cells after transplantation. *J. Cereb. Blood Flow Metab.* **2002**, *22*, 899–907.

60 S. A. Schmitz, M. Taupitz, S. Wagner, K. J. Wolf, D. Beyersdorff, B. Hamm, Magnetic resonance imaging of atherosclerotic plaques using superparamagnetic iron oxide particles. *J. Magn. Reson. Imaging* **2001**, *14*, 355–361.

61 M. E. Kooi, V. C. Cappendijk, K. B. Cleutjens, A. G. Kessels, P. J. Kitslaar, M. Borgers, P. M. Frederik, M. J. Daemen, J. M. van Engelshoven, Accumulation of ultrasmall superparamagnetic particles of iron oxide in human atherosclerotic plaques can be detected by in vivo magnetic resonance imaging. *Circulation* **2003**, *107*, 2453–2458.

62 S. A. Schmitz, S. E. Coupland, R. Gust, S. Winterhalter, S. Wagner, M. Kresse, W. Semmler, K. J. Wolf, Superparamagnetic iron oxide-enhanced MRI of atherosclerotic plaques in Watanabe hereditable hyperlipidemic rabbits. *Invest. Radiol.* **2000**, *35*, 460–471.

63 P. Constantinides, Plaque fissuring in human coronary thrombosis. *J. Atheroscler. Res.* **1966**, *6*, 1–17.

64 L. Oltrona, C. M. Speidel, D. Recchia, S. A. Wickline, P. R. Eisenberg, D. R. Abendschein, Inhibition of tissue factor-mediated coagulation markedly attenuates stenosis after balloon-induced arterial injury in minipigs. *Circulation* **1997**, *96*, 646–652.

65 J. S. Kerr, S. A. Mousa, A. M. Slee, $\alpha_v\beta_3$-Integrin in angiogenesis and restenosis. *Drug News Perspect.* **2001**, *14*, 143–150.

66 G. G. Bishop, J. A. McPherson, J. M. Sanders, S. E. Hesselbacher, M. J. Feldman, C. A. McNamara, L. W. Gimple, E. R. Powers, S. A. Mousa, I. J. Sarembock, Selective $\alpha_v\beta_3$-receptor blockade reduces macrophage infiltration and restenosis after balloon angioplasty in the atherosclerotic rabbit. *Circulation* **2001**, *103*, 1906–1911.

67 M. H. Corjay, S. M. Diamond, K. L. Schlingmann, S. K. Gibbs, J. K. Stoltenborg, A. L. Racanelli, $\alpha_v\beta_3$, $\alpha_v\beta_5$, and osteopontin are coordinately upregulated at early time points in a rabbit model of neointima formation. *J. Cell Biochem.* **1999**, *75*, 492–504.

68 P. C. Brooks, S. Stromblad, R. Klemke, D. Visscher, F. H. Sarkar, D. A. Cheresh, Antiintegrin $\alpha_v\beta_3$ blocks human breast cancer growth and angiogenesis in human skin. *J. Clin. Invest.* **1995**, *96*, 1815–1822.

69 S. A. Anderson, R. K. Rader, W. F. Westlin, C. Null, D. Jackson, G. M. Lanza, S. A. Wickline, J. J. Kotyk, Magnetic resonance contrast enhancement of neovasculature with $\alpha_v\beta_3$–targeted nanoparticles. *Magn. Reson. Med.* **2000**, *44*, 433–539.

70 P. M. Winter, S. D. Caruthers, A. Kassner, T. D. Harris, L. K. Chinen, J. S. Allen, E. K. Lacy, H. Zhang, J. D. Robertson, S. A. Wickline, G. M. Lanza, Molecular imaging of angiogenesis in nascent Vx-2 rabbit tumors using a novel $\alpha_v\beta_3$–targeted nanoparticle and 1.5 tesla magnetic resonance imaging. *Cancer Res.* **2003**, *63*, 5838–5843.

71 A. N. Tenaglia, K. G. Peters, M. H. Sketch Jr., B. H. Annex, Neovascularization in atherectomy specimens from patients with unstable angina: implications for pathogenesis of unstable angina. *Am. Heart J.* **1998**, *135*, 10–14.

72 Y. Zhang, W. J. Cliff, G. I. Schoefl, G. Higgins, Immunohistochemical study of intimal microvessels in coronary atherosclerosis. *Am. J. Pathol.* **1993**, *143*, 164–172.

73 K. S. Moulton, K. Vakili, D. Zurakowski, M. Soliman, C. Butterfield, E. Sylvin, K. M. Lo, S. Gillies, K. Javaherian, J. Folkman, Inhibition of plaque neovascularization reduces macrophage accumulation and progression of advanced atherosclerosis. *Proc. Natl. Acad. Sci. U. S. A.* **2003**, *100*, 4736–4741.

74 W. Cacheris, T. Richard, R. Grabiak, A. Lee, Paramagnetic complexes of N-alkyl-N-hydroxylamides of organic acids and emulsions containing same for magnetic resonance imaging. United States: HemaGen/PFC; **1997**.

75 C. W. Grant, S. Karlik, E. Florio, A liposomal MRI contrast agent: phosphatidylethanolamine-DTPA. *Magn. Reson. Med.* **1989**, *11*, 236–243.

76 A. M. Morawski, P. M. Winter, K. C. Crowder, S. D. Caruthers, R. W. Fuhrhop, M. J. Scott, J. D. Robertson, D. R. Abendschein, G. M. Lanza, S. A. Wickline, Targeted nanoparticles for quantitative imaging of sparse molecular epitopes with MRI. *Magn. Reson. Med.* **2004**, *51*, 480–486.

77 T. M. Button, R. J. Fiel, Isointense model for the evaluation of tumor-specific MRI contrast agents. *Magn. Reson. Imaging* **1988**, *6*, 275–280.

78 E. T. Ahrens, U. Rothbacher, R. E. Jacobs, S. E. Fraser, A model for MRI contrast enhancement using T1 agents. *Proc. Natl. Acad. Sci. U. S. A.* **1998**, *95*, 8443–8448.

79 D. A. Sipkins, D. A. Cheresh, M. R. Kazemi, L. M. Nevin, M. D. Bednarski, K. C. Li, Detection of tumor angiogenesis in vivo by $\alpha_v\beta_3$–targeted magnetic resonance imaging. *Nat. Med.* **1998**, *4*, 623–626.

80 D. Artemov, N. Mori, R. Ravi, Z. M. Bhujwalla, Magnetic resonance molecular imaging of the HER-2/neu receptor. *Cancer Res.* **2003**, *63*, 2723–2727.

III

Application of Nanotechnology in Biomedical Research

Nanofabrication Towards Biomedical Applications. C. S. S. R. Kumar, J. Hormes, C. Leuschner (Eds.)
Copyright © 2005 WILEY-VCH Verlag GmbH & Co. KGaA, Weinheim
ISBN 3-527-31115-7

have been used as gene delivery vehicles by many investigators [5]. In addition to their ability to transfer the genes, viral particles interfere with the immune system by triggering the production of antibodies and other proteins [6, 7]. In many cases, the production of undesired proteins has triggered serious side effects, including deaths in clinical trials [7]. Therefore, several academic and industrial laboratories are involved in the development of nonviral gene delivery vehicles.

Development of nonviral gene delivery vehicles involves the interaction of negatively charged DNA phosphate groups with agents that carry multiple positively charged groups and/or polymeric chains. This process, known as DNA condensation, has been the subject of intense research during the past three decades [1, 8–12]. In many cases, the cationic molecules are derived from the chemical structures of the natural polyamines putrescine $(H_2N(CH_2)_4NH_2)$, spermidine $(H_2N(CH_2)_3NH(CH_2)_4NH_2)$, and spermine $(H_2N(CH_2)_3NH(CH_2)_4NH(CH_2)_3NH_2)$. Under physiological ionic and pH conditions, these molecules are positively charged, and hence a dominant force in their interaction with DNA is electrostatic, although site-specific interactions have also been implicated as playing a secondary role [13–15]. An inorganic cation, cobalt hexamine $(Co(NH_3)_6^{3+})$ has also been used by many investigators to study the mechanism of DNA condensation [9, 14–16]. In general, condensation results in morphologically distinct DNA nanoparticles, as evidenced by electron and atomic force microscopic techniques [17, 18]. However, DNA complexes formed with these small cationic molecules are labile to dissociation under physiological ionic conditions. Therefore, higher-valency polyamines and derivatives of spermidine and spermine have been synthesized as gene delivery vehicles [1, 19]. The synthesized polycations condense DNA into nanoparticles and facilitate its cellular uptake. The cationic polyamines and polymers can provoke the condensation of large DNA molecules into compact particles, permitting the incorporation of gene-regulatory regions [20]. Nonviral vectors are also flexible to incorporate functional groups so that cell-specific targeting and nuclear localization can be facilitated [21]. Stable gene delivery vehicles that are nontoxic and biodegradable with the ability to protect DNA from degradation are of utmost interest [20–22]. In addition, these vehicles may facilitate cellular uptake through membrane receptors and allow endosomal release of the DNA [23–26].

The first step in the cellular transport of DNA nanoparticles is the electrostatic interaction between the carrier/DNA complex and the anionic plasma membrane [1, 27, 28]. Complex formation between DNA and cationic molecules results in positively charged nanoparticles, as measured by ζ potential. A positively charged structure facilitates adherence to the negatively charged cell surface receptors and endocytosis. The amount of negative charges on the cell surface and the size of the carrier/DNA complex appear to be the determinants of successful gene delivery [29]. Depending on the size of the carrier/DNA complex, receptor-mediated endocytosis, pinocytosis, or phagocytosis may occur. In the cytoplasm, endosomes are destabilized, leading to the release of the DNA. A schematic diagram of DNA uptake in cells is shown in Fig. 10.1 [1]. The common agents used for DNA condensation and gene delivery applications are listed below.

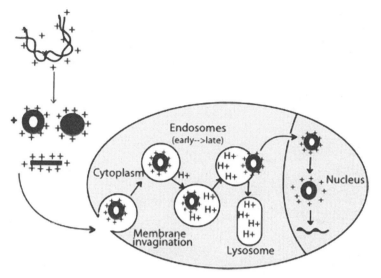

Figure 10.1. DNA uptake by mammalian cells. DNA is compacted in the presence of polycations into ordered structures such as toroids, rods, and spheroids. These particles interact with the anionic proteoglycans at the cell surface and are transported by endocytosis. The cationic agents accumulate in the acidic vesicles and raise the pH of the endosomes, inhibiting the degradation of DNA by lysosomal enzymes. They also sustain a protein influx, which destabilizes the endosome and releases DNA. The DNA is then translocated to the nucleus either through the nuclear pore or with the aid of nuclear localization signals, and decondenses after separation from the cationic delivery vehicle. (Reproduced with permission from Ref. [1].)

10.2
Agents That Provoke DNA Nanoparticle Formation

10.2.1
Polyamines

The polyamines spermidine and spermine and their synthetic analogues are excellent promoters of DNA nanoparticle formation [8, 9, 14, 18, 30–33]. The ability of polyamines and their analogues (Fig. 10.2) to compact DNA has been studied by several investigators as a model system to understand the mechanism of DNA compaction in phage head [8, 9]. Bloomfield and colleagues [8, 9, 13, 14, 32] have used the counterion condensation theory developed by Manning [34] and Record et al. [35] to calculate DNA charge neutralization in the presence of spermidine and spermine, and found that DNA collapse occurs at >89% charge neutralization. Earlier work considers cationic polyamines as point charges without any defined structure. However, two trivalent cations, cobalt hexamine^{3+} and spermidine^{3+}, differ in their ability to condense DNA, suggesting the importance of site-specific interactions in addition to the overriding electrostatic interactions [14]. Recent studies with a series of isovalent spermine homologues further show the importance of polyamine structure in DNA condensation (Fig. 10.3) [30]. There are also significant differences in the relative affinity of these homologues to DNA, as measured by the ethidium bromide displacement assay (Tab. 10.1). Although tetravalent polyamines are not very efficient in DNA transport in cells, higher-valency analogues, including the hexamines facilitated the transport of a 37-nucleotide (nt) oligonucleotide in breast cancer cells (Fig. 10.4) [19]. The hexamines can condense a plasmid DNA to toroidal nanoparticles, as measured by atomic force microscopy (AFM) (Fig. 10.5; Tab. 10.2).

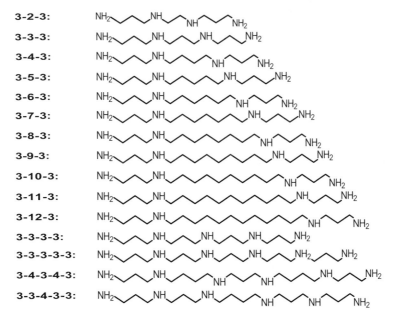

Figure 10.2. Chemical structures of the natural polyamine, spermine (3–4-3) and its analogues. The polyamine analogues are abbreviated by a number system that represents the number of methylene bridging groups between the primary and secondary amino groups.

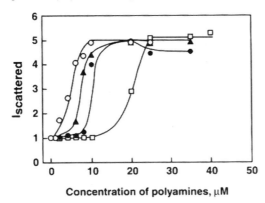

Figure 10.3. Typical plots of the relative intensity of scattered light at 90° plotted against the concentrations of spermine (O), 3–10–3 (▲), 3–11–3 (●, and 3–12–3 (□). The λ-DNA solution had a concentration of 1.5 μM DNA phosphate, dissolved in 10 mM Na cacodylate buffer, pH 7.4. (Reproduced with permission from Ref. [30].)

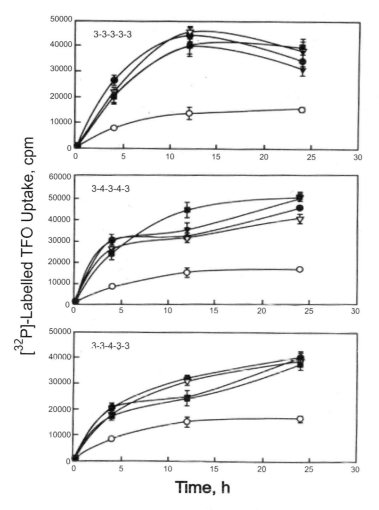

Figure 10.4. Time-course of uptake of a 37-nt triplex-forming oligonucleotide (TFO) in the presence of hexamines, 3–3-3–3-3, 3–4-3–4-3 and 3–3-4–3-3. MCF-7 breast cancer cells (5 x 10^5 per well) were plated in six well plates and allowed to adhere for 24 h. Triplicate wells were treated with a 250,000 cpm level of ^{32}P-labeled TFO in 0.5 ml prewarmed (37 °C) medium. Cells were harvested at the indicated time points. Cell lysate was prepared and radioactivity determined by scintillation counting. The concentrations of hexamines used in these experiments were: 0 (O), 0.25 (◆), 0.5 (■), 1 (▽), and 2.5 μM (●). Data shown are mean ± SD from triplicate experiments. (Reproduced with permission from Ref. [19].)

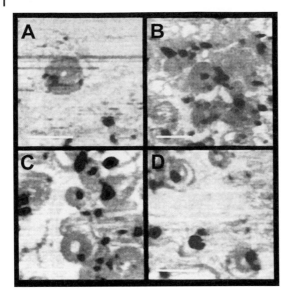

Figure 10.5. Atomic force microscopy images showing the toroid structures of pGL3 plasmid DNA formed by incubation with 25 μM spermine (A), 5 μM 3–3-3–3 (B), 2 μM 3–4-3–4-3 (C), and the partly formed toroids observed in the presence of 2 μM 3–4-3–4-3 (D). D provides evidence for spooling of DNA to form toroids. Scale bar is 200 nm. (Reproduced with permission from *Nucleic Acids Research* **2004**, *32*, 127–134. Copyright 2004 Oxford University Press Ref. [33].)

Table 10.1 Relative binding constants of polyamine analogues for calf thymus DNA, as measured by the ethidium bromide competition method.

Polyamine homologue	Relative binding constant[a]
3–2-3	0.4
3–3-3	0.6
3–4-3 (spermine)	1.0
3–5-3	1.3
3–6-3	1.1
3–7-3	1.0
3–8-3	0.7
3–9-3	0.5
3–10–3	0.5
3–11–3	0.4

[a] The binding constants were calculated as the reciprocal of the 50% concentration of polyamine homologues required to displace 50% ethidium bromide bound to λ-DNA. The reproducibility with these results was within 3% in repeated measurements. The results are normalized with respect to spermine. (Reproduced with permission from Ref. [30].)

Table 10.2 AFM measurement of the outer diameter and height of toroidal DNA nanoparticles formed in the presence of polyamine analogues.

Polyamine	Outer diameter (nm)	Mean height (nm)
3–4-3 (Spermine)	191 ± 12^a	2.61 ± 0.77^b
3–3-3–3	168 ± 5.4	
3–3-3–3-3	117 ± 8.8	
3–4-3–4-3	118 ± 10.8	

a Mean ± S.E.M. of 5–7 toroids measured in each case.
b The toroid height given here is the average value for 28 toroids, formed in the presence of polyamines shown in column 1 of this table.
(Reproduced with permission from Ref. [33].)

10.2.2
Cationic Lipids

A group of cationic molecules that has attracted much attention in preclinical and clinical gene therapy trials is the cationic lipids [26, 27, 36, 37]. These molecules possess a hydrophobic group and a polar group (Fig. 10.6). A large number of lipids have been developed to transfect DNA; the best known are DOTAP (N-1 (-2,3-dio-leoyloxy) propyl)-N, N, N-trimethylammoniumethyl sulfate), DOTMA (N-1- (2,3-dio-lcoyloxy) propyl)-N, N, N-trimethylammonium chloride), DOGS (dioctadecylamido glycylspermine-4-trifluoroacetic acid), and DOSPA (2,3-dioleoyloxy-N-[2-(spermine-carboxamido)-ethyl]N,N-dimethyl-propan-1-aminium trifluoroacetate). These are available as commercial DNA transfection agents. These agents can form DNA nanoparticles by electrostatic interactions between the positively charged polar head group of the lipid and the negatively charged phosphate group of DNA [36]. Increasing the number of amino groups per molecule and the distance between amino group and hydrophobic group favors the transfection efficiency of cationic lipids. Nanoparticle formation by complexation with lipids not only improves the transport of DNA through the cellular barriers, but also protects it from enzymatic degradation in cell culture media [38–41]. The size of the particles ranges from 50 to 1000 nm. The large particles of liposome/DNA complexes are found to transfect cells *in vitro*. Liposome DNA complexes are taken up by endocytosis, and after internalization DNA is released from the endocytic vehicle, leaving behind the liposome [42]. However, the transfection efficiency of liposome/DNA complex is relatively low *in vivo*. In addition, many available cationic lipids are reported to be toxic [43]. Studies in mice and macaques have shown that exposure to high doses or repeated doses results in microscopic pathology and gross lung pathology.

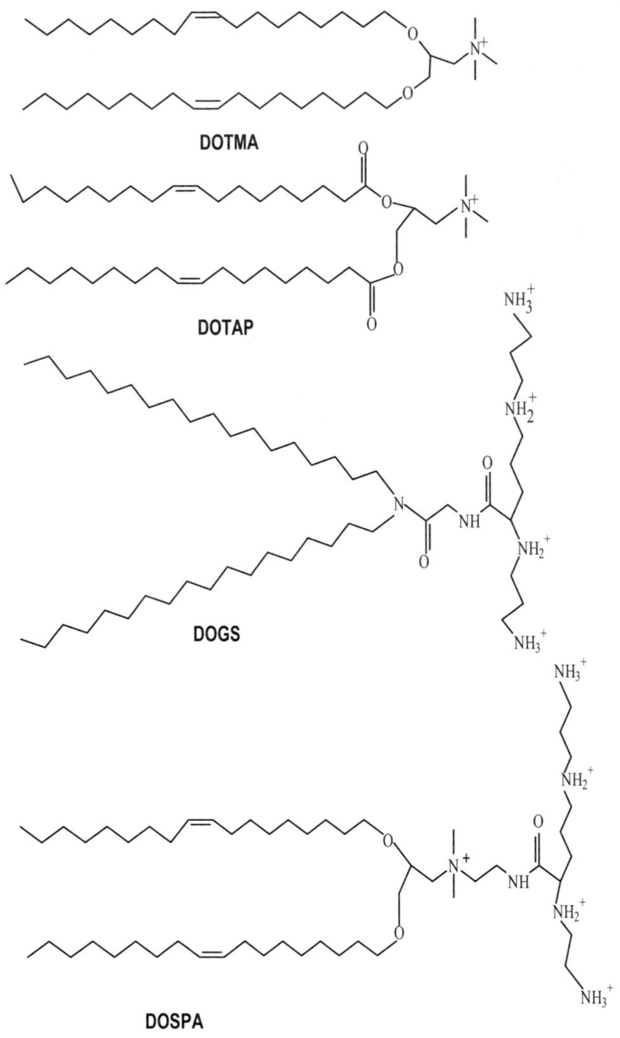

DOTMA

DOTAP

DOGS

DOSPA

Figure 10.6. Chemical structures of commonly used cationic lipids. The abbreviations are defined in the text.

10.2.3
Polyethylenimine

Polyethylenimine (PEI) is a group of synthetic polymers that are known to be efficient in the transport of oligonucleotides and plasmid DNA in a variety of cell types and animal models [44]. Linear and branched PEI (Fig. 10.7) can induce the condensation of DNA to nanoparticles. The linear and branched nature of PEI as well as its molecular weight play important roles in DNA condensation and transfection efficiency [45]. Branched PEI contains primary, secondary, and tertiary amino groups, and acts as a proton sponge at the endosomal pH [1, 23]. The buffering capacity of PEI is believed to contribute to its ability to deliver DNA within cells without degradation. Unlike other polymers, PEI possesses intrinsic endosomolytic activity and does not need any endosomolytic agent to escape from endosome [23, 45, 46]. A confocal microscopic investigation shows that PEI DNA transport occurs without the dissociation of the complex [47]. Poor solubility of the DNA/PEI complex at physiological pH and toxicity of PEI in animal model studies are the major limitations in using this polymer as a gene carrier. Derivatization of PEI with poly(ethylene glycol) (PEG) is reported to increase the solubility and reduce the cytotoxicity of the PEI [47–50]. PEGylation improves the stability of nanoparticles and increases their *in vivo* circulation time. A terpolymer of polylysine, PEG, and PEI has also been investigated as an agent for DNA nanoparticle formation [51].

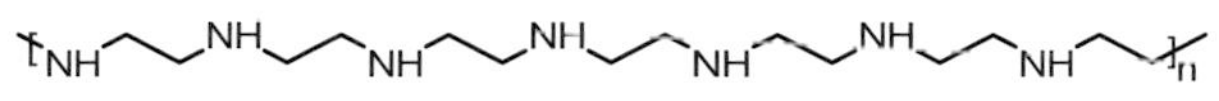

Polyethylenimine (linear)

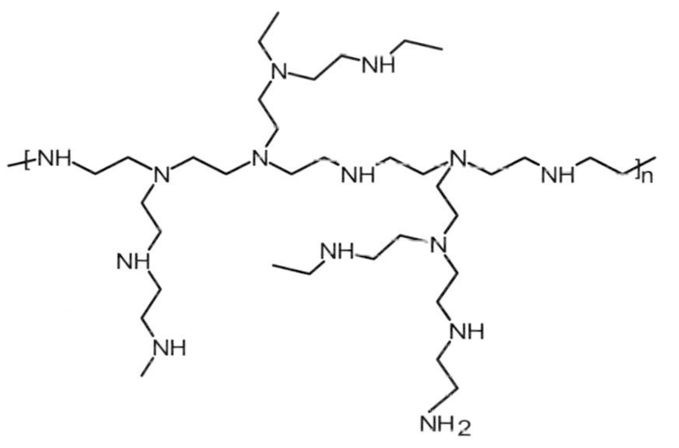

Polyethylenimine (branched)

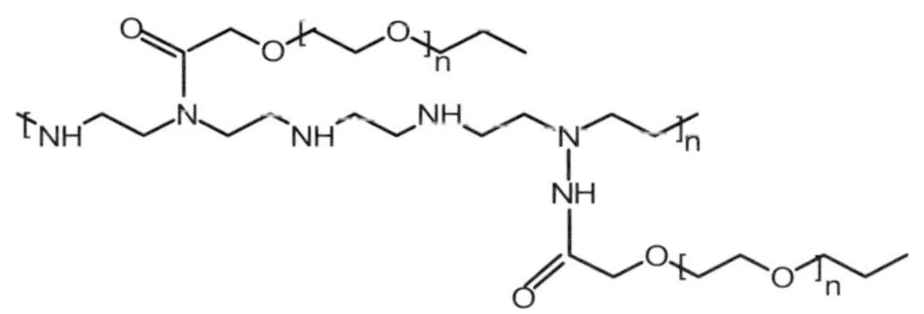

PEGylated Polyethylenimine

Figure 10.7. Chemical structures of linear, branched, and polyethylene glycol (PEG) derivatized polyethylenimines.

10.2.4
Dendrimers

Polyamidoamine (PAMAM) and polypropylenimine (PPI, Fig. 10.8) dendrimers have been studied for their ability to provoke DNA nanoparticle formation and facilitate DNA transport [52–57]. Tomalia et al. [52] reported the first preparation of an entire series of dendrimers, possessing trigonal, 1->2 N-based, branching centers. Compared to linear and branched polymers, the monodispersity and

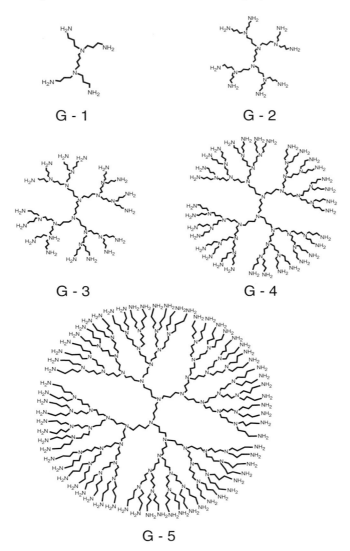

G - 1

G - 2

G - 3

G - 4

G - 5

Figure 10.8. Chemical structures of five generations of polypropylenimine dendrimers. The complexity of the dendrimers is represented by generation numbers (G1 to G5).

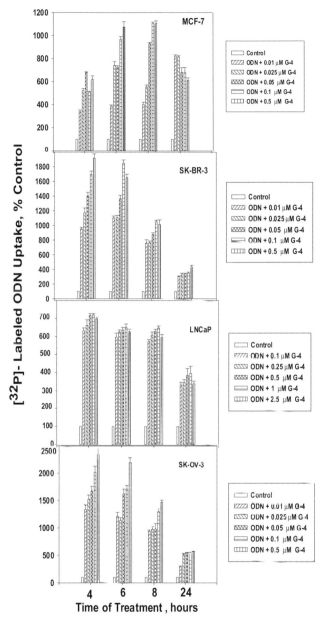

Figure 10.9. Cellular uptake of a ^{32}P-labeled 31-nt oligonucleotide (ODN) in the presence of polypropylenimine generation 4 (G-4) dendrimer. Different cell lines (MCF-7 and SK-BR-3 breast cancer, LNCaP prostate cancer, and SK-OV-3 ovarian cancer) were treated with a 250,000 cpm level (0.4 nM) of ^{32}P-labeled ODN after complexing with G-4 dendrimer in six-well plates. At the indicated times, the medium was removed, cells washed three times with cold phosphate buffered saline, lysed, and cell-associated radioactivity quantified by scintillation counting. Control indicates the use of labeled ODN in the absence of dendrimer. The percentage increase in uptake is calculated with reference to the control group. Results presented are the mean of three separate triplicate measurements. Error bars indicate S.E.M. (Reproduced with permission from Ref. [56].)

domains that can bind with the phosphodiester backbone of DNA. Protamine-mediated DNA condensation causes adjacent arginine residues to interlock both strands of the DNA helix, making it transcriptionally inactive [60, 61]. There have been several studies using protamine and related proteins to induce and stabilize DNA nanoparticles for gene delivery.

The cationic polypeptide, poly-L-lysine (PLL) (Fig. 10.11) contains protonated amine groups, which can interact with DNA and provoke nanoparticle formation [17, 18, 62, 63]. The size of the nanoparticles thus formed depends on the nature of the PLL (linear versus branched as well as ionic conditions of the medium) (Fig. 10.12) [17]. Although PLL is not a very efficient agent for transfecting DNA in

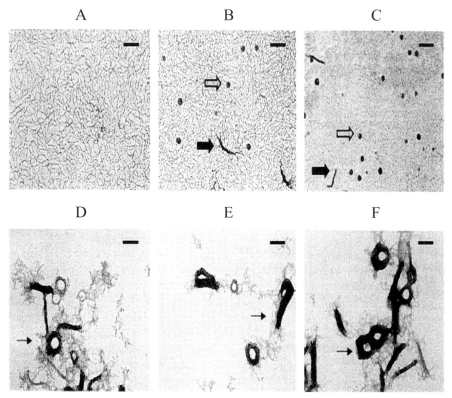

Figure 10.12. Electron micrographs of the structures of condensed DNA complexes prepared at different ionic strengths and poly-L-lysine/DNA ratios. Electron micrographs of 0.04% uranyl acetate-stained DNA complexes were prepared using plasmids pCMV-Luc and GalPLL256 [17]. Samples were stained within 30 min after their preparation. (A–C) DNA samples prepared at 1 M NaCl with GalPLL256. (D–F) DNA samples prepared at 0 M NaCl with GalPLL256. r values (poly-L-lysine to DNA ratio) are as follows: (A, D) 0.25; (B, E) 0.5; (C, F) 0.75. There was no adjustment of salt concentration before processing samples. Open arrows, spherical complexes; solid arrows, rod-like complexes (major diameters of ~100 nm); thin arrows, aggregated structure including large rods and toroids complexes. The bars in all panels represent 100 nm. (Reproduced with permission from Ref. [17].)

cells, conjugation to a receptor enhances its ability as a gene delivery vehicle [64–67]. Complex formation between PLL and DNA can also protect the DNA from enzymatic degradation within the cellular environment [68].

Low-molecular-weight peptides are also reported to condense DNA to nanoparticles, but the *in vivo* efficiency of these complexes for gene delivery is poor because of the low affinity of the peptides for DNA. To improve the DNA binding and transfection efficiency, stable cross-links were introduced to the condensates using bifunctional agents such as glutaraldehyde or sulfhydryl groups [9, 69–71]. Short peptides possessing multiple cysteine residues form interpeptide disulfide bonds when bound to DNA and after internalization. The half-life of DNA in mouse liver is extended by a formulation of sulfhydryl-linked gene delivery system [72].

10.2.6
Polymers

Chitosan is a natural polymer consisting of two subunits, D-glucosamine and *N*-acetyl-D-glucosamine linked together by glycosidic bonds (Fig. 10.13). It is comparatively nontoxic and has a high transfection efficiency after nanoparticle formation [73–77]. Chemical modification is feasible in this system because of the availability of a large number of surface functional groups. Chitosan and its derivatives were found to be promising vectors for gene delivery [73].

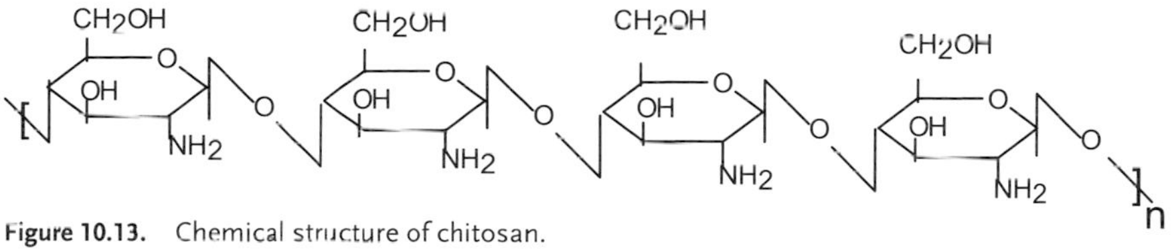

Figure 10.13. Chemical structure of chitosan.

Neutral polymers such as PEG and poly(ethylene oxide) cause DNA condensation at high ionic strength [78]. These uncharged flexible polymers act as crowding agents and provoke DNA condensation through an excluded volume mechanism. PEG is being used as part of the graft polymers in conjunction with PEI, folate receptors, and polylysine to improve the stability of the DNA condensate and transfection efficiency [79–81].

In summary, a large number of agents that can interact with DNA and provoke nanoparticle formation are being studied in academic and industrial settings for the development of nonviral gene delivery vehicles.

10.3
Characterization of DNA Nanoparticles

The following three methods are commonly used for the characterization of DNA nanoparticles in gene therapy applications: (i) laser light scattering; (ii) electron microscopy; and (iii) atomic force microscopy.

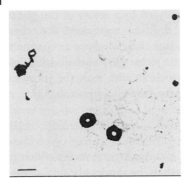

Figure 10.15. Electron micrograph of DNA nano-particles formed in the presence of spermidine. Poly(dA-dT).poly(dA-dT) solution (3 μM DNA phosphate) in 10 mM NaCl, 1 mM sodium cacodylate, and 200 μM spermidine was placed on a carbon-coated grid and counterstained with 1% uranyl acetate. The bar indicates 100 nm.

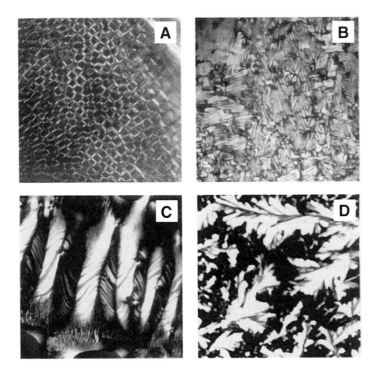

Figure 10.16. Effects of spermine and N^1-acetylspermine on the liquid crystalline phase transitions of calf thymus DNA. (A) DNA (25 mM in Na cacodylate buffer) was mixed with 1 mM spermine and incubated on a glass slide at 22 °C for 15 min (100×). A planar cholesteric phase with a three-dimensional network is observed. (B) The glass slide in (A) was incubated for 12 h at 37 °C and viewed through the plate under crossed polars (200×). Fingerprint texture with antiparallel grain boundaries is found. (C) Large pitch cholesteric phase was observed when the glass slide was further incubated for 24 h at 37 °C (180×). (D) DNA was mixed with 1 mM N^1-acetylspermine and incubated for 48 h at 37 °C (45×). A crystalline phase is obtained. (Reproduced with permission from Ref. [59].)

The coverslips are sealed with a solution of polystyrene and plasticizer in xylene to prevent dehydration. Textures are allowed to stabilize for a few minutes to a few hours. Recent studies indicate the formation of two types of liquid crystalline phase: a cholesteric phase and a columnar hexagonal phase in DNA treated with a natural polyamine (Fig. 10.16) [59]. The thermodynamic stability of cholesteric phases formed from DNA can be determined by monitoring different phases under a microscope as a function of temperature, salt concentration, and DNA dilution. It is of interest to examine whether the ability of DNA to form different liquid crystalline phases is related to the efficacy of different agents to facilitate the cellular transport of DNA and oligonucleotides.

10.3.3
Atomic Force Microscopy

Atomic force microscopy (AFM) is the preferred technique for visualizing DNA nanoparticles and making quantitative three-dimensional measurements. AFM, which was invented in 1986, expanded the application of scanning tunneling microscopy to nonconductive, soft, and living biological samples [85–88]. AFM has several capabilities, including imaging topographic details of surfaces from the submolecular level to the cellular level [89], monitoring the dynamic processes of single molecules in physiologically relevant solutions [90], measuring molecular interactions [91], and characterizing the mechanical properties of single molecules or single nanostructures [92]. In general, mixing DNA and condensing agent in solution forms DNA nanoparticles. However, nanoparticles have also been prepared on a solid surface for imaging. When the nanoparticles are formed in solution, they can be immobilized onto a freshly cleaved mica surface and AFM imaging done in air or in solution. In general, imaging in air is much easier than imaging under solution conditions, and can provide valuable structural and morphological information about DNA nanoparticles in three dimensions. For example, the size, height, shape, and volume of the particles can be easily obtained for spherical or rod-like structures, after obtaining the AFM images. For toroidal structures, which are commonly formed during DNA condensation, the toroidal radius and volume can be determined by AFM software as described by Rackstraw et al. [92] (Fig. 10.17). Cross sections of the ring-like particles can be taken at 45° separations, and toroidal radii determined from the cross sections at half-maximum height in order to minimize tip convolution effects. The volume of single nanoparticles can be calculated using the equation:

$$\text{Volume} = 2\pi^2 r^2 R, \tag{3}$$

where the values of r and R are as defined in the legend to Fig. 10.17.

In the second case, the condensates can be formed directly on the substrate and imaged by AFM directly in air or under solution. There are many attractive features with regard to imaging under solution. The most obvious one is the ability to follow the dynamic condensation process in physiologically relevant buffers in real time.

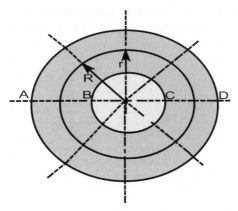

Figure 10.17. Representation of a toroidal structure used to calculate the condensate volume using Eq. (3). In this figure, $r = \{(A–D)(B–C)\}/4$, and $R = \{(B–C)/2\} + r$.

Compared to electron microscopy, the ability to image in solution under native conditions is the most striking feature of AFM. AFM can be operated in both tapping mode and contact mode to image the DNA nanoparticles. Tapping mode is more commonly utilized, because the destructive effect of the AFM tip on the DNA nanoparticles (which are usually soft) is much less in tapping mode than in contact mode imaging.

10.3.3.1 DNA Nanoparticle Studies by AFM

Examples of DNA nanoparticle formation in the presence of PPI dendrimer (generation 4) and spermidine are shown in Fig. 10.18. The DNA structure gradually changes from fully uncondensed plasmid structures to partially condensed flower-like structures in the presence of $2.5\,\mu$M PPI dendrimer. In the presence of $10\,\mu$M spermidine, compact, fully condensed nanoparticles are observed. Using AFM, Iwataki et al. [93] have observed competition between compaction of single chains and bundling of multiple chains in giant DNA molecules with spermidine as a condensing agent. When the concentration of DNA was <1 μM in base-pair units, individual DNA molecules condensed into a compact structure, whereas a thick fiber-like assembly of multiple chains occurs when the concentration of DNA is increased to $10\,\mu$M. In other studies, different morphologies, including flowers, disks, toroids, and branch-like structures are observed by AFM at different stages of DNA condensation or under different condensation conditions [94–96].

DNA nanoparticle formation in the presence of poly-L-lysine (PLL), modified PLL, spermidine, and PEI has been extensively studied using AFM [18, 96–101]. Hansma and collaborators [18, 96, 101] employed AFM to study the extent of DNA condensation in approximately 100 different complexes of DNA with PLL or PLL attached to the glycoproteins, asialoorosomucoid (AsOR), or orosomucoid. They found that DNA condensation is efficient with 10-kDa PLL covalently attached to AsOR at a lysine:nucleotide (Lys:nt) ratio of 5:1 or higher. Large numbers of toroids and short

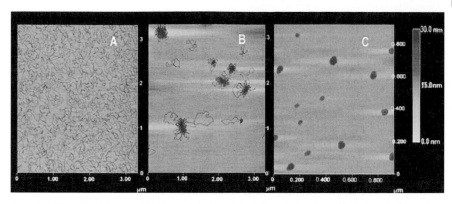

Figure 10.18. AFM images of DNA nanoparticle formation from plasmid pGL3 in the presence of PPI dendrimer and spermidine. (A) Plasmid DNA on mica surface; (B) partially condensed plasmid DNA in the presence of 2.5 μM generation-4 PPI dendrimer; (C) plasmid DNA condensed in the presence of 10 μM spermidine.

rods with contour lengths of 300–400 nm are produced under these conditions. Furthermore, it has been found that PLL-AsOR enhanced gene expression in the mouse liver approximately 10- to 50-fold as compared with PLL alone. The study by Wolfert and Seymour [99] on the influence of the molecular weight of PLL on the size of DNA nanoparticles shows that the smallest PLL (MW: 3970) produce more homogeneous complexes with diameters ranging from 20 to 30 nm as compared to the large complexes (120–300 nm) formed in the presence of high-molecular-weight PLL (MW: 224,500). DNA nanoparticles formed with low-molecular-weight PLL have significantly lower cytotoxicity than those formed with high-molecular-weight PLL. Interestingly, PEI is more efficient than PLL for nanoparticle formation for gene delivery applications [45, 94, 100, 101]. AFM studies show DNA aggregation and partially condensed nanoparticles in the presence of PLL, whereas PEI can provoke nanoparticles of 30–60 nm. Rings, extended linear structures, toroids, rods, and other intermediates are found in AFM studies of nanoparticles provoked by PLL polymer conjugates.

Han et al. [102] show condensation of plasmid DNA by a water-soluble lipopolymer into spherical particles of approximately 50 nm diameter. Lim et al. [103] confirm by AFM the formation of self-assembling biodegradable complexes between DNA and poly[α-(4-aminobutyl)-L-glycolic acid], a non-toxic polymeric gene carrier.

In addition to cationic DNA condensing agents, neutral amphiphiles that possess a DNA molecular recognition unit with a hydrophobic region and a PEG unit have also been shown to condense plasmid DNA successfully [78, 79]. AFM studies show a range of collapsed states from partial collapse to the condensed supramolecular toroidal structures, which have an outer and inner diameter of approximately 100 nm (range 106±22 nm) and 20 nm (range 21±11 nm), respectively. These amphiphiles represent the first step toward identifying the design requirements for a selective DNA binding/condensation agent. Isobe et al. [104] have studied the condensation of plasmid DNA with functionalized fullerenes by AFM. Gallyamov et al. [105]

have investigated the process of compaction of high-molecular-weight T4 DNA directly in an AFM liquid cell. AFM images of globules formed by compaction of DNA in water–alcohol environments at high isopropanol concentration (80%) have also been obtained. By using the technique of deconvolution of the AFM images, it is shown that the globule contains only one closely packed DNA molecule. Our recent studies demonstrate toroid formation of a plasmid DNA in the presence of polyamine analogues (Fig. 10.5) [33].

DNA condensation occurring on surfaces has also been studied [106–111]. Dimitriadis et al. [106] used AFM to study the conformation of DNA adsorbed from aqueous solutions containing monovalent (Na^+), divalent (Ca^{2+}), and trivalent (spermidine^{3+}) cations on different solid substrates (hydrophobic, moderately hydrophilic, and hydrophilic surfaces). It is shown that the substrate surface energy mediated DNA/DNA interaction, leading to the formation of a variety of structures. It has also been shown that the condensed DNA structures are formed on hydrophobic surfaces even in the absence of multivalent cations. The study by Allen et al. [108] on DNA–protamine complexes bound to mica by AFM shows that the morphology of the structures varies depending on the sample preparation method. For example, interstrand, side-by-side fasciculation of DNA and toroidal-like structures are observed for complexes formed in solution, following direct mixing of the DNA and protamine. Large aggregates of DNA are also observed occasionally. However, if the DNA is first attached to the mica surface, prior to addition of the protamine, only well-defined toroidal structures are formed, without DNA fasciculation or aggregation. The mean diameter of the toroids is 39.4 nm. The structures indicate that the condensed DNA is stacked vertically by four to five turns with each coil containing as little as 360–370 bp of B-form DNA. Fang et al. [110, 111] showed the two-dimensional condensation of DNA molecules on cationic lipid membrane by AFM. It was found that the fluidity of cationic membrane promoted the close packaging of both linear and circular DNA molecules, independent of the length of the DNA. The average interhelical distance for closely packed DNA is over twice the average diameter of DNA.

It is noteworthy that different nanoparticle structures are formed with DNA, depending on whether the condensation process is performed in solution or on a solid surface. For example, Allen et al. [108] show distinct toroidal structures when protamine is added to DNA on a mica surface. However, when DNA and protamine are mixed and incubated in solution before transfer onto mica surface, a network of DNA with a few thin toroids is found. Fang et al. [110] reported similar results when they used silanes as condensing agents. When silanes and DNA are preincubated in solution, looped structures including flower and sausage shapes are observed, whereas on a silicon surface silanes condense DNA into distinct toroids, whose size depends on the length of the DNA.

10.3.3.2 Limitation of AFM Technique

Compared to transmission electron microscopy, AFM not only provides direct three-dimensional images of DNA nanoparticles, but also has the capability of operating in solution, thereby providing a new avenue to determine the structures of biological specimens at high resolution in physiologically relevant ionic and pH conditions. However, like any physical technique, AFM is not perfect, and has limitations based on the sample preparation method. As previously discussed, most DNA nanoparticles for nonviral gene delivery are prepared in solution. In order to image these nanoparticles by AFM, they have to be first transferred and immobilized onto a substrate, which in most cases is mica, and then the substrate fastened to the AFM instrument. A common sample preparation follows a two-step procedure:

1. Deposit an aliquot of the solution of DNA nanoparticles on a freshly cleaved mica surface (modified or unmodified).
2. Rinse and blow dry the sample on the mica surface after a predetermined incubation period.

While following this basic procedure, details of sample preparation differ from laboratory to laboratory. For example, Sun et al. [112] use a filter paper to remove the residual solution after depositing the condensation solution onto the mica surface with an incubation period of 1 min. This procedure may interfere less with the immobilization of the DNA condensates on the surface than rinsing with water, which can either wash away the DNA condensates or induce decondensation. Fang et al. [111] find that rinsing with water prior to drying the mica surface fails due to the rapid decondensation of the condensates. To avoid decondensation, they dry the samples with compressed air after a 5-minute incubation, and then conduct the AFM imaging without rinsing. This procedure can prevent decondensation of the condensates on the mica surface, with the limitation that a high concentration of condensing agents may interfere with the AFM imaging. In most cases, immobilization of DNA condensates is mainly due to the electrostatic interaction between the DNA condensates and the mica surface. Therefore, when the DNA condensates are overall positively charged, the condensates can usually adsorb and adhere to the unmodified mica with overall negative charge. However, when the DNA condensates are overall negatively charged or the interaction between positively charged DNA condensates and unmodified mica surface is too weak to immobilize the DNA condensates, an appropriate procedure must be adopted to obtain successful immobilization. A common procedure employed is to modify the mica surface. For example, Dunlap et al. [94] have modified mica to be positively charged (poly-L-ornithine-coated mica) to immobilize the incomplete condensates with overall negative charge while bare mica with negative charge is used to immobilize the completely saturated condensates, which are positively charged. Furthermore, to enhance the adhesion of some condensates to the surface, Dunlap et al. [94] use another procedure, in which they first submerge the prepared mica substrate in the diluted condensate solution, centrifuge it for 10 min, and then rinse the substrate for imaging. By this procedure, the condensates and substrate surface are forced to be in close contact with each other, resulting in stronger interaction.

It is evident from the above examples that the development of an appropriate procedure to prepare the DNA condensates on the mica surface for imaging is key to the success of the AFM imaging. Without successful immobilization of the DNA condensates on the mica surface, it is misleading to conclude that no condensates are formed, based on the result that no condensates are observed by AFM. It may also be wrong to conclude that some condensates are formed in solution only, based on the AFM results. As previously discussed, uncondensed DNA or partially condensed DNA may get fully condensed on the mica surface [104–108].

AFM can be used for the characterization of particles of a wide size range, from 1 to 8000 nm. This technique also allows resolution of single double-stranded DNA molecules on the mica surface.

10.4
Mechanistic Considerations in DNA Nanoparticle Formation

DNA is a highly charged molecule in solution and interacts with solvent and other solutes over a potentially long distance. The negatively charged phosphate groups of DNA molecule in solution repel each other, resulting in a worm-like chain with persistence length of approximately 50 nm [113]. The conformation of DNA in solution depends on solvent–polymer interaction and on interaction between the different segments of the polymer. Condensation occurs as a result of the interplay of different interactive forces in the presence of a condensing agent. In the presence of cationic molecules, DNA condensation is predominantly governed by electrostatic interaction between the cations and negatively charged phosphate groups. DNA condensation in solution can be achieved *in vitro* by simple addition of multivalent cations, cationic polymers, or neutral crowding molecules [9].

Theories based on electrostatic interaction of positively charged ions with DNA postulate ion competition, and hence the driving force for multivalent ion binding to DNA derives from a net gain in entropy by the release of DNA-bound monovalent ions, such as Na^+, into solution:

$$\text{Multivalent ion}^{m+} + \text{DNA.nNa}^+ \rightarrow \text{DNA.(n-m)Na}^+\cdot \text{multivalent ion}^+ + m\text{Na}^+$$

When the bulk Na^+ concentration is increased in the medium, the net gain in entropy on releasing the bound Na^+ to the solution decreases, and an increased concentration of multivalent ions is required to compete with Na^+ to collapse DNA to compact structures, such as toroids.

The dependence of the EC_{50} (concentration of condensing agent at 50% DNA condensation) value of multivalent ions on the Na^+ concentration is usually represented by a plot of $\ln[EC_{50}]$ against $\ln[Na^+]$. The slope of this plot is a measure of the binding affinity of counterions with DNA according to the counterion condensation theory [8, 30, 32, 34, 35, 114]:

$$1 + \ln(1000\Theta_1/c_1 v_{p1}) = -2z_1\xi(1-z_1\Theta_1-z_2\Theta_2)\ln(1-e^{-\kappa b}) \tag{4}$$

$$\ln(\Theta_2/c_2) = \ln(v_{p2}/1000e) + (z_2/z_1)\ln(1000\Theta_1e/c_1 v_{p1}) \tag{5}$$

where c_1 and c_2 are the concentrations of counterions of charges z_1 and z_2 contributing to fractional charge neutralization of Θ_1 and Θ_2 and occupying volumes v_{p1} and

v_{p2}, respectively, when they are bound to DNA and κ is the Debye screening parameter. $\xi = q_p^2/\varepsilon k Tb$ where q_p is the charge of the proton, ε is the bulk dielectric constant, and b is the average linear charge spacing of the polyelectrolyte in the absence of any associated ions. In other words, the parameter ξ is given by the ratio between the Bjerrum length and the average axial charge spacing, that is, the contour length divided by the number of charge groups. For double-helical B-DNA, $\xi = 4.2$, while for the single-stranded DNA, $\xi = 1.8$. Manning [34] and Record et al. [35] calculated these values, while Olson and Manning [115] provided a configurational interpretation of this result. κ is given by the equation:

$$\kappa = 3.29 z^{1/2} c_1^{1/2} \, (\text{nm}^{-1}) \tag{6}$$

Using this value of κ in Eq. (4) and introducing the first term of Eq. (4) into Eq. (5), the following equation can be derived after rearrangement:

$$\ln c_2 = [\ln \Theta_2 - \ln(v_{p2}/1000e) + 2 z_2 \xi (1-r) \ln(3.29b\zeta^{1/2})] + z_2 \xi (1-r) \ln c_1 \tag{7}$$

In Eq. (7), $r = z_1 \Theta_1 + z_2 \Theta_2$, and the approximation of $\kappa b \sim (1 - e^{-\kappa b})$ has been introduced in Eq. (4). The maximum extent of charge neutralization of DNA by a combination of Na^+ and multivalent ion has been calculated to be ~91% [8]. Substituting this value for r, the slope of a plot of $\ln c_2$ against $\ln c_1$ can be calculated to be 1.5 from Eq (7). Experimentally determined values of the slope are very close to this value for trivalent and tetravalent polyamines. However, this theory breaks down when counterions of >4 positive charges are considered because of the high affinity of these ions for DNA at low Na^+ concentration [33]. A theory for calculating the charge neutralization of DNA by polycations, such as PEI and PLL, is yet to evolve.

Counterion fluctuations and hydration forces play an important role in DNA condensation. Thermal fluctuation in the condensed counterion charge density along the DNA chain can lead to nonuniformity. Interaction with a positive polyelectrolyte can lead to inversion of the charge distribution on DNA [116]. At high ionic strength, increased screening of repulsion between the charges causes the configuration of DNA to be more compact. Fenley et al. [117] have studied the condensation of counterions on different conformations of DNA chains, in linear DNA, DNA circles, and closed circular supercoiled DNA configuration. They found that counterion condensation is correlated to the conformation of DNA. More compact DNA condenses more counterions.

Reorganization of water molecules between DNA duplexes in the presence of condensing agents is also an important aspect of DNA condensation. Kankia et al. [118] studied the hydration effect in the condensation of DNA in the presence of $Co(NH_3)_6^{3+}$ using ultrasonic and densitometric techniques. The binding was found to be accompanied by dehydration from the hydration shells of DNA and $Co(NH_3)_6^{3+}$, corresponding to the direct contact between the molecules. Conformational change in DNA structure toward a C-form was observed in the presence of Mn^{2+} ions by electronic, vibrational, and circular dichroism studies [119]. The ability of Mn^{2+} to disturb the structure of DNA might be via destabilization of hydration layers on the double helix and partial breakage of the hydrogen bonds between the base pairs [120, 121].

Two major theories have been advanced to explain the formation of toroidal structures during DNA condensation (Fig. 10.19). Circumferential winding of DNA to form toroids has been considered for toroid formation in earlier studies [9, 32, 33, 120, 121]. Some electron microscope and AFM studies support this model. Hud et al. [122] proposed a kinetic mechanism in which the DNA molecule in solution is coiled with a constant radius into a series of equally sized loops, which form toroids in the presence of condensing agents. This model is in good agreement with several observations in the literature. According to this model, spontaneous formation of DNA loops in the presence of condensing agents provides a nucleation event for toroid formation. This is kinetically favored because of random polymer fluctuation. The size of the initial loop is suggested to be a factor in determining the size of the condensate [122]. During DNA toroid formation, the growth of the loop proceeds inward and outward equally, keeping the toroid diameter the same as the nucleation loop. The most probable diameter of the loop that spontaneously forms from a DNA molecule is calculated to be 50 nm under conditions at which the ionic strength of the medium is >1 mM. Conwell et al. [83] have compared the condensates formed at low ionic strength and in the presence of additional salt and shown that toroid thickness is dependent on salt concentration. In the presence of added salt, secondary growth of the toroid to the outside occurs, increasing the size of the toroid from the initial loop diameter.

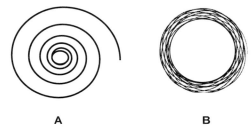

Figure 10.19. Schematic representation of (A) circumferential winding and (B) constant loop models for the formation of toroidal nanoparticles of DNA.

An alternate mechanism of coil–globule transition has also been suggested as the cause of condensation [33, 123]. The theory based on the coil-to-globule transition of flexible polymer in poor solvents is valid only at low DNA concentrations [124, 125]. Since the DNA concentration is very high inside the condensates, this theory cannot fully explain the actual process of condensation. In the condensed state, DNA exists as liquid crystals with lattice spacings close to the bulk DNA liquid crystals at the same osmotic pressure [59, 123]. Condensates can be considered as small liquid crystals with free energy contributed from the bulk and due to the finite size [126]. The contribution from the finite size includes surface energy and elastic energy. The compromise between these energies is suggested to be a factor in determining the shape of the condensate.

Crowding-induced DNA condensation occurs in the presence of neutral polymers such as polyethylene glycol and polyethylene oxide. DNA compaction in bacterial cells is reported to be crowding-induced [126]. This form of condensation is favored at high ionic strength, where electrostatic repulsion among the chain is screened to favor condensation. The ionic strength dependence of crowding-induced DNA condensation shows the existence of two states. At high ionic strength, the concentration of condensing agent required for condensation is independent of the ionic strength. However, at low ionic strength, the critical concentration of condensing agent is sensitive to ionic strength [127].

10.5
Systemic Gene Therapy Applications

While most of the nonviral gene delivery vehicles are being developed using *in vitro* cell culture models, systemic gene delivery poses numerous challenges. Overcoming serum endonuclease-mediated degradation of plasmid DNA, prevention of the aggregation of nanoparticles in the presence of serum proteins, and the need for tissue-specific targeting are major requirements for administering plasmid constructs for therapeutic purposes [128]. In addition to the use of liposomes, cationic lipids, polymers, and peptides/proteins, attempts have been made to hydrophobize DNA and assemble it into nanometer- or micron-sized particles with substantial improvement in transgene expression [129]. Encapsulation of plasmid DNA in lipid envelope and coating with PEG allows extended circulation after intravenous administration and results in accumulation of 10% of the injected dose in distal tumor [130]. Chenosy et al. [131] describe plasmid DNA encapsulation in cationic lipids and stabilization with PEG with significant decrease in plasma clearance. Cui and Mumper [132] have reported entrapping plasmid DNA using emulsifying wax and a cationic surfactant. The stable, uniform nanoparticles (100–160 nm) are produced and plasmid DNA coated on the surface of these preformed cationic nanoparticles or entrapped inside. Tail vein injections in mice show that 40% of the entrapped plasmid DNA remains in circulation, compared to 16% of naked DNA, 30 min after injection.

An example of successful nonviral vehicle used in gene therapy is leuvectin [133]. Leuvectin is a plasmid DNA/lipid complex composed of an expression vector encoding human interleukin-2 (IL-2) in a vehicle containing 5:1 mass ratio with DMRIE/DOPE lipid (1,2-dimyristyloxypropyl-3-dimethylhydroxyethyl ammonium bromide/dioleoylphosphatidyl ethanolamine). *In vitro* transfection of IL-2 plasmid DNA/DMRIE/DOPE complex results in the expression of sustained levels of biologically active IL-2 [134]. In a murine model of renal cell carcinoma, direct intratumoral administration of an IL-2 plasmid DNA/DMRIE/DOPE complex has resulted in complete tumor regression in the majority of mice. Leuvectin has been tested in phase I and II clinical trials [135]. Results of these studies suggest its potential use in prostate cancer, renal cell carcinoma, and melanoma patients. Other promising vectors include DOTAP:cholesterol cationic liposome, which has been used to deliver p53 and FHIT genes to primary and metastatic lung cancers [136].

PEI-based aerosol formulations of gene therapy hold promise for pulmonary dysfunctions such as cystic fibrosis, α_1-antitrypsin deficiency, pulmonary hypertension, asthma, and lung cancer [137]. PEI-based aerosol is stable during nebulization and results in highly efficient transfection and therapeutic response in several animal lung tumor models in conjunction with p53 and IL-12 genes. However, activation of the immune and inflammatory system is a real problem in the administration of nonviral vectors to the lung [138]. Current research is focused on improvement of gene delivery and expression by modification of delivery vehicles as well as by addition of physical techniques such as electroporation and ultrasound [139, 140].

10.6
Future Directions

Nanotechnology is a multidisciplinary field undergoing rapid expansion with the promise of new developments in medicine, genomics, communications, and robotics. At the nanometer scale, the self-ordering forces and properties of materials seem to be different from those at the macroscale. Theoretical understanding of the properties of DNA at the nanoscale and the techniques to deliver DNA as nanoparticles to specific cell types and disease states need to be perfected. Innovative ideas such as DNA self-assembly on metal carbonate nanoparticles and dissolution of the metal carbonate by acids to form DNA microcapsules are also under development [141]. In current AFM studies of DNA condensation, mainly topographic information is explored, although a variety of information including mechanical properties, magnetic properties, and thermodynamic information can be obtained from AFM. Experiments involving pulling on single DNA molecules and condensing agents and elasticity measurements by AFM could provide valuable insights into the dynamic changes in DNA morphology under different conditions. Two-dimensional mapping of sample elasticity could be achieved by recording force curves while the AFM tip scans across the sample [142]. Surprisingly, reports on the mechanical properties of DNA condensates, such as elasticity and stiffness, are scant in the literature. It has been accepted that the condensed DNA is transported into cells by endocytosis and escapes into the cytoplasm from endosomes. Subsequent transfer from the cytoplasm to the nucleus has been thought to be a critical barrier because the condensates often change their structure after endocytosis. It would be valuable to investigate the morphological and mechanical properties of the DNA condensates in relation to the stability of the condensates after endocytosis, and finally in relation to the gene transfection efficiency.

Acknowledgments

This work was supported by NIH grants CA042439, CA073058, and CA080163 from the National Cancer Institute, and a grant from the Susan G. Komen Breast Cancer Research Foundation.

Abbreviations

AAV	adeno-associated virus
AFM	atomic force microscopy
DMRIE	1,2-dimyristyloxypropyl-3-dimethylhydroxyethyl ammonium bromide
DOGS	dioctadecylamidoglycylspermine-4-trifluoroacetic acid
DOPE	dioleoylphosphatidyl ethanolamine
DOSPA	2,3-dioleoyloxy-N-[2-(sperminecarboxamido)-ethyl]N,N-dimethyl-propan-1 aminium trifluoroacetate
DOTAP	N-1 (-2,3-dioleoyloxy) propyl)-N, N, N-trimethylammoniumethyl sulphate
DOTMA	N-1- (2,3-dioleoyloxy) propyl)-N, N, N-trimethylammonium chloride
EM	electron microscope
ODN	oligodeoxyribonucleotide
PAMAM	polyamidoamine
PEG	polyethylene glycol
PEI	polyethylenimine
PLL	poly-L-lysine
PPI	polypropylenimine

References

1 Vijayanathan, V., Thomas, T., Thomas, T. J. (2002) DNA nanoparticles and development of DNA delivery vehicles for gene therapy. *Biochemistry 41*, 14085–14094.

2 Huang, L., Hung, M.-C., Wagner, E. (1999) *Nonviral Vectors for Gene Therapy.* Academic Press, San Diego, CA.

3 Varmus, H. (1988) Retroviruses. *Science 240*, 1427–1435.

4 Bukrinsky, M. I., Haggerty, S., Dempsey, M. P., Sharova, N., Adzhubel, A., Spitz, L., Lewis, P., Goldfarb, D., Emerman, M., Stevenson, M. (1993) A nuclear localization signal within HIV-1 matrix protein that governs infection of non-dividing cells. *Nature 365*, 666–669.

5 Amalfitano, A. (2003) Use of multiply deleted adenovirus vectors to probe adenovirus vector performance and toxicities. *Curr. Opin. Mol. Ther. 5*, 362–366.

6 Chuah, M. K., Collen, D., Van den Driessche, T. (2003) Biosafety of adenoviral vectors. *Curr. Gene Ther. 3*, 527–543.

7 Reid, T., Warren, R., Kirn, D. (2002) Intravascular adenoviral agents in cancer patients: lessons from clinical trials. *Cancer Gene Ther. 9*, 979–986.

8 Wilson, R. W., Bloomfield, V. A. (1979) Counterion-induced condensation of deoxyribonucleic acid. A light scattering study. *Biochemistry 18*, 2192–2196.

9 Bloomfield, V. A. (1996) DNA condensation, *Curr. Opin. Struct. Biol. 6*, 334–341.

10 Blessing, T., Remy, J. S., Behr, J. P. (1998) Monomolecular collapse of plasmid DNA into stable virus-like particles, *Proc. Natl. Acad. Sci. U.S.A. 95*, 1427–1431.

11 Montign, W. J., Houchens, C. R., Illenye, S., Gilbert, J., Coonrod, E., Chang, Y. C., Heintz, N. H. (2001) Condensation by DNA looping facilitates transfer of large DNA molecules into mammalian cells. *Nucleic Acids Res. 29*, 1982–1988.

12 Blagbrough, I. S., Geall, A. J., Neal, A. P. (2003) Polyamines and novel polyamine conjugates interact with DNA in ways that can be exploited in non-viral gene therapy. *Biochem. Soc. Trans. 31*, 397–406.

13 Bloomfield, V.A. (1991) Condensation of DNA by multivalent cations: considerations on mechanism. *Biopolymers 31*, 1471–1481.

14 Thomas, T. J., Bloomfield, V. A. (1983) Collapse of DNA caused by trivalent cations: pH and ionic specificity effects. *Biopolymers* 22, 1097–1106.

15 Widom, J., Baldwin, R. L. (1983) Monomolecular condensation of λ-DNA induced by cobalt hexamine. *Biopolymers 22*, 1595–1620.

16 Thomas, T. J., Bloomfield, V. A. (1985) Quasielastic laser light scattering and electron microscopy studies of the conformational transitions and condensation of poly(dA-dT).poly (dA-dT). *Biopolymers 24*, 2185–2194.

17 Liu, G., Molas, M., Grossmann, G. A., Pasumarthy, M., Perales, J. C., Cooper, M. J., Hanson, R. W. (2001) Biological properties of poly-L-lysine-DNA complexes generated by cooperative binding of the polycation. *J. Biol. Chem. 276*, 34379–34387.

18 Golan, R., Pietrasanta, L. I., Hsieh, W., Hansma, H. G. (1999) DNA toroids: stages in condensation. *Biochemistry 38*, 14069–14076.

19 Thomas, R. M., Thomas, T., Wada, M., Sigal, L. H., Shirahata, A., Thomas, T. J. (1999) Facilitation of the cellular uptake of a triplex-forming oligonucleotide by novel polyamine analogues: structure-activity relationships, *Biochemistry 38*, 13328–13337.

20 Wu, C. H., Wilson, J. M., Wu, G. Y. (1989) Targeting genes: delivery and persistent expression of a foreign gene driven by mammalian regulatory elements *in vivo. J. Biol. Chem. 264*, 16985–16987.

21 Zauner, W., Ogris, M., Wagner, E. (1998) Polylysine-based transfection systems utilizing receptor-mediated delivery. *Adv. Drug Deliv. Rev. 30*, 97–113.

22 Lim, Y. B., Choi, Y. H., Park, J. S. (1999) A self-destroying polycationic polymer: biodegradable poly(4-hydroxy-L-proline ester). *J. Am. Chem. Soc. 121, 5633–5639.*

23 Zuber, G., Dauty, E., Nothisen, M., Belguise, P., Behr, J. P. (2001) Towards synthetic viruses, *Adv. Drug Deliv. Rev. 52*, 245–253.

24 Luo, D., Aaltzman, W.M. (2000) Synthetic DNA delivery systems. *Nat. Biotechnol. 18*, 33–37.

25 Rudolph, C., Muller, R. H., Rosenecker, J. (2002) Jet nebulization of PEI/DNA polyplexes: physical stability and in vitro gene delivery efficiency. *J. Gene Med. 4*, 66–74.

26 Wagner, E., Zenke, M., Cotten, M., Beug, H., Birnstiel, M. L. (1990) Transferrin-polycation conjugates as carriers for DNA uptake into cells. *Proc. Natl. Acad. Sci. U.S.A. 78*, 3410–3414.

27 Felgner, P. L., Gadek, T. R., Holm, M., Roman, R., Chan, H. W., Wenz, M., Northrop, J. P., Ringold, G. M., Danielsen, M. (1987) Lipofection: a highly efficient, lipid-mediated DNA-transfection procedure. *Proc. Natl. Acad. Sci. U.S.A. 84*, 7413–7417.

28 Kumar, V. V., Singh, R. S., Chaudhuri, A. (2003) Cationic transfection lipids in gene therapy: successes, set-backs, challenges and promises. *Curr. Med. Chem. 10*, 1297–306.

29 Matsui, H., Johnson, L. G., Randell, S. H., Boucher, R. C. (1997) Loss of binding and entry of liposome-DNA complexes decreases transfection efficiency in differentiated airway epithelial cells. *J. Biol. Chem. 272*, 1117–1126.

30 Vijayanathan, V., Thomas, T., Shirahata, A., Thomas, T. J. (2001) DNA condensation by polyamines: a laser light scattering study of structural effects. *Biochemistry 40*, 13644–13651.

31 Böttcher, C., Endisch, C., Fuhrhop, J. H., Catterall, C., Eaton, M. (1998) High yield preparation of oligomeric C-type DNA toroids and their characterization by cryoelectron microscopy. *J. Am. Chem. Soc. 120*, 12–17.

32 Bloomfield, V. A. (1997) DNA condensation by multivalent cations. *Biopolymers 44*, 269–282.

33 Vijayanathan, V., Thomas, T., Antony, T., Shirahata, A., Thomas, T.J. (2004) Formation of DNA nanoparticles in the presence of novel polyamine analogues: a laser light scattering and atomic force microscopic study. *Nucleic AcidsRes. 32*, 127–134.

34 Manning, G. S. (1978) The molecular theory of polyelectrolyte solutions with applications to the electrostatic properties of polynucleotides. *Q. Rev. Biophys. 11*, 179–246.

35 Record, M. T. Jr., Anderson, C. F., Lohman, T. M. (1978) Thermodynamic analysis of ion effects on the binding and conformational equilibria of proteins and nucleic acids: the roles of ion association or release, screening, and ion effects on water activity. *Q. Rev. Biophys. 11*, 103–178.

36 Pedroso de Lima, M. C., Simoes, S., Pires, P., Faneca, H., Duzgunes, N. (2001) Cationic lipid-DNA complexes in gene delivery: from biophysics to biological applications. *Adv. Drug Deliv. Rev. 47*, 277–294.

37 Sugiyama, M., Matsuura, M., Takeuchi, Y., Kosaka, J., Nango, M., Oku, N. (2004) Possible mechanism of polycation liposome (PCL)-mediated gene transfer. *Biochim. Biophys. Acta 1660*, 24–30.

38 Brown, M. D., Schatzlein, A. G., Uchegbu, I. F. (2001) Gene delivery with synthetic (non viral) carriers. *Int. J. Pharm. 229*, 1–21.

39 Templeton, N. S., Lasic, D. D., Frederik, P. M., Stray, H. H., Roberts, D. D., Pavlakis, G. N. (1997) Improved DNA: liposome complexes for increased systemic delivery and gene expression, *Nat. Biotechnol. 15*, 647–652.

40 Juliano R. L., Akhtar, S. (1992) Liposomes as a drug delivery system for antisense oligonucleotides. *Antisense Res. Dev. 2*, 165–176.

41 Zelphati, O., Szoka, Jr., F. C. (1996) Mechanism of oligonucleotide release from cationic liposomes, *Proc. Natl. Acad. Sci. U.S.A. 93*, 11493–11498.

42 Simoes, S., Moreira, J. N., Fonseca, C., Duzgunes, N., De Lima Pedroso, M. C. (2004) On the formulation of pH-sensitive liposomes with long circulation times. *Adv. Drug Deliv. Rev. 56*, 947–965.

43 Filion, M.C., Philip, N. C. (1998) Major limitations in the use of cationic liposomes for DNA delivery. *Int. J. Pharm. 162*, 159–170.

44 Boussif, O., Lezoualch, F., Zanta, M.A., Mergny, M.D., Scherman, D., Demeneix, B., Behr, J.P. (1995) A versatile vector for gene and oligonucleotide transfer into cells in culture and in-vivo-polyethylenimine. *Proc. Natl. Acad. Sci. U.S.A. 92*, 7297–7301.

45 Kichler, A., Leborgne, C., Coeytaux, E., Danos, O. (2001) Polyethylenimine-mediated gene delivery: a mechanistic study. *J. Gene Med. 3*, 135–144.

46 Klemm, A. R., Young, D., Lloyd, J. B. (1998) Effects of polyethyleneimine on endocytosis and lysosome stability. *Biochem. Pharmacol. 56*, 41–46.

47 Godbey, W. T., Wu, K. K., Mikos, A. G. (1999) Tracking the intracellular path of poly(ethylenimine)/DNA complexes for gene delivery. *Proc. Natl. Acad. Sci. U.S.A. 96*, 5177–5181.

48 Kircheis, R., Wightman, L., Wagner, E. (2001) Design and gene delivery activity of modified polyethylenimines. *Adv. Drug Deliv. Rev. 53*, 341–358.

49 Ogris, M., Brunner, S., Schuller, S., Kircheis, R., Wagner, E. (1999) Pegylated DNA/transferrin- PEI complexes: reduced interaction with blood components, extended circulation in blood and potential for systemic gene delivery. *Gene Ther. 6*, 595–605.

50 Petersen, H., Fechner, P. M., Martin, A. L., Kunath, K., Stolnik, S., Roberts, C. J., Fischer, D., Davies, M. C., Kissel, T. (2002) Polyethylenimine-graft-poly(ethylene glycol) copolymers: influence of copolymer block structure on DNA complexation and biological activities as gene delivery system. *Bioconjug. Chem. 13*, 845–854.

51 Oupicky, D., Ogris, M., Howard, K. A, Dash, P. R., Ulbrich, K., Seymour, L. W. (2002) Importance of lateral and steric stabilization of polyelectrolyte gene delivery vectors for extended systemic circulation. *Mol. Ther. 5*, 463–472.

52 Tomalia, D. A., Baker, H., Dewald, J., Hall, M., Kallos, G., Martin, S., Roeck, J., Ryder, J., Smith, P. (1985) A new class of polymers: starbust-dendritic macromolecules. *Polym. J. (Tokyo) 17*, 117–132.

53 Haensler, J., Szöka, F. (1993) Polyamidoamine cascade polymers mediate efficient transfection of cells in culture. *Bioconjug. Chem. 4*, 372–379.

54 Zinselmeyer, B. H., Mackay, S. P., Schatzlein, A. G., Uchegbu, I. F. (2002) The lower generation poly(propylenimine) dendrimers are effective gene transfer agents. *Pharm. Res. 19*, 960–967.

55 Choi, Y., Mecke, A., Orr, B. G., Holl, M. M. B., Baker, J. R., Jr. (2004) DNA-directed synthesis of generation 7 and 5 PMAM dendrimer nanoclusters. *Nano Lett. 4*, 391 397.

56 Santhakumaran, L. M., Thomas, T., Thomas, T. J. (2004) Enhanced cellular uptake of a triplex forming oligonucleotide by nanoparticle formation in the presence of polypropylenimine dendrimers. *Nucleic Acids Res. 32*, 2102–2112.

99 Wolfert, M. A., Seymour, L. W. (1996) Atomic force microscopic analysis of the influence of the molecular weight of poly-L-lysine on the size of polyelectrolyte complexes formed with DNA. *Gene Ther. 3*, 269–273.

100 Marschall, P. Malik, N., Larin, Z. (1999) Transfer of YACs up to 2.3 Mb intact into human cells with polyethylenimine. *Gene Ther. 6*, 1634–1637.

101 Sitko, J. C., Mateescu, E. M., Hansma, H. G. (2003) Sequence-dependent DNA condensation and the electrostatic zipper. *Biophys. J. 84*, 419–431.

102 Han, S. Mahato, R. I., Kim, S. W. (2001) Water-soluble lipopolymer for gene delivery. *Bioconjug. Chem. 12*, 337–345.

103 Lim, Y.-B. Han, S.-O., Kong, H.-U., Lee, Y., Park, J.-S., Jeong, B. K., Sung, W. (2000) Biodegradable polyester, poly[-(4-aminobutyl)-L-glycolic acid], as a non-toxic gene carrier. *Pharm. Res. 17*, 811–816.

104 Isobe, H., Sugiyama, S., Fukui, K., Iwasawa, Y., Nakamura, E. (2001) Atomic force microscope studies on condensation of plasmid DNA with functionalized fullerenes. *Angew. Chem. (Int. Ed.) 40*, 3364–3367.

105 Gallyamov, M. O., Pyshkina, O. A., Sergeyev, V. G., Yaminsky, I. V. (2000) T4 DNA condensation in water-alcohol media. *Poverkhnost 7*, 88–91.

106 Dimitriadis, E. K., Pascual, J., Horkay, F. (2004) Morphology of DNA adsorbed from solutions on hydrophobic surfaces. 227th ACS National Meeting, Anaheim, CA.

107 Koltover, I., Wagner K., Safinya, CR. (2001) DNA condensation in two dimensions. *Proc. Natl. Acad. Sci. U.S.A. 97*, 14046–14051.

108 Allen, M. J., Bradbury, E. M., Balhorn, R. (1997) AFM analysis of DNA-protamine complexes bound to mica. *Nucleic Acids Res. 25*, 2221–2226.

109 Martin, A. L., Davies, M.C., Rackstraw, B. J., Roberts, C. J., Stolnik, S., Tendler, S. J. B., Williams, P. M. (2000) Observation of DNA-polymer condensate formation in real time at a molecular level. *FEBS Lett. 480*, 106–112.

110 Fang, Y., Yang, J. (1997) Two-dimensional condensation of DNA molecules on cationic lipid membranes. *J. Phys. Chem. 101*, 441–449.

111 Fang, Y., Spisz, T. S., Hoh, J. H. (1999) Ethanol-induced structural transitions of DNA on mica. *Nucleic Acid Res. 27*, 1943–1949.

112 Sun, X.-G., Cao, E.-H, Zhang, X.-Y., Liu, D., Bai, C., (2002) The divalent cation-induced DNA condensation studied by atomic force microscopy and spectra analysis. *Inorg. Chem. Commun. 5*, 181–186.

113 Bloomfield, V. A., Crothers, D. M., Tinoco, I. Jr. (1999) *Nucleic Acids Structures, Properties, and Functions.* University Science Books, Sausalito, CA.

114 Manning, G. S., Ray, J (1998) Counterion condensation revisited. *J. Biomol. Struct. Dyn. 16*, 461–476.

115 Olson, W. K., Manning, G. S. (1976) A configurational interpretation of the axial phosphate spacing in polynucleotide helices and random coils. *Biopolymers 15*, 2391–2405.

116 Nguyen, T. T., Shklovskii, B. I. (2002) Model of inversion of DNA charge by a positive polymer: fractionalization of the polymer charge. *Phys. Rev. Lett. 89*, 018101.

117 Fenley, M. O., Manning, G. S., Marky, N. L., Olson, W. K. (1998) Excess counterion binding and ionic stability of kinked and branched DNA. *Biophys. Chem. 74*, 135–152.

118 Kankia, B. I., Buckin, V., Bloomfield, V. A. (2001) Hexammine cobalt(III)-induced condensation of calf thymus DNA: circular dichroism and hydration measurements. *Nucleic Acids Res. 29*, 2795–2801.

119 Polyanichko, A. M., Andrushchenko, V. V., Chikhirzhina, E. V., Vorob'ev, V. I., Wieser, H. (2004) The effect of manganese (II) on DNA structure: electronic and vibrational circular dichroism studies. *Nucleic Acids Res. 32*, 989–996.

120 Marx, K. A., Reynolds, T. C. (1983) Ion competition and micrococcal nuclease digestion studies of spermidine-condensed calf thymus DNA. Evidence for torus organization by circumferential DNA wrapping. *Biochim. Biophys. Acta 741*, 279–287.

121 Kornyshev, A. A., Leikin, S. (1998) Symmetry laws for interaction between helical macromolecules. *Biophys. J. 75*, 2513–2519.

122 Hud, N. V., Downing, K. H., Balhorn, R. (1995) A constant radius of curvature model for the organization of DNA in toroidal condensates. *Proc. Natl. Acad. Sci U.S.A. 92*, 3581–3585.

123 Pelta, J. Jr., Durand, D., Doucet, J., Livolant, F. (1996) DNA mesophases induced by spermidine: structural properties and biological implications. *Biophys. J. 71*, 48–63.

124 Yoshikawa, K., Takahashi, M., Vasilevskaya, V. V., Khokhlov, A. R. (1996) Large discrete transition in a single DNA molecule appears continuous in the ensemble. *Phys. Rev. Lett. 76*, 3029–3031.

125 Ubbink, J., Odijk, T. (1995) Polymer- and salt-induced toroids of hexagonal DNA. *Biophys. J. 68*, 54–61.

126 Cunha, S., Woldringh, C. L., Odijk T. (2001) Polymer-mediated compaction and internal dynamics of isolated *Escherichia coli* nucleoids. *J. Struct. Biol. 136*, 53–66.

127 de Vries R. (2001) Flexible polymer-induced condensation and bundle formation of DNA and F-actin filaments. *Biophys. J. 80*, 1186–1194.

128 Chesnoy, S., Huang, L. (2000) Structure and function of lipid-DNA complexes for gene delivery. *Annu. Rev. Biophys. Biomol. Struct. 29*, 27–47.

129 Hara, T., Tan, Y., Huang, L. (1997) *In vivo* gene delivery to the liver using reconstituted chylomicron remnants as a novel nonviral vector. *Proc. Natl. Acad. Sci. U.S.A. 94*, 14547 14552.

130 Monck, M. A., Mori, A., Lee, D., Tam, P., Wheeler, J. J., Cullis, P. R., Scherrer, P. (2000) Stabilized plasmid-lipid particles: pharmacokinetics and plasmid delivery to distal tumors following intravenous injection. *J. Drug Target. 7*, 439–452.

131 Chesnoy, S., Durand, D., Doucet, J., Stolz, D. B., Huang, L. (2001) Improved DNA/emulsion complex stabilized by poly (ethylene glycol) conjugated phospholipid. *Pharm. Res. 18*, 1480–1484.

132 Cui, Z., Mumper, R. J. (2002) Plasmid DNA-entrapped nanoparticles engineered from microemulsion precursors: in vitro and in vivo evaluation. *Bioconjug. Chem. 13*, 1319–1327.

133 Kaushik, A. (2001) Leuvectin Vical Inc. *Curr. Opin. Investig. Drugs. 2*, 976–981.

134 Hoffman, D. M., Figlin, R. A. (2000) Intratumoral interleukin 2 for renal-cell carcinoma by direct gene transfer of a plasmid DNA/DMRIE/DOPE lipid complex. *World J. Urol. 18*, 152–156.

135 Galanis, E., Hersh, E. M., Stopeck, A. T., Gonzalez, R., Burch, P., Spier, C., Akporiaye, E. T., Rinehart, J. J., Edmonson, J., Sobol, R. E., Forscher, C., Sondak, V. K., Lewis, B. D., Unger, E. C., O'Driscoll, M., Selk, L., Rubin, J. (1999) Immunotherapy of advanced malignancy by direct gene transfer of an interleukin-2 DNA/DMRIE/DOPE lipid complex: phase I/II experience. *J. Clin. Oncol. 17*, 3313–3323.

136 Ramesh, R., Saeki, T., Templeton, N. S., Ji, L., Stephens, L. C., Ito, I., Wilson, D. R., Wu, Z., Branch, C. D., Minna, J. D., Roth, J. A. (2001) Successful treatment of primary and disseminated human lung cancers by systemic delivery of tumor suppressor genes using an improved liposome vector. *Mol. Ther. 3*, 337–50.

137 Densmore, C. L. (2003) Polyethyleneimine-based gene therapy by inhalation. *Expert Opin. Biol. Ther. 3*, 1083–1092.

138 Ferrari, S., Griesenbach, U., Geddes, D. M., Alton, E. (2003) Immunological hurdles to lung gene therapy. *Clin. Exp. Immunol. 132*, 1–8.

139 Fratantoni, J. C., Dzekunov, S., Singh, V., and Liu, L. N. (2003) A non-viral gene delivery system designed for clinical use. *Cytotherapy 5*, 208–210.

140 Hosseinkhani, H., Tabata, Y. (2004) PEGylation enhances tumor targeting of plasmid DNA by an artificial cationized protein with repeated RGD sequences, Pronectin(R). *J. Control. Release 97*, 157–171.

141 Shchukin, D. G., Patel, A. A., Sukhorukov, G. B., Lvov, Y. M. (2004) Nanoassembly of biodegradable microcapsules for DNA encasing. *J. Am. Chem. Soc. 126*, 3374 3375.

142 Thompson, J. B., Hansma, H. G., Hansma, P. K., Plaxco, K. W. (2002) The backbone conformational entropy of protein folding: experimental measures from atomic force microscopy. *J. Mol. Biol. 322*, 645–652.

11
Nanoparticles for Cancer Drug Delivery

Carola Leuschner and Challa Kumar

11.1
Introduction

Nanotechnology involves materials and systems that measure between 1 and 100 nm in at least one dimension. Nanomaterials form a bridge between the realm of individual atoms and molecules and the macroworld. Nanoparticles are too small to simply obey classical physics and too large to be described by straightforward application of quantum mechanics. They show very often unforeseen properties and these properties can very often be modified by varying "just" the size of the particle. Nature has been utilizing nanostructures for billions of years. These two properties, (i) being about the size of "typical" biological objects and (ii) the possibility of tailoring their properties by changing their size, make nanoparticles attractive for biomedical applications.

Breathtaking developments in the area of nanotechnology (nanoparticles, nanofilms, nanotubes, etc.) over the last years have opened up avenues to dramatic changes in the way devices, materials, and systems are fabricated. This new phenomenon spanning various branches of science including chemistry, physics, biology, medicine, and engineering is expected to raise quality of life by several orders of magnitude [1]. Among the various forms of nanomaterials, there is a huge potential for the application of nanoparticles in the field of biomedicine as demonstrated by several publications over the last few years. Some of the biomedical applications for functionalized nanoparticles that are under intensive investigation by several research groups are: (i) sensors for detection of biomolecular interactions [2]; (ii) cancer treatment using magnetic drug targeting [3]; (iii) contrast agents in magnetic resonance imaging [4]; and (iv) magnetic carriers for cell separation, immobilization of enzymes, and extracorporeal blood purification [5]. Magnetic nanoparticles form the basis for all these applications. The other type of application depends on functionalization of the nanoparticles.

A thorough literature search has indicated that even though there are several research publications on the application of nanoparticles in cancer therapy and diagnosis, to the best of the present authors' knowledge there is no complete review of the existing literature. This review focuses on how nanoparticles are being used as vehicles in the chemotherapy of cancer. Literature relating to the use of liposomes,

Nanofabrication Towards Biomedical Applications. C. S. S. R. Kumar, J. Hormes, C. Leuschner (Eds.)
Copyright © 2005 WILEY-VCH Verlag GmbH & Co. KGaA, Weinheim
ISBN 3-527-31115-7

micelles, or dendrimers is beyond the scope of this review, as is the application of nanoparticles in cancer diagnosis.[FettU]

11.2
Cancer: A Fatal Disease and Current Approaches to Its Cure

Cancer is the second leading cause of death in the USA≪1, claiming a total of 553,768 lives per year. Despite new discoveries of drugs and treatment combinations for cancer, the mortality rate has not changed in the past 53 years. However, the overall survival rate has been increased by 13% since 1974, due to more sensitive diagnostic techniques and early detection methods. Of the 699,560 men and 668,470 women predicted to be diagnosed with cancer in the year 2004, 64% are expected to survive [6].

The second leading cause of cancer death is prostatic adenocarcinoma in men and mammary carcinoma in women. Both these cancers develop mixed populations of hormone (estrogen/androgen)-dependent and hormone-independent cells, which are poorly to well differentiated and have variable proliferation rates. In cases of organ-confined disease, radical prostatectomy is the preferred treatment in men and mastectomy in women. Initially patients are treated with radiation and/or chemotherapy (cyclophosphamide, doxorubicin, 5-fluorouracil, [7, 8], in addition to hormone ablation. Although androgen ablation in prostate cancer patients (LHRH agonists, leuprolide [9]) leads to a reduction of the primary tumor and to its partial regression, within 2 years the disease can re-emerge in a poorly differentiated, androgen-independent form, after which no therapy is available to prolong the life of the patient [10]. In breast cancer patients estrogen ablation is achieved by tamoxifen treatment [11] or ovariectomy. Surgical removal of the primary tumor is invasive and has side effects depending on the area of resection and on the tumor (whether encapsulated or invasive). Often, removal of the primary tumor can lead to increased proliferation of metastases, single metastatic cells, and dormant cancer cells.

Up to 70% of prostate cancer patients [12] and 40% of breast cancer patients have occult metastases at the time of diagnosis [13]. Bone and lymph node metastases occur in 26% of patients with prostate adenocarcinoma [14] and in 23% of breast cancer patients [13]. More than 70% of patients die from skeletal metastases [15] (Fig. 11.1).

Current treatments can only prolong the life of the patients, but do not cure. Because of the high toxicity and poor specificity of currently used drugs, increasing the chemotherapeutic dosages is not possible. Often, patients in relapse do not respond to chemotherapy due to an acquired drug resistance which decreases the efficacy of chemotherapy. Dormant metastatic disease cannot be targeted by chemotherapy, and proliferation of occult breast cancer cells in the bone marrow is responsible for a relapse of at least one-third of the patients [13, 16, 17].

Currently used chemotherapeutic drugs are systemically active and cannot target single dormant cancer cells, tumors, or slow-growing tumors [13, 16, 18], as most

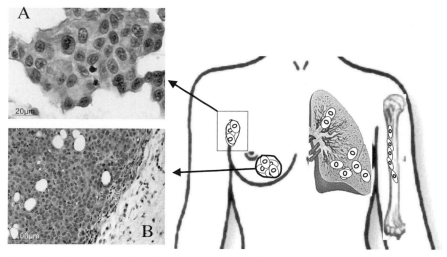

Figure 11.1. Single cell seeding from primary tumors leads to lymph node and bone metastases. Histological sections from (A) lymph node metastasis from primary breast cancer tumor and (B) primary breast cancer tumor.

chemotherapeutic drugs are not selective for cancer cells. To reach therapeutically effective concentration at the tumor site, therefore, high systemic dosages are required, which cause severe side effects and peripheral tissue damage. Among these side effects are destruction of bone marrow cells (which impairs the production of erythrocytes), cardiotoxicity, nephrotoxicity, hepatotoxicity, and hematotoxicity. Bone marrow cells produce erythrocytes for oxygen transport, so patients become anemic as a result of bone marrow destruction. Hematotoxicity includes the destruction of platelets needed for blood clotting and leukocytes to fight infections. Immediate side effects are nausea, alopecia, and fatigue. Longer-lasting effects are the increased susceptibility to infections, which persists until the immune system has recovered from the chemotherapy (4–6 weeks).

Most of the chemotherapeutic drugs interfere with the proliferation machinery of the cancer cells. Cyclophosphamide (Cytoxan) destroys genetic material which controls tumor cell growth. Methotrexate and 5-fluorouracil (5FU) are antimetabolites which interfere with cancer cell division. Antimicrotubule reagents prevent cell division by acting on the microtubules; among these are paclitaxel (Taxol), docetaxel (Taxotere), vincristine (Oncovin), and vinblastine (Velban). Doxorubicin (Adriamycin) is a tumor antibiotic. In order to be effective these drugs need to accumulate in the tumor cells.

High-dose chemotherapy does not necessarily cure the patient. Many patients relapse. Recurrence of the cancer occurs after high-dose chemotherapy, often manifested by the lack of response to further chemotherapy, which in turn leads to terminal disease even after several years of apparent freedom from disease. This phenomenon is defined as multidrug resistance and is attributed to multiple mechanisms [19, 20].

Most cancers, such as colon, kidney, breast, ovarian, prostatic, and lung cancers, overexpress the p-glycoprotein gene (also known as the multidrug resistance gene). Its gene product is the protein Pgp, which is located in membranes, Golgi apparatus, and nucleus [21, 22]. The p-glycoprotein (Pgp) is a transmembrane efflux pump that actively excretes cytotoxic drugs of different molecular structure through an ATPase mechanism [23]. decreasing intracellular drug concentrations. Modulators for Pgp pumps have been reported and include verapamil, a compound which competes with the drug efflux pump. During differentiation and progression, cancer cells can acquire the ability to remove the administered drug through such a pump mechanism. Other mechanisms include alteration of enzymatic activities such as topoisomerase or glutathione S-reductase activity, altered apoptosis regulation, altered transport, or alteration of intracellular drug distribution due to increased sequestration of cytoplasmic vesicles [24].

The severity of side effects during administration of chemotherapeutic drugs and the occurrence of multidrug resistance emerges as nonresponsive or refractory disease to chemotherapeutic drugs.

11.3
Characteristics of Tumor Tissues

Tumor tissue consists of various structures and areas of which the actual cancer cells can occupy less than 50%, the vasculature 1–10%. The remaining structure consists of a collagen-rich matrix. Histological evaluation of excised tumors revealed rapidly growing regions interspersed with necrotic regions [25]. Tumors develop a tortuous, chaotic capillary network that distinguishes it from normal vasculature, which follows a hierarchic branching pattern [26]. The capillary vasculature in tumors is often accompanied by occlusions, caused by rapidly proliferating cancer cells. Compression of the vasculature [27] causes hypoxia and eventually necrosis of viable tumor cells. In contrast to normal tissue, tumors often lack a functional lymphatic system. In addition to a high proportion of proliferating endothelial cells, tumor vasculature shows aberrant basement membranes. Tumor blood vessels differ from normal vasculature in being up to 3–10 times more permeable [28], to counterbalance the high oxygen and nutrient requirements of the fast-growing tumor [29] (Fig. 11.2).

Macromolecules and drugs are transported into the tumor cells through interendothelial junctions and vesicular vacuolar organelles and fenestrations. The range of pore cutoff size in tumor tissue has been reported at between 100 and 780 nm [30, 31]. This range has been confirmed by measurements made *in vivo* through fluorescence microscopy [32]. The pore size can be increased up to 800 nm by perfusion with low dosages (10 µg/ml) of vascular endothelial growth factor (VEGF) in human colon tumors [33]. In a mouse model bearing human colon cancer xenografts, the extravasation of polyethyleneglycol-stabilized liposomes of diameters from 100 to 400 nm was examined. VEGF perfusion increased the frequency of 400-nm pores in the tumor vasculature and increased the number of transvascular pathways, open

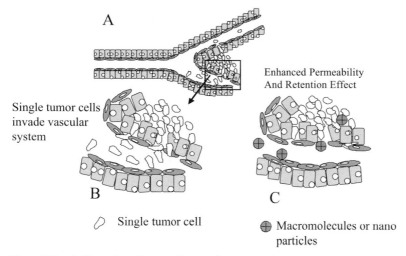

A

Single tumor cells
invade vascular
system

Enhanced Permeability
And Retention Effect

B

C

⬭ Single tumor cell

⊕ Macromolecules or nano
particles

Figure 11.2. A. Normal vaculature with part of cancerous tissue
(box). B. Single tumor cells pass through the hypermpermeable
vasculature. C. Enhanced permeability and retention effect. Grey
circles represent nanoparticles or macromolecules.

junctions, and fenestrae. For comparison, normal vasculature endothelium diameter is less than 2 nm, 6 nm in postcapillary venules, 40–60 nm in kidney glomeruli and 150 nm in sinusoidal epithelium of liver and spleen.

11.4
Drug Delivery to Tumors

The interstitial compartment of a tumor contains a network of collagen and elastic fiber, which is immersed by hyaluronate and proteoglycan-containing fluid. The interstitial pressure within the tumor tissue is elevated due to the lack of a lymphatic drainage system [34, 35]. Increased interstitial pressure and rapid aberrant cell growth are believed to be responsible for the compression and occlusion of blood and lymphatic vessels in solid tumors [25] and for hindering the extravasations and accumulation of the drugs in the tumor tissue.

The transport of a drug into the tumor area is dependent on the interstitial pressure as well on its composition, charge, and the characteristics of the drug (hydrophobicity, size) [36] as the interstitial pressure is higher than the pressure within the blood vessels [37]. In addition the dense packing of tumor cells limits the movement of molecules from the vessel into the interstitial compartment [38]. For example, colloidal particles larger than 50 kDa enter the interstitial compartment through leaky vessels and accumulate in the tumor tissue. This phenomenon is called the *enhanced permeability and retention* (EPR) effect. The EPR effect is characterized by an accumulation of a drug in the interstitium, exceeding the drug concentration in the plasma [39, 40], which then in turn hinders diffusion of the drug into the interstitial compartment.

The lack of tumor response to a drug (multidrug resistance) can be due to poorly vascularized tumor regions and lack of drug accumulation in the tumor cells at a therapeutically effective concentration. Interaction of macromolecules (drug) with plasma can inactivate the drug by degradation or hydrolysis. The acidic environment of the tumor can inactivate basic molecules by ionization and thereby prevent their diffusion across the cell membrane. Altered drugs can loose their therapeutic efficacy. This cannot be avoided by increasing the dosage of an administered drug to reach the therapeutic effective plasma concentration, because most chemotherapeutics are highly toxic and lead to severe systemic side effects.

It is apparent that the required therapeutic effective drug concentration in the tumor tissue is difficult to achieve and maintain because of vascular and lymphatic characteristics of the tumor tissue. The combination of chemotherapeutics and nanoparticle technology could reduce some of the currently observed problems encountered with systemic treatment regimens, e.g., protecting the drug from degradation, increasing the solubility of lipophilic drugs, increasing the specificity of drugs, and circumventing multidrug resistance.

11.5
Physicochemical Properties of Nanoparticles in Cancer Therapy

Nanoparticles are spherical particles, whether naked or functionalized, measuring less than 100 nm in at least one dimension. It is not surprising that nanoparticles have implications in biological systems in general and as drug delivery systems in particular since most cells are 10,000–20,000 nm in diameter. Nanoparticles can enter cells, even nuclear compartments, and interact with DNA and cellular proteins. They can be fabricated at different sizes and surface modifications, which determines their properties in biological systems. Particles less than 100 nm have longer circulation times, large effective surface areas, low sedimentation rates, and they show enhanced diffusion potential and are easily internalized in tumor cells through the membrane pores. Agglomeration of nanoparticles has to be avoided to prevent thrombosis [41]. Small particles have access to capillaries and are more resistant to the macrophage uptake of the reticuloendothelial system [42–44].

Nanoparticles applicable for cancer treatments need to counter the adverse effects of the current chemotherapeutic approaches described in the previous section. In general, nanoparticles have the potential to improve cancer drug delivery and provide tools to

- deliver the pharmacologically required concentration of the drug
- increase drug concentration at the target site through extended or controlled release
- overcome multidrug resistance
- eradicate side effects to vital organs by reducing systemic exposure
- avoid immune response and hematopoietic toxicity
- destroy malignant cells specifically, sparing normal cells
- kill primary tumors inaccessible to surgery

- destroy seeded cancer and dormant cells and metastases
- protect the active drug from alteration and inactivation
- detect cancers at a early stage.

Figure 11.3 summarizes different types of nanoparticles that have the potential to improve various aspects of cancer treatment. Nanoparticles currently under development for cancer treatment can be classified into four groups:

- Magnetic nanoparticles
- Polymeric nanoparticles
- Fluorescent nanoparticles or quantum dots
- Silica nanoparticles

Some important elements of cancer treatment that are affected by nanoparticles can be broadly classified as:

- Site-specific delivery of drugs
- Controlled release of drugs
- Protection of anticancer agents or drugs
- Treatment using hyperthermia
- Prevention of multi-drug resistance

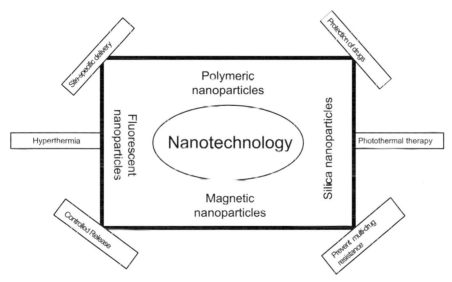

Figure 11.3. Applications of nanotechnology for cancer drug delivery and treatment.

Factors important for *in vivo* application of nanoparticles that should be taken into account in the selection of materials and fabrication of particles are:

- Biocompatibility of particles and coatings
- Particle size

- Immunogenicity
- Surface properties
- Degradation properties
- Drug loading capacities and release
- Stability of the drug during encapsulation
- Storage and use of the fabricated nanoparticles

11.5.1
In Vivo Circulation Pathways of Nanoparticles

Nanoparticles injected into biological systems are rapidly coated with plasma proteins such as immunoglobulins and fibronectin and build aggregates. This process is called opsonization. Opsonized particles are recognized by the reticuloendothelial system (RES) or mononuclear phagocytic system (MPS), which is comprised of macrophages related to liver (Kupffer cells), spleen, lymph nodes (perivascular macrophages), nervous system (microglia), and bones (osteoclasts) [45, 46]. The RES is a defense system and comprises highly phagocytotic cells derived from bone marrow. These cells travel in the vascular system as monocytes and reside in their particular tissues.

These macrophages [47] internalize the opsonized nanoparticles through phagocytosis and deliver them to the liver, spleen, kidney, lymph node, and bone marrow (Fig. 11.4). This clearance can occur within 0.5–5 min, [48], thus removing the active nanoparticles from the circulation and prevent their access to the tumor tissue. Coating the nanoparticles and reducing their size to less than 100 nm can mask them that they are no longer recognized by the MPS and remain in circulation for longer [49]. Rather than normal tissue, such nanoparticles preferentially access tumors through their hyperpermeable vasculature [50], which can be regulated by

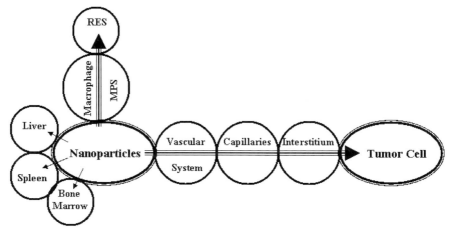

Figure 11.4. Distribution and routes of nanoparticles after injection.

low dosages of VEGF [33]. Nanoparticles accumulate in the interstitial compartment of the tumors. The retention of nanoparticles is mainly due to the lack of lymphatic clearance in the tumor tissue (the EPR effect) [39, 40, 51]. If nanoparticles are biodegraded in the interstitium, the drug can be released and enter the tumor cells through diffusion.

Cellular uptake mechanisms for nanoparticles and macromolecules are pinocytosis [52], endocytosis [53, 54], and receptor-mediated endocytosis [55] (Fig. 11.5). Macromolecules and DNA, which are susceptible to lysosomal degradation, can be delivered by nanoparticles which escape lysosomal degradation. Polylactide glycolic acid (PLGA) particles are nonspecifically transported into the cells by fluid phase pinocytosis, a process which is mediated by clathrin (Fig. 11.5). At pH 7–7.5 (physiological pH), negatively charged PLGA nanoparticles are transported to primary endosomes, become positively charged, and enter the acidic secondary endosomes and lysosomes, from which the nanoparticles escape into the cytosol. Nanoparticles transported to early or primary endosomes enter the secondary endosomes and become cationic. Local interaction with the membrane releases the particles into the cytoplasm, escaping the lysosomal compartment [56].

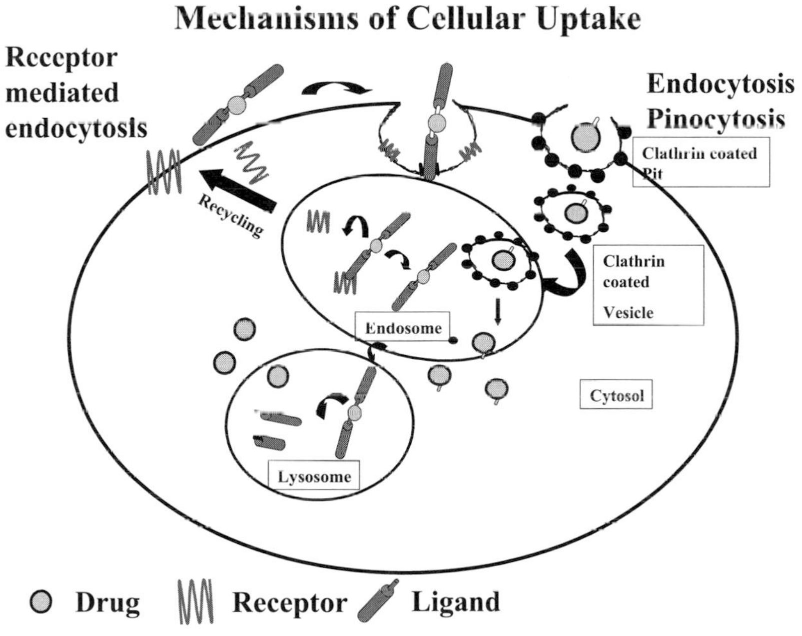

Figure 11.5. Mechanisms of drug uptake into tumor cells: phagocytosis, clathrin-coated pits, endocytosis, and targeted delivery via receptor mediated endocytosis.

11.5.2
Surface Treatment or Coating of Nanoparticles

Through coating with biodegradable matrices, nanoparticles become "invisible" to macrophages. The choice of hydrophilic or hydrophobic matrices for coating determines the fate of the nanoparticles. Hydrophilic coating prevents interaction of nanoparticles with macrophages of the RES, reduces their removal from the circulation, and increases their circulation half-life [57–59]. Hydrophobic coatings are applied to increase opsonization, leading to copious interaction with macrophages, and the nanoparticles are therefore rapidly removed from the circulation. The latter approach is applied for targeted delivery of nanoparticles to the RES of liver and spleen.

Hydrophilic coatings are dextran [60], polyethylene glycol (PEG) [61], polyethylene oxide (PEO), poloxamers and poloxamines [62], and silicones [59, 60, 63]. PEG coating resulted in enhanced circulation time of the particles [64] and reduced opsonization [65, 66]. Ishiwata et al. showed that PEG coating resulted in suppression of macrophage interaction [67].

The concept of particle coating seems promising, although *in vivo* applications need to be optimized. The choice of the coating polymer can prolong the half-life of the particle in circulation. Polymethylmethacrylate (PMMA) nanospheres coated with polyoxamer 407 or poloxamine 908 showed increased retention and accumulation in B16 and Mtu tumors in mice [68]. However, PMMA is not biodegradable and has no further application *in vivo*.

Amphiphilic copolymers like polylactic acid, poly ε-caprolactone, and polycyanoacrylate were coupled with PEG [69–71], appear to have high circulation profiles, and have not yet been used for tumor targeting.

11.5.3
Polymers for Encapsulation

Nanoparticles as drug carriers can be prepared in two different approaches. One is as a drug reservoir, which consists of an oily core as vehicle, which carries the drug, and a polymeric outer core layer with a coating. Vehicles are vegetable oil, triglycerides, or cotton seed oil. In the other approach, the particles are nanospheres in which the drug is dispersed in a polymeric matrix. Nanoparticles and nanospheres can be prepared from synthetic biodegradable polymers such polyvinylpyrrolidone (PVP) [72], chitosan [73], or polyalkylcyanoacrylates and polylactides [74–77] such as polyisohexylcyanoacrylate (PIHCA), polyethylcyanoacrylate, and polyisobutylcyanoacrylate (PIBCA). PLGA, which has been approved by the Food and Drug Administration for human application, degrades slowly, releasing the drug, and is therefore used for controlled release. Breakdown products are lactic and glycolic acids, which are metabolized through the Krebs cycle. A nonbiodegradable polymer is PMMA, which is not suitable for *in vivo* injection. Preparations of nanocapsules have been described and characterized by Puisieux and Couvreur et al. [75–77]. Depending on the preparation method, nanocapsules of different sizes have been prepared. An advantage of synthetic biodegradable polymers lies in the fact that their degradation rate can be predicted and manipulated by the choice of co-polymer composition.

One of the most important characteristics of nanoparticles is their ability to encapsulate drugs. This feature can reduce systemic exposure and deliver pharmaceutically effective concentrations of a drug to the target. Typically, drugs against cancer are administered systemically at high dosages to ensure a therapeutically effective concentration at the tumor site. Commonly, dosages of chemotherapeutics are in the 100-μg range; sometimes they are as high as 1 g per day. In the past 30 years a number of drug delivery devices have been developed, ranging from macro-sized (1 mm) to micro- (100–0.1 μm) and nano-sized [78, 79]

Macromolecules or drugs can become altered during plasma exposure and thus lose their potency. Systemic exposure leads to severe side effects as the drugs destroy nonmalignant tissue. Encapsulation of the drug offers a solution for drug protection. Drug encapsulation in a lipophilic environment may even enhance the solubility of a lipophilic drug and reduce severe systemic side effects. Liposomes or polymeric nanoparticles or nanocapsules have been tested *in vitro* and *in vivo* for their ability to house drugs or DNA, and their ability to deliver the drugs has been investigated.

11.6
Site-Specific Delivery of Chemotherapeutic Agents Using Nanoparticles

Using nanoparticles to deliver drugs to tumors offers an attractive possibility to avoid obstacles that occur during conventional systemic drug administration. However, new obstacles come into play when nanoparticles are introduced into the system. The route of administration, size of the particles, degradable coatings, biologically acceptable coatings, endocytotic properties, composition of particles, and stability in physiological salinity all determine the fate of nanoparticles. Nanoparticles injected into biological systems should not agglomerate, in order to avoid macrophage uptake and thrombosis. Enhanced accumulation of nanoparticles at the target can be achieved by attaching ligands to the surface of the nanoparticles or, in the case of magnetic nanoparticles, by using an external magnetic field to concentrate them in the area of the tumors.

Nanoparticles can be directed to the tumors by passive or active targeting. Passive targeting includes manipulation of the size and/or hydrophobicity or other physicochemical characteristics and can be applied to target the RES; active targeting involves the direction of magnetic particles by using an external magnetic field or by using ligand-conjugated nanoparticles. The following outline provides examples for the *in vivo* application of nanoparticles with various coatings and various materials. Polymer-complexed micelles are not discussed in this review.

11.6.1
Passive Targeting

Biodegradable nanoparticle systems have been developed to target the lymphatic system; this can be exploited to deliver drugs to the lymph nodes and destroy lymph node metastases and prevent further metastatic progression. Targeting lymphatic systems (to avoid hepatic drug degradation) can be achieved by using different routes of administration: intramuscular, subcutaneous, intraperitoneal, or oral. Coating nanoparticles is advantageous to enhance circulation time and avoid their recognition by the MPS, in turn increasing their uptake and accumulation in tumor. Therefore, long-circulating nanoparticles can reach targets outside the MPS [80].

The importance and effects of different coating materials in passive targeting are indicated in the following examples.

11.6.1.1 Targeting Lymph Nodes with Nanoparticles

Hydrophobic coating further enhances lymph node accumulation, because the nanoparticles are incorporated by the MPS and delivered to the lymph nodes. Poly-acrylcyanoacrylate particles loaded with insulin have been orally administered to rats and were incorporated into the lymphatic system via the Peyer's patches in the intestinal lining [81].

The behavior and biodistribution of nanoparticles and egg phosphatidyl-choline (EPC) or phospholipid (PL) emulsions as drug carrier have been compared and revealed that EPC and PL were cleared faster from the injection site: the lymphatic retention time was 17–24 h compared to 131 h in the case of the nanocapsules. Nanocapsules coated with PIBCA were more stable in plasma than the emulsion particles, because the nanoparticles were protected from lipolysis. PIBCA nanoparticles were largely incorporated into lymphocytes, whereas only a small portion of the emulsion particles were detectable in lymphocytes [82]

Nanoparticles of different materials can result in various biodistribution of the drug and therefore alter drug efficacies. Cucurbitacin BE polylactic acid nanoparticles (47–120 nm) were developed for delivery to cervical lymph nodes. The nanoparticles were loaded with 23% of the drug. Since polylactic acid is biodegradable, the release in vivo was slow and the acute toxicity of the cucurbitacin was lowered by 50% [82, 83].

11.6.1.2 Increasing Bioavailability of a Compound

Highly hydrophilic compounds are rapidly excreted after systemic injection; their efficacy can be diminished and they may even lose their potency.

Gadolinium neutron capture therapy (GdNCT) is a two-step radiotherapy. Gd-157 emits gamma radiation as a result of its neutron capture reaction from an external neutron source, which inactivates the tumor tissue. Commercially available Gd formulas such as Magnevist® – a dimeglumine gadopentetate aqueous solution – are highly hydrophilic, poorly retained in tumors, and rapidly excreted even after intra-tumoral injection. In order to be effective, Gd has to be delivered and retained at the tumor site at high concentrations. Nanoparticles loaded with gadopentaacetic acid were fabricated from chitosan poly[β-(1–4)-2amino-2-deoxy-D-glucopyranose]. As a naturally occurring polysaccharide, chitosan is biodegradable, bioadhesive, and biocompatible. The Gd-nanoCP = Gadoliniumpentetic acid in chitosan nanoparticles were fabricated by emulsion droplet coalescence technique, with a particle size of 426 nm, and contained 9.3% Gd. Gd nanoCP particles entered tumor cells through endocytosis and in a comparison to Magnevist® significantly enhanced Gd accumulation and retention, by a factor of 200 in B16F10 melanoma cells and SCC-VII squamous cell carcinoma *in vitro* [84]. Mice bearing subcutaneous B16F10 melanoma were injected intratumorally with the Gd-nanoCP nanoparticles containing 1200 µg natural gadolinium. After thermal neutron irradiation was performed at the tumor site, tumor growth in the nanoparticle-administered group was significantly suppressed compared to that in the gadopentetate solution-administered group. However, a complete treatment was not achieved due to the uneven distribution of Gd in the tumor tissue. This study demonstrated the potential usefulness of Gd-NCT using Gd-loaded nanoparticles [85]. Watanabe et al. tested Gd lipid nanoemulsions (Gd-nanoLE) synthesized from hydrogenated phosphatidyl choline, and gadolinium diethylenetetraaminepentaacetic acid (Gd-DTPA), which were surface-treated with polyoxyethylene to create a hydrophilic moiety. The Gd-nanoLE Gd-nanoLE = Gd lipid nanoemulsion; LE = lipid emulsionparticles (70–90 nm) were injected intravenously and intraperitoneally into melanoma-bearing hamsters (1.5 mg ml^{-1} Gd). Intravenous injection was advantageous over intraperitoneal injection with respect to bioavailability, tumor retention, and accumulation. When injected intraperitoneally the bioavailability was reduced to 57% compared to intravenous injection. Tumor accumulation was 49.7 µg Gd per gram of tumor at 24 h compared to 21 µg Gd per gram of tumor at 12 h in the intraperitoneally treated groups. However, intravenous administration resulted in higher Gd accumulation in liver, spleen, lung, and kidney compared to intraperitoneally injected groups. Repeated injection with a two-fold enriched formulation led to 100 µg Gd per gram of tissue [86]. The increased efficacy was due to prolonged circulation of the particles, reduced interaction with the RES, reduced excretion of the compound Gd, and increased retention in the tumor tissue.

Polymeric nanoparticles prepared through the polymer–metal complex formation between cisplatinum (CDDP) and poly(ethylene glycol)-poly(glutamic acid) block copolymers were tested for their efficacy in delivering cisplatinum to tumors. The nanoparticles (CDDP/m) had a size of 28 nm and exhibited sustained release *in vitro*. Lewis lung carcinoma-bearing mice were injected intravenously with free CDDP or CDDP/m 4 mg kg^{-1}. CDDP/m had prolonged blood circulation time, and accumulated in the tumors at a 20-fold higher rate compared to free CDDP, whereas the accumulation in normal tissue was reduced. Complete tumor regression was observed only in mice treated with CDDP/m; no change of body weight occurred during treatment. Free CDDP treatment at the same dosage caused 20% body weight loss and retained tumor survival [87]. These data clearly show the advantages of drug encapsulation in respect of increased bioavailability, increased target accumulation, and reduced accumulation in normal organs.

PIBCA nanospheres containing mitoxantrone coated with poloamine [88] accumulated in tumors of melanoma-bearing mice when injected intravenously. In this study encapsulation of mitoxantrone did not change the biodistribution of the drug compared to free mitoxantrone. The plasma doxorubicin concentrations in rats injected intravenously with polysorbate-80-coated PIBCA nanoparticles loaded with doxorubicin was $0.1\,\mu g\,g^{-1}$ compared to $6\,\mu g\,g^{-1}$ in the brain 2–4 h after intravenous administration. The nanoparticles were able to pass the blood–brain barrier which is possible through apolipoprotein E adsorption and low-density lipoprotein-receptor-mediated transport [89, 90]. PEG-coated hexadecylcyanoacrylate particles accumulated three-fold compared to uncoated particles in the rat brain [91].

Irinotecan was encapsulated in polylactic nanoparticles from PEG-PPG-PEG block polymers. The drug content of the nanoparticles was 4.5%, the size distribution 80–210 nm. These particles were injected into sarcoma-bearing mice and rats. Irinotecan was antitumorigenic only in the encapsulated form when injected intravenously or subcutaneously. The plasma concentration of irinotecan in nanoparticle injections was increased compared to free CPT 11 = Irinotecan [92]. The small size of particles increased the tumor uptake and therefore improved the chemotherapy.

Taxol is one of the most potent antineoplastic drugs. However, its therapeutic effect has been limited due to poor solubility. Passive targeting by size selection has been tested *in vivo* on B16F10 melanoma-bearing mice, which were injected intravenously with PVP nanospheres (60 nm) containing paclitaxel (Taxol). Repeated injection resulted in increased survival and reduction of tumor mass compared to treatments with free paclitaxel [72]. This approach is advantageous as paclitaxel shows poor aqueous solubility. Prolonged therapeutic concentration is required to achieve clinical efficacy. Paclitaxel-loaded PLGA nanoparticles with polymer coatings are biodegradable. When intravenously injected the drug was released in a biphasic pattern, with an initial fast release followed by a slower continous release. The fast release may be due to dissolution diffusion of the drug and poor entrapment, whereas the slower release can be attributed to diffusion through the PLGA core. *In vitro* toxicity was enhanced 10-fold in the PLGA particle approach compared to free paclitaxel in a human small cell lung cancer cell line. Since paclitaxel was more active with cells in the G2/M cell cycle, longer incubation times are more effective. The use of different copolymers in the nanoparticle preparation resulted in prolonged release time and increased the efficacy of the treatment [93].

Paclitaxel has been encapsulated in gelatine nanoparticles of diameter range 600–1000 nm. The large size was chosen to ensure residence of the particles in the urinary bladder, where they were implanted and rapidly released the encapsulated drug. The IC_{50} was comparable to that of free paclitaxel [94]. This approach prevented systemic drug exposure and therefore prevented the side effects.

A commercially available formulation for i.v. administration is a formulation of paclitaxel in Cremophor EL and alcohol (50:50). The solvent is incompatible with PVC infusion tubing and causes severe side effects such as nephrotoxicity and cardiotoxicity; it could only be administered after steroid and antihistamine premedication [95–97]. The administration was long, requiring up to 3 h infusion time.

In a new approach Cremophor was replaced by biodegradable PLGA nanoparticles containing d-α-tocopherylpolyethyleneglycol 1000 succinate (TPGS) as emulsifier/stabilizer to dissolve paclitaxel for oral administration. The drug encapsulation efficiency was 100% and the release kinetic controllable. The particle size varied from 300 to 1000 nm [98].

Tocosol, a vitamin-E-based paclitaxel emulsion, has been tested in a phase II clinical trial. Taxane-naïve patients with ovarian cancer, bladder cancer, colorectal cancer, or non-small cell lung cancer received a 15-min infusion of Tocosol once a week. The treatment was well tolerated and steroid pretreatment was no longer required. A response was seen in 4–26% of patients; half of the patients had stable disease [99].

In a phase I clinical trial using ABI-007, an albumin nanoparticle formulation, encapsulated paclitaxel was administered intravenously to patients. The infusion rate was increased six-fold, no premedication was required, and a higher maximum tolerated dosage (MTD) was achieved during encapsulation compared to the free drug [100]. Hypersensitivity reactions were absent, there was a lower incidence of myelosuppression, and mild hematologic toxicity was observed compared to the free drug. As of 2003, ABI-007 is in phase III clinical trials for metastatic breast cancer.

ABI-007 has also been tested for pulmonary delivery in rats through intratracheal administration. The administered dosage was 5 mg paclitaxel per kilogram body weight. Biodistribution was determined using tritium-labeled paclitaxel/ABI-007. Maximum drug concentration was achieved within 5 min after administration, mean absorption half-life was 0.01 h, and elimination time was 4.7 h. After 10 min, 28% of the drug was discovered in the lungs, less than 1% was detected in other tissue [101].

Oral administration of ABI-007 was tested in rats and resulted in 45% accumulation; this was increased to 100% accumulation in the presence of cyclosporin A. ABI-007/cyclosponrin A/paclitaxel reached maximum concentration six times faster (within 30 min) compared to free paclitaxel. Interestingly, during oral administration saturation was not achieved. Rapid oral uptake and increased bioavailability (100% vs. 38%) suggests a novel mechanism of oral drug uptake [102].

The pharmacologically effective drug concentration in the tumor is a prerequisite for successful treatment, yet difficult to achieve in systemic treatments. Delivery of tamoxifen to estrogen-receptor-positive breast cancer cells was conducted by poly ε-caprolactone nanoparticles. *In vitro* the intracellular distribution of tamoxifen containing poly ε-caprolactone nanoparticles was found in the perinuclear region in MCF-7 cells after 1 h. The intracellular concentration of tamoxifen was saturable [103].

The antiestrogen RU 58668 was encapsulated in biodegradable nanocapsules (110 nm) and nanospheres (250 nm) consisting of PEG-coated polylactic acid nanoparticles. These particles were intravenously injected into MCF-7 xenograft-bearing mice. Coating with PEG prolonged the antiestrogenic potency of RU 58668 compared to noncoated nanoparticles. The antitumoral activity of the encapsulated drug with PEGylated nanocapsules was stronger than from the nanosphere formulation. The particles accumulated in liver, spleen, and tumors [104]

11.6.2
Active Targeting

11.6.2.1 Magnetically Directed Targeting to Tumor Tissue[FettU]
Magnetic nanoparticles were first reported by Gilchrist et al. in 1957. These authors studied the hyperthermic effect of exposure to a magnetic field (1.2 MHz) on tissue, which were immersed in ferrofluids of 20–100 nm (γ-Fe_2O_3) [105]. Magnetic nano-

particles can be composed of iron oxide, magnetite, or nickel, cobalt, or neodymium–iron boron oxides. All of these particles are magnetic or superparamagnetic, and range in size between 10 and 200 nm. Magnetite (Fe_3O_4) and maghemite (γ-Fe_2O_3) are preferentially used as they are biocompatible and not toxic to humans. Nickel and cobalt, although highly magnetic, are not suitable for administration to humans due to their toxicity [106].

Iron oxide is easily degradable and therefore useful for *in vivo* applications. Ferrous nanoparticles are biologically safe. They are metabolized into elemental iron and oxygen by hydrolytic enzymes, the iron joins the normal body stores and is subsequently incorporated into hemoglobin. Iron homeostasis is controlled by absorption, excretion, and storage. Acute toxicity has not been observed; in rats the administered iron was excreted over a period of 4 weeks. In human clinical trials no toxicity was observed [107]. Renal function, hepatic parameters, serum electrolytes, and lactate dehydrogenase remained unchanged from baseline parameters after treatment with ferrofluids [107]. The elevation of serum iron levels persisted for a maximum of 48 h and caused no symptoms. In rats, iron particles (250 mg kg^{-1}) injected intravenously were without any side effects [108]. In patients 2.6 mg iron particles were without any side effects [109].

Numerous studies are in progress to target tumors *in vivo* for therapeutic applications such as drug delivery and hyperthermic destruction of tumor. Other studies are pursuing the use of nanoparticles for diagnostic applications, such as their use as contrast agents in NMR and MRI. Upon application of an external magnetic field (EMF) the particles can be guided to the tumor tissue, and they should be immobilized when the EMF is then removed. Coating the surface of magnetic nanoparticles increases their circulation time and facilitates the binding of ligands or other molecules on the surface for targeting and receptor-mediated endocytosis.

The efficacy of magnetic therapy is dependent on the applied field strength as well as on the volumetric and magnetic properties of the particles. Magnetic nanoparticles have to meet the following requirements: high magnetization to be controlled by an external magnetic field and exceed linear blood flow rates of 10 cm s^{-1} in arteries and 0.05 cm s^{-1} in capillaries, biocompatibility, long circulation time, size of less than 200 nm, and they should not agglomerate. A magnetic field of 0.8 T is sufficient to exceed linear blood flow rates in carriers containing 20% magnetite [110]. For most carriers, flux densities must be 0.2 T with field gradients of 8 T m^{-1} for femoral arteries. Tissue depth penetration has been observed for the range of 8–12 cm [111].

Increased tissue depth reduces the efficacy of magnetic targeting. Magnetic targeting and delivery are not readily applicable to tumors located deep in the body.

Active targeting with anticancer drugs has been tested in many *in vivo* studies in animals and humans. Adriamycin has limited application in humans due to its high cardiotoxicity. An approach to avoiding high toxicity in vital organs is to guide magnetic-particle-bearing drugs to the tumors.

Luebbe et al. fabricated ferrofluids coated with anhydroglucose, reversibly linked to 4-epirubicin. These particles, 100 nm in size, were injected intravenously into tumor-bearing rats and mice. Upon exposure to a magnetic field of 0.2–0.5 T, epirubicin ferrofluid accumulated in the tumor region. Most of the ferrofluids were detected in liver and spleen, reaching moderate levels after 18 days, whereas only a little iron deposition occurred in lung, kidney, and heart [112]. Tumor reduction occurred only if an external magnetic field was applied after ferrofluid 4-epirubicin injection. In contrast, free epirubicin treatment was not followed by tumor response. Only high doses of epirubicin resulted in tumor response, but those were highly toxic, even lethal. Ferrofluid 4-epirubicin complex alone at higher concentrations was as toxic as epirubicin alone [112].

The first phase I clinical trial was published in 1996 by Luebbe et al [107]. 4-Epirubicin was reversibly bound to magnetic ferrofluids of a particle size of 100 nm and was injected intravenously into 14 patients with squamous carcinoma of the breast or head and neck. Table 11.1 shows the characteristics of the ferrofluids used in this study. The dosages ranged from 6 mg ml^{-1} in infusion volumes of 20–70 ml. The dosages translated to 5–100 mg m^{-2} of magnetic fluid. During the time of infusion the magnetic field (0.5–0.8 T) was applied for 60–120 min in the tumor area. The 4-epirubicin ferrofluids accumulated in the tumor area in six patients and were well tolerated. The systemic effect of 4-epirubicin was reduced within 60 min after the injection. Hematologic toxicity was seen as leukopenia or thrombocytopenia, from which the patients recovered within 21 days, and was lower than after 4-epirubicin treatment alone. The transient elevated serum iron levels lasted for 24–48 h. No changes in renal function or hepatic parameters were observed. Iron uptake into the tumors was visible by discoloration of the tumor area. The amount of particles delivered into liver or spleen could not be determined [107]. Evidently the patients with tumors close to the skin showed a greater response in accumulation of iron particles in the tumor site. This result might be due to the smaller depth of tissue to be penetrated by the magnetic field.

Table 11.1 Characteristics of magnetic fluids. (From Ref. [112].)

Particle Size	100 nm
Magnetites	1.5% of total weight
Iron content (wt/wt)	60%
pH	7.4
Color	Black
Odor	Neutral
Iron content	6 mg/ml
Carbohydrate content	5 mg/ml
No. of particles	10^8/ml
Weight by volume	10 mg/ml
Epirubicin desorption within 30 min	

The distribution of magnetic particles is dependent on the route of administration. Different routes were compared when mitoxantrone-bound ferrofluids (100 nm) were injected intravenously and intraarterially into squamous cell carcinoma-bearing rabbits with exposure to an external magnetic field (1.7 T). Treatment efficacies were compared with respect to magnetic targeting to the tumors. In this model, tumor regression after 12 days was achieved only with intraarterial injection. Intravenous injections were cytostatic, suggesting the intraarterial administration was more successful than intravenous administration. Free mitoxantrone administration (intraarterial) was less effective, causing alopecia, with tumor regression seen after 24 days. Arterial administration resulted in increased drug concentration at the tumor site. Histological examination of the tumor tissue showed that the ferrofluids were concentrated in the intraluminal space and deposited in the endothelium, the tumor interstitium and surrounding tissues. No pathological changes were observed in the liver, kidney, spleen, lung, or brain This study suggests that intravenous injection might be less effective as a route to target tissue through magnetic targeting, because the reticuloendothelial system (RES) removed the injected nanoparticles from the system during the hepatic passage. The macrophages of the RES transport the phagocytosed particles to liver, spleen, bone marrow, and lymph nodes. These organs appear to be a more appropriate target for this approach. Intraarterial injection, however, was successful in transporting the nanoparticles to tumors located in the hind leg of rabbits, by avoiding the liver passage [113].

The application of magnetic force was essential in concentrating the ferrofluids at the tumor site. Linking the mitoxantrone to ferrofluids reduced the circulating mitoxantrone concentration by 50–80% compared to when free mitoxantrone was given [114].

When long-circulating dextran-coated iron oxide particles were injected intravenously into a gliosarcoma rodent model, the particles accumulated in the tumor tissue and were detected by MRI. The particle distribution in the tumor was 19% interstitial, 49% intracellular in glioma cells, and 21% incorporated in tumor-associated macrophages. The nanoparticles crossed the blood–brain barrier [115].

Doxorubicin delivery with magnetic nanoparticles has been conducted in swine [116, 117], rabbits [113], and rats [118, 119]. In all cases the effective doxorubicin concentration could be reduced by a factor of 10. Implanted magnets in the tumor area were used to guide magnetite containing liposome/doxorubicin particles to the tumor tissue. This resulted in an increase of antitumor activity compared to intravenously injected doxorubicin and side effects were eliminated [120, 121].

11.6.2.2 Ligand-Directed Active Targeting

The functionalization of nanoparticles with targeting agents such as ligands or antibodies, which bind to receptors on the tumor cell membranes, facilitate specific and increased uptake into the target cells through receptor-mediated endocytosis and thus can increase the *in vivo* specificity of the drug. *In vitro* receptor-mediated nanoparticle accumulation has been studied.

Gadolinium hexanedione-containing nanoparticles were fabricated from PEG 400 (<125 nm). A folate ligand was bound to the nanoparticles by linking folic acid to distearoylphosphatidylethanolamine via PEG spacer. Folate-coated nanoparticles specifically targeted the folate receptor expressing human nasopharyngeal epidermal carcinoma cells (KB); MCF-7 cells served as a folate-receptor-negative control.

Folate-coated nanoparticles were incorporated into KB cells by receptor-mediated endocytosis and their concentration was 10-fold higher compared to uncoated particles. Cell death occurred through gadolinium neutron capture therapy [122] in KB cells but not in the receptor-negative MCF-7 cells. Cell death was observed only after gadolinium–folate nanoparticles had accumulated in the KB cells [122]. Similar results were obtained when folate-receptor-positive breast cancer cells and macrophages were incubated with folate-decorated superparamagnetic nanoparticles. Macrophages did not accumulate the folate-decorated particles [123]. These findings show the high specificity of the receptor-mediated process.

Vitamin H receptors are overexpressed in cancer cells compared to normal cells. Pullulan acetate (PA) and vitamin H were self-aggregated to form particles of 80–125 nm diameter and were loaded with doxorubicin (Adriamycin). Using these nanoparticles, tumor targeting, internalization, and controlled drug release were tested on HepG2 cells. The loading efficiency decreased with higher vitamin H content, suggesting a controlled delivery might be feasible. Vitamin-H-decorated nanoparticles showed strong interaction with the HepG2 cells, whereas the PA particles alone showed poor interaction, suggesting receptor-mediated endocytosis. The release of doxorubicin was saturable within 60 min and depended on the loading capacity of the nanoparticles. The release of doxorubicin could be controlled by the vitamin H content of the PA nanoparticles [124].

11.6.2.3 Targeted Drug Delivery Using Magnetic Guidance

Due to the large surface area of nanoparticles, drugs or ligands can be covalently bound to the magnetic nanoparticles themselves or to their polymer coating. Exposure to an alternating electric field accumulates the particles in the target tissue. When the forces of the magnetic field exceed the linear blood flow rate ($0.05 \, \text{cm}^{-1}$ in capillaries), the magnetic particles are retained in the interstitial compartment through the EPR effect. The particles can be internalized into the tumors by ligand-controlled endocytosis or diffusion and can deliver the drug at concentration required for therapy. Recently, silica-coated magnetic holospheres were synthesized [125, 126] with an average size of 150 nm. Carriers have been designed in the form of magnetic particle cores coated with polymers or porous polymer in which magnetic nanoparticles are located inside the pores [127].

11.7
Nonviral Gene Therapy with Nanoparticles

Up to 70% of the European clinical trials use viral gene therapy. Tumor cells frequently have mutations or overexpression of genes which regulate apoptosis and proliferation pathways. These alterations in the genome cause the tumor cells to grow exponentially. Overexpression of certain proteins like Bcl-2 in cancer cells prevents programmed cell death and results in chemoresistance. Other targets for gene therapy are p53 (tumor suppressor) gene transfer, c-myc, and cytokines such as IL-12 to enhance antitumor activity. Antisense oligonucleotides introduced into cancer cells can prevent overexpression of certain proteins which prevent the tumor cells from apoptosis, resulting in chemoresistance.

Viral gene therapy encounters problems which can limit the transfection and transcription rate of the gene/DNA. The transfection rate is lowered because the negatively charged tumor cell environment hinders the uptake of negatively charged antisense DNA. Other obstacles are the degradation of viral DNA in circulation and after transfection by endo-/exonucleases, which is caused by lysosomal degradation. Viral therapy can initiate immune response in the patient. Manufacturing viral DNA constructs is costly and requires high safety measures and precautions [128–130].

Many of the problems encountered by viral gene delivery can be avoided by encapsulation of DNA in nanoparticles, which opens a new era in gene therapy [131]. Nonviral gene delivery protects DNA from endonuclear/exonuclear degradation, and avoids viral-initiated immune response. Consequently, nanoparticle delivery of DNA increases the transfection rate and has major advantages for manufacturing and safety reasons [128–130]. To enhance transfection rates, nanoparticles for delivery of DNA into tumor cells have to meet the following requirements: small-size DNA packages (<25 nm) to facilitate DNA delivery to the nucleus; protection from serum endo-/exonucleases; stabilization during the uptake process; and they must bypass the endocytic pathway/lysosomal degradation [132]. Macromolecules which enter the cellular compartment through endocytosis are degraded by lysosomes and hence become inactivated. Escape from endolysosomal degradation has been shown for PLGA nanoparticles. Encapsulated DNA was slowly released, resulting in sustained gene expression [133].

When nondividing cells, growth-arrested neuroblastoma, and hepatoma cells were transfected with DNA liposome mixtures, the transfection increased 7000-fold compared to naked DNA. The size limit of the liposomes for transfection was 25 nm, which is the nuclear pore size. Particles >25 nm showed decreased transfection capability. Gene delivery through viral vectors linked to magnetic coating surface is under development for gene therapy. The advantage of this approach is the extended contact of the viral vector with the target cell, increasing the transfection rate and the expression rate of the introduced gene [134, 135].

Targeted gene delivery has been demonstrated to inhibit angiogenesis in tumor tissue. Cationic polymerized lipid-based nanoparticles (40–50 nm) were covalently coupled to integrin $\alpha_V\beta_3$-ligands, which target angiogenic blood vessels [136]. *In vitro*, these particles delivered the green fluorescence protein (GFP)-encoding gene specifically to human melanoma cells expressing the integrin receptor. The plasmids were coupled on the surface of the particles via electrostatic attraction. Importantly, nanoparticles without the integrin $\alpha_V\beta_3$-ligands were not incorporated into the cells, hence GFP was not delivered. This approach was tested *in vivo* on human melanoma xenograft-bearing mice. The mice were injected intravenously with nanoparticles containing the luciferase gene and the $\alpha_V\beta_3$-ligand ($\alpha_V\beta_3$-NP-luciferase) to determine whether $\alpha_V\beta_3$ can deliver genes to angiogenic-tumor-associated blood vessels. Maximal luciferase activity was detected in the tumor area after 24 h but none in vessels of lung, liver, brain, kidney, or skeletal muscle. The tumor-associated blood vessels were specifically targeted as the gene delivery was blocked in the presence of $\alpha_V\beta_3$ inhibitor [136].

Mice bearing subcutaneous M21-L melanoma ($\alpha_V\beta_3$-negative) were injected intravenously with $\alpha_V\beta_3$-NP-Raf(–) to block endothelial signaling and angiogenesis. Treatment resulted in apoptosis of tumor-associated endothelium leading to tumor cell apoptosis regression of primary tumor and metastases after 6 days [136].

Most DNA delivery constructs attach the DNA to the surface of the particles. *In vivo* systemic delivery of p53 with Transferrin-Lip-p53 complex improved the transfection rate compared to untargeted construct. Complete tumor regression of DU145 human prostate cancer xenografts were observed in mice treated with radiation [137].

DNA delivery through nanotechnological approaches has important future applications in treating diseases of any kind as it facilitates:

- Increased transfection and transcription rates
- Sustained transcription through slow release
- Decreased biohazard risk in production and application
- Decreased immune response in the patient
- Protection of DNA and biosystem
- Specific targeting.

Nonviral gene delivery may be superior to viral gene delivery in the future.

11.8
Hyperthermia

Tumor cells are more sensitive to temperatures above 42 °C than normal cells [138, 139] and can be destroyed by increasing the temperature locally to 41–42 °C for 30 min [140–143]. Magnetic particles generate heat under an alternating magnetic field (AMF) by hysteresis loss [143, 144].

Heat production occurs during the thermal loss resulting from reorientation of the magnetism of the magnetic material with low electrical conductivity [144]. The heating potential is directly correlated to the size of the magnetic particle, which can be controlled by appropriate synthesis methods. For example, nanoparticles require less AC power than microparticles, and increased dispersion of nanoparticles requires less AC power [145, 146]. Specific absorption power rates (SARs) have been determined in suspensions of magnetite nanoparticles of various size and coating. SARs of dextran ferrites were 180–210 W g^{-1} Fe (120 nm), sonicated dextran ferrites ranged from 12 to 240 W g^{-1} Fe, uncoated ferrosuspension from 0 to 45 W g^{-1} Fe (6–10 nm). Ultrasonification caused dispersion of agglomerated particles and can in part destroy the dextran coating. Carboxymethyldextran magnetite particles (130 nm) had a SAR of 90 W g^{-1} Fe. SAR data allow an estimate of the amount of particles necessary to heat human tissues [147, 148].

To test the efficacy of hyperthermia treatment, magnetic fluids with a particle size of 3 nm coated with dextran were injected into C3H mammary tumors of mice.

After 30 min the whole bodies of the mice were exposed to an AMF of 6–12 kA m^{-1} at 520 kHz for 30 min. The intratumoral temperature rose to 46 °C and ferrofluid accumulation increased in the tumor tissue from 15% to 45%. Ferrofluid injected alone did not change the tumor histology. Tumor necrosis was observed in mice with ferrofluid and exposure to the AMF. However, tumor growth after treatment was heterogeneous, with a response of 44%, suggesting that the tumor inhomogeneity coincides with ferrofluid distribution [148].

After intravenous injection of 100 mg dextran magnetite (400 mg kg^{-1}) into rats an accumulation of nanoparticles in the mammary tumors was observed. Exposure to an AMF (12 min, 450 kHz) resulted in shrinking and necrosis of the tumor tissue and coagulation of the blood vessels [149]. In mice, iron concentration was increased in liver and spleen after intratumoral injection due to RES uptake. Iron contents in tumors were initially 15% and decreased to 2% after 52 h. Application of AMF resulted in a 2.5-fold increase of iron specifically in the tumor tissue over a time frame of 30 min. Except for the region around the tumor no systemic heating occurred, suggesting that the iron content in the tumors was sufficient to cause hyperthermia. In order to effectively destroy tumors through hyperthermia, the required amount of iron inside the tumor tissue has been suggested to be 5–10 mg cm^{-3} [150].

To further increase the accumulation of iron content in the tumors, magnetoliposomes (ML) were conjugated with an antibody which recognizes renal tumor cells. Mice bearing renal implanted tumors were injected with G250F-ML.

The incorporation of iron in the renal tumors was 27-fold higher when G250F-ML was injected compared to the ML particles alone. Exposure to AMF (5 kW, 118 kHz, 30.6 kA m^{-1}) arrested tumor growth within 2 weeks [151]. Necrosis was observed in tumors but not in livers after AMF exposure, although the liver accumulated ML. Total destruction of tumor tissue was achieved after three AMF exposures, suggesting that repeated exposures are effective. The iron distribution was 50% tumor, 35% liver, 33% blood.

Magnetic nanoparticles injected intratumorally into human breast cancer-bearing mice resulted in partial release of the particles from the tumors and accumulated in liver, spleen, and lung. These organs may be damaged upon body exposure to AMF [145].

A combined hyperthermia therapy approach has been suggested for U251-SP xenograft in nude mice to increase treatment efficacy. TNF-α gene therapy driven by a heat-inducible promoter was combined with hyperthermia using magnetite cationic liposomes. When mice were injected intratumorally with magnetic cationic liposomes and exposed to AMF 118 kHz, 30.6 kA m^{-1}), tumor cell death was induced within 3 min. TNF-α expression increased three-fold during AMF heat induction even in peripheral areas which were not affected by hyperthermia [152].

Specific targeting of cancer cells and incorporation of magnetic nanoparticles conjugated to luteinizing-hormone-releasing hormone (LHRH) showed *in vitro* accumulation of magnetic nanoparticles through receptor-mediated endocytosis. LHRH-receptor-expressing MCF-7 human breast cancer cells were incubated with LHRH magnetic nanoparticles coated with silica with a final particle size of

20–50 nm. These particles accumulated on the surface and inside LHRH-receptor-expressing cells dependent on the LHRH receptor capacities, but not in UCI 107 cells, which do not express LHRH receptor [153]. LHRH-decorated nanoparticles were eight times more potent in destroying MCF-7 cells. The number of lysed cells was linearly dependent on the magnetic field exposure time and the concentration of nanoparticles. UCI 107 cells were not lysed under the same conditions. These results suggest that specific targeting can increase the efficacy of hyperthermic therapy and keep other organs unaffected.

An alternative to AMF-exposure-related thermal damage to tumor tissue has been shown with near-infrared light (NIR). The fabricated gold nanoshells consisted of silica as core particles, which were surrounded by a thin gold shell coated with PEG. Such particles possess a tunable plasmon resonance through light exposure. The relative thickness of the core and shell layers can be controlled and results in various optical absorption from near-UV to mid-IR. NIR light showed minimal absorption in tissue with optimal penetration. The nanoshells were injected into the tumors of xenograft-bearing mice and exposed to NIR light (820 nm, 4 W cm^{-2}). Within 4–6 min a temperature increase of 37 °C in the tumor region occurred [154], resulting in necrosis. Gold shell silica nanoparticles show different absorption spectra, dependent on the thickness of the shell: 60-nm core radius particles with a 20-nm shell have an absorption maximum at λ=680 nm versus a 5-nm shell which gives λ=1000 nm [155]. These fabricated particles are unique. They are tunable because their absorption properties can be designed during the fabrication process, making them highly suitable for imaging (primarily light scattering) or the photothermal-based therapy (primarily absorbing). *In vitro* exposure of carcinoma cells to NIR-absorbing gold–silica nanoparticles followed by NIR laser irradiation led to photothermal destruction of the cells. *In vivo* when nanoshells were injected into the tumor (canine TVT cells), irradiation with NIR light for 6 min (820 nm, 4 W cm^{-2}) resulted in tumor cell necrosis within 4–6 min, spanning a zone of thermal damage of 4 mm. The heating occurred in two phases: initial rapid heating and gradual heating. The nanoshells diffused throughout the tumor tissue and adjacent tissue and caused coagulation and cell shrinkage in adjacent areas which were infiltrated by the nanoshell. Further surrounding tissue stayed intact [154].

In yet another approach, nanoscale metallic particles were used to target specific biological structures and tissues through controlled laser-induced breakdown (LIB) process.

So far there have been no reports of successful application of hyperthermia treatment in humans, although the treatment had been very effective in animal studies. In humans it is difficult to acquire adequate quantities of the magnetic nanoparticles in the target tissue without exceeding the tolerable amount, since in humans the applied external magnetic field needs to be stronger than the tolerable field strength. Heating of the tissue is opposed by blood circulation, which creates a cooling effect. The volume of the tumor tissue which needs to be heated is a limiting factor and should not exceed 300 mm^3 [150, 156]. The maximum tolerated dose for humans has been reported to be H × f = 4.85 × 10^8 Am^{-1} s^{-1} [157]. In addition, the tissue depth causes loss of field strength. With a particle size of 0.5–5 μm in a swine model, a targeting depth of 8–12 cm was achieved [116].

11.9
Controlled Delivery of Chemotherapeutic Drugs Using Nanoparticles

Controlled release systems continue to draw the attention of several research groups owing to their application in a broad range of areas such as drug delivery, paper, pesticide, printing, cosmetics, and so on [158, 159]. Most of the work so far has focused on achieving controlled release of active ingredients, encapsulated in polymeric matrices, in response to specific stimuli [159–161] or on the use of microfabrication technology [162]. Advances in microelectromechanical systems (MEMS) technology has resulted in further improvements in microchip-based implantable drug delivery systems [136, 158, 163]. In the field of drug delivery, the method of delivery has a tremendous impact on the therapeutic efficacy [159, 164]. More complicated delivery systems are also being designed for targeted delivery of genes using viral vectors, liposomes, and cationic nanoparticles [164].

Of the aforementioned approaches, responsive polymers and microchips appear to be the more versatile methods for controlled release of drugs. Polymer-based drug delivery systems for controlled release are well known and have been in vogue for the delivery of a variety of drugs [165–167]. They have also been designed to respond to specific stimuli such as ultrasound, light, enzymes, temperature, pH, and electric and magnetic fields [164]. Of these stimuli, magnetic stimulus is particularly attractive as it is a soft approach. Reproducible regulation of drug release from polymers was demonstrated in cases where small magnetic spheres or cylindrical magnets were embedded in a polymeric matrix containing drug [168, 169].

The majority of the controlled drug delivery systems being developed are for clinical applications where systemic exposure and release of a drug was desired. Some of these are: delivery of insulin [165, 166], antiarrhythmics [166], gastric acid inhibitors [170], contraception [171, 172], general hormone replacement [173, 174], and immunization [175]. Only recently has there been a growing interest in the development of controlled drug delivery systems for cancer chemotherapy [176, 177]. The approaches can be broadly classified into the following categories:

- Sustained release through polymer degradation
- Enzymatically controlled release
- Controlled release through use of thermosensitive polymers
- Photochemically controlled release
- pH-responsive release systems
- Laser-induced breakdown (LIB)
- Ultrasound-mediated release.

Sustained release using biodegradable polymeric nanoparticles is one of the most popular approaches for controlled release of cancer chemotherapeutic agents. Sustained release depends on the chemical nature of the polymer [178–182] and the method of preparation [98, 183, 184] and surface modification(s) of the polymer–drug nanoparticles [185]. In an interesting study by Yoo and Park, doxorubicin was chemically conjugated to a terminal end group of PLGA by an ester linkage. The doxorubicin–PLGA conjugate and doxorubicin were formulated into nanoparticles and sustained release profiles (over a 1-month period) with minimal initial bursts were observed [186]. In contrast, when unconjugated free doxorubicin was incorporated into nanoparticles, the initial burst was absent and the drug release was complete within 5 days. Conjugation of anticancer drugs to the polymeric nanoparticles resulted in linear drug-release profiles over an extended period [187, 188].

Examples of approaches have been reported where physiological properties of cancer cells have been taken into consideration while designing the controlled release systems for anticancer agents, e.g., angiogenesis, which is a specific physiological property of cancer cells. A controlled release system for an anticancer drug, all-*trans*-retinoic acid (atRA), was demonstrated based on enzymatic degradation [189]. In this study, PEGylated gelatin nanoparticles carrying atRA released the anticancer drug very sensitively by the action of the enzyme collagenase IV, one of the major metalloproteases involved in angiogenesis. Similarly, pH values in the tumor interstitium are lower than those of normal cells and this difference can be exploited to develop pH-sensitive polymeric nanoparticles for controlled release of anticancer agents. A new class of pH-responsive polymers carrying the anticancer agent doxorubicin (Adriamycin), was prepared using sulfadimethoxine and succinylated pullulan acetate. The nanoparticles were found to be very stable at physiological pH of 7.4 but showed degradation to release doxorubicin at a lower pH close to that of tumor cells [190].

Another approach for controlled release of cancer chemotherapeutics taking advantage of thermal and photochemical properties of polymeric nanoparticles has been investigated. Yoshinobu et al. [191] have reported the design of a novel acrylic composite nanoparticle with a hydrophobic core and a thermosensitively swellable shell demonstrating a thermosensitive mode of drug release. The microcapsules exhibited an exceptionally rapid on–off response of drug release in a thermosensitive manner. Sershen et al. [192] have also demonstrated photothermally modulated drug delivery using gold nanoshells.

11.10
Nanoparticles to Circumvent MDR

Lack of response to chemotherapy is defined as multidrug resistance (MDR) which involves a mechanism to evade chemotherapy. Pgp recognizes drugs when located in the plasma membrane, but not in the cytosol of the lysosomal compartment [193, 194]. Doxorubicin is a Pgp substrate and has been shown to be exported from the cytosol in multidrug-resistant cancer cell lines. Nanoparticles circumventing MDR via the Pgp mechanism have to enter the tumor cells in order to be effective. Two approaches using PIHCA and PIBCA doxorubicin-loaded nanospheres have been reported in an effort to avoid MDR. In the case of PIHCA doxorubicin nanospheres the intracellular doxorubicin concentration was lower in doxorubicin-resistant glioblastoma cells compared to that seen with free doxorubicin [194]. The doxorubicin nanospheres were only efficient if the MDR was based on Pgp. When a fast-degrading polymer was used for encapsulation of doxorubicin in a doxorubicin-resistant murine leukemia cell line overexpressing Pgp, the intracellular doxorubicin concentration was high and the doxorubicin efflux was comparable to that seen with the free doxorubicin group.

PIBCA nanospheres were degraded before they entered the cellular compartment [195]. Thus doxorubicin at high concentration was associated with the tumor cell membrane being extravasated through the Pgp efflux pump. The coupling of doxorubicin with polymethacrylate resulted in internalization of the particles through endocytosis in U937 doxorubicin-resistant monocyte-like cancer cells. These particles showed sustained slow doxorubicin release, resulting in a significant higher cytotoxicity than free doxorubicin [196].

Several other attempts have been conducted to increase doxorubicin delivery in resistant cancer cells. Doxorubicin incorporated into gelatin nanospheres showed low cytotoxicity in a mouse colon carcinoma xenograft; even increased cardiotoxicity was observed. The different effects were due to slow dissociation of the nanosphere doxorubicin complexes, slow diffusion rate across the cell membrane, and the failure of the complex to enter cells, causing elevated drug release in the cell membrane. MDR can only be avoided when close contact with polycyanoacrylate is achieved [197].

Folate-receptor-mediated uptake of polyethyleneglycol diastearoyl phosphatidylethanolamine liposomes (70–100 nm) loaded with doxorubicin resulted in increased cytosol uptake and release of doxorubicin into the cytoplasm. The drug was released within 2 h *in vitro* in M109-HiFR multidrug resistant cancer cells. Folate-targeted liposome drug uptake was increased 10-fold compared to free doxorubicin and was more toxic *in vivo* than free doxorubicin [198].

Doxorubicin delivered by folate-targeted liposomes did not avoid the Pgp efflux system, which was explained by the aggregation form of doxorubicin in the folate-targeted liposome nanoparticle. It has been suggested that encapsulated doxorubicin is dimeric [199]. Doxorubicin loaded into nanospheres (300 nm) consisting of polycyanoacrylate were injected into leukemia cells (P388ADR-) of xenograft-bearing mice intraperitoneally. The treatment prolonged the survival of the mice compared to free doxorubicin administration, which was ineffective. However, no cure was obtained whether nanoparticles or free doxorubicin was administered. IC_{50} *in vitro* was 4.3 µM for free dox versus 0.08 µM for doxorubicin-loaded nanospheres. The multidrug-resistant cell lines were 30- to 250-fold more sensitive to doxorubicin-loaded nanospheres than to free doxorubicin, and even complete reversion of MDR was observed in some cell lines. PIHCA nanospheres loaded with doxorubicin were biodegradable; nanospheres of a size of 200 nm they entered cells by endocytosis and delivered doxorubicin into lysosomes. Doxorubicin-loaded nanospheres were not recognized by Pgp, thus circumventing the MDR.

11.11
Potential Problems in Using Nanoparticles for Cancer Treatment

Potential problems associated with nanoparticle use *in vivo* include the risk of thrombosis through agglomerating particles or their breakdown products. Most experiments have been conducted in rodents and are difficult to scale up to humans with respect to distances and volume in circulation, tissue densities, and the amount of particles required. This problem needs to be taken into account if the drug release is uncontrollable with a magnetic field. There is always a potential toxicity of magnetic carrier or coating polymers and their breakdown products, which may occur after long-time exposure or injections. Also, the application of an external magnetic field can lead to cardiac arrhythmia, muscle stimulation, and seizures and needs to be closely monitored. Some patients may not be eligible for these particular applications.

11.12
Future Outlook

Development and research in the field of nanotechnology in applications like biomedicine requires the understanding and interaction of experts from physics, mathematics, the biomedical sciences, and from chemists and medical personnel. The potential for the use of nanoparticles in the task of drug delivery and disease detection is huge and may change the way of currently used drug delivery systems. Nanoparticles offer a broad application range, and represent enough flexibility to custom-design a treatment regimen in the future.

Future emphasis will move on to metastases detection and treatment with a view to eventually eradicating cancer as a disease.

Acknowledgements

The authors would like to thank Janice Keener and Eric Guilbeau of the Pennington Biomedical Research Center for their help in preparing the manuscript and some of the figures.

Abbreviations

AMF – alternating magnetic field
Apo E – apolipoprotein E
CPT – Irinotecan
DNA – desoxyribonucleic acid
EMF – external magnetic field
EPC – Egg Phosphatidylcholine

EPR – enhanced permeability and Retention
5 FU – 5-fluorouracil
Gd-DTPA – gadolinium-diethylenetetraminc penta acetic acid
GdNCT – gadolinium neutron capture therapy
GFP – green fluorescence protein
IC_{50} – half maximum inhibitory concentration
LHRH – luteinizing hormone releasing hormone
MDR – multidrug resistance
MPS – mononuclear phagocytic system
MTD – maximum tolerated dosage
NIR – near-infrared
NP – nanoparticles
NS – nanospheres
PA – pullulan acetate
PCL – polycaprolactone
PEG – polyethylene glycol
PEO – polyethylene oxide
Pgp – p-glycoprotein
PL – Phospholipid
PIBCA – polyisobutylcyanoacrylate
PIHCA – polyisohexylcyanoacrylate
PLGA – polylactide glycolic acid
PMMA – poly(methyl methacrylate)
PVP – polyvinylpyrrolidone
RES – reticuloendothelial system
SAR – specific abruption rate
TNF-α – tumor necrosis factor-α
VEGF – vascular endothelial growth factor

References

1 NSF report on "Societal implications of Nanoscience and Nanotechnology", March 2001.

2 J. M. Nam, S. Park, C. A. Mirkin, Bio-barcodes based on oligonucleotide-modified nanoparticles, *J. Am. Chem. Soc.* **2002**, *124*, 3820–3821.

3 A. S. Lubbe, C. Alexiou, C. Bergermann, Clinical applications of magnetic drug targeting, *J. Surg. Res.* **2001**, *95*, 200–206.

4 L. Illum, A. E. Church, M. D. Butterworth, A. Arien, J. Whetstone, S. S. Davis, Development of systems for targeting the regional lymph nodes for diagnostic imaging: in vivo behavior of colloidal PEG-coated magnetite nanospheres in the rat following interstitial administration, *Pharm. Res.* **2001**, *18*, 640–645

5 L. K. Komissarova, A. A. Kuznetsov, N. P. Gluchoedov, M. V. Kutushov, M. A. Pluzan, Absorptive capacity of iron-based magnetic carriers for blood detoxification, *J. Magn. Magn. Mater.* **2001**, *225*, 197–201

6 Cancer Statistics, American Cancer Society, 2004.

7 M. De Lena, C. Brambilla, A. Morabito, G. Bonadonna, Adriamycin plus vincristine compared to and combined with cyclophosphamide, methotrexate and 5-fluorouracil for advanced breast cancer, *Cancer* **1975**, *35*, 1108–1115.

8 J. M. Bull, D. C. Tormey, S. H. Li, P. P. Carbone, G. Falkson, J. Blom, E. Perlin, R. Simon, A randomized comparative trial of adriamycin versus methotrexate in combination drug therapy, *Cancer* **1978**, *41*, 1649–1657.

9 R. J. Santen, Endocrine treatment of prostate cancer, *J. Clin. Endocrinol. Metab.* **1992**, *75*, 685–689.

10 J. L. Emmet, L. F. Greene, A. Papantoniou, Endocrine therapy in carcinoma of the prostate gland: 10-year survival studies, *J. Urol.* **1960**, *83*, 471–484.

11 R. M. O'Regan, V. C. Jordan, Tamoxifen to raloxifene and beyond, *Semin Oncol* **2001**, *28*, 260 273.

12 S. A. Eccles, G. Box, W. Court, J. Sandle, C. J. Dean, Preclinical models for the evaluation of targeted therapies of metastatic disease, *Cell Biophys.* **1994**, *24–25*, 279–291.

13 K. Pantel, M. Otte, Occult micrometastasis: enrichment, identification and characterization of single disseminated tumour cells, *Semin. Cancer Biol.* **2001**, *11*, 327–337.

14 C. J. Mettlin, G. P. Murphy, R. Ho, H. R. Menck, The National Cancer Database report on longitudinal observations on prostate cancer, *Cancer* **1996**, *77*, 2162–2166.

15 C. S. B. Galasko, The anatomy and pathways of skeletal metastases, in L. Weiss, A. H. Gilbert (eds.) Bone metastases, Boston, GK Hall, **1981**, pp. 49–63.

16 K. Pantel, R. J. Cote, O. Fodstad, Detection and clinical importance of micrometastatic disease, *J. Nat. Cancer Inst.* **1999**, *91*, 1113–1124.

17 S. Honig, Hormonal therapy and chemotherapy, in J. R. Harris, M. E. Lippman, M. Morrow, S. Hellman (eds.) Diseases of the breast, Philadelphia, Lippincott–Raven Publishers, **1996**, pp. 669–734.

18 S. Braun, and K. Pantel, Biological characteristics in micrometastatic cancer cells in bone marrow, *Cancer Metastasis Rev.* **1999**, *18*, 75–90.

19 M. Kavallaris, D. Y. Kuo, C. A. Burkhart, D. L. Regl, M. D. Norris, M. Haber, S. B. Horwitz, Taxol-resistant epithelial ovarian tumors are associated with altered expression of specific beta-tubulin isotypes, *J. Clin. Invest.* **1997**, *100*, 1282–1293.

20 M. Lehnert, S. Emerson, W. S. Dalton, R. de Giuli, S. E. Salmon, In vitro evaluation of chemosensitizers for clinical reversal of P-glycoprotein-associated Taxol resistance, *J. Natl. Cancer Inst. Mongr.* **1993**, *15*, 63–67.

21 N. Baldini, K. Scotlandi, M. Serra, T. Shikita, N. Zini, A. Ognibene, S. Santi, R. Ferracini, N. M. Maraldi, Nuclear immunolocalization of P-glycoprotein in multidrug-resistant cell lines showing similar mechanisms of doxorubicin distribution, *Eur. J. Cell Biol.* **1995**, *68*, 226 239.

22 A. Molinari, A. Calcabrini, S. Meschini, A. Stringaro, M. Del Bufalo, M. Cianfriglia, G. Arancia, Detection of P-glycoprotein in the Golgi apparatus of drug-untreated human melanoma cells, *Int. J. Cancer* **1998**, *75*, 885–893.

23 J. Hamada, T. Tsuruo, Characterization of the ATPase activity of the 170 to 180 kilodalton membrane glycoprotein associated with multidrug resistance: the 170 to 180 kilodalton membrane glycoprotein is an ATPase, *J. Biol. Chem.* **1988**, *263*, 1454–1458.

24 R. Krishna, L. D. Mayer, Multidrug resistance (MDR) in cancer – mechanisms reversal using modulators of MDR and the role of MDR modulators in influencing the pharmacokinetics of anticancer drugs, *Eur. J. Cancer Sci.* **2000**, *11*, 265–283.

25 J. C. Murray, J. Carmichael, Targeting solid tumours: challenges, disappointments and opportunities, *Adv. Drug. Del. Rev.* **1995**, *17*, 117–127.

26 R. K. Jain, Molecular regulation of vessel maturation, *Nat. Med.* **2003**, *9*, 685–93.

27. T. P. Padera, B. R. Stoll, J. B. Tooredman, D. Capen, E. di Tomaso, R. K. Jain, Pathology: cancer cells compress intratumour vessels, *Nature* **2004**, *427*, 695.

28 K. Weindel, J. R. Moringlane, D. Marme, H. A. Weich, Detection and quantification of vascular endothelial growth factor/vascular permeability factor in brain tumor tissue and cyst fluid: the key to angiogenesis? *Neurosurgery* **1994**, *35*, 439–449.

29 S. P. Olesen, Rapid increase in blood brain barrier permeability during severe hypoxia and metabolic inhibition, *Brain Res.* **1986**, *368*, 24–29.

30 S. K. Hobbs, W. L. Monsky, F. Yuan, W. G. Roberts, L. Griffith, V. P. Torchilin, R. K. Jain, Regulation of transport pathways in tumor vessels: role of tumor type and microenvironment, *Proc. Natl. Acad. Sci. USA* **1998**, *95*, 4607–4612.

31 F. Yuan, M. Dellian, D. Fukumura, M. Leunig, D. A. Berk, V. P. Torchilin, R. K. Jain, Vasular permeability in human tumor xenograft: molecular size dependence and cut-off size, *Cancer Res.* **1995**, *55*, 3752–3756.

32 S. Unezaki, K. Maruyama, J. I. Hosoda, I. Nagae, Y. Koyanagi, M. Nakata, O. Ishida, M. Iwatsuru, S. Tsuchiya, Direct measurement of the extravasation of polyethylenegly-col-coated liposomes into solid tumor tissue by in vivo fluorescence microscopy, *Int. J. Pharm.* **1996**, *144*, 11–17.

33 W. L. Monsky , D. Fukumura, T. Gohongi, M. Ancukiewcz, H. A. Weich, V. P. Torchilin, R. K. Jain, Augmentation of transvascular transport of macromolecules and nanoparticles in tumors using vascular endothelial growth factor, *Cancer Res.* **1999**, *59*, 4129–4135.

34 Y. Boucher, M. Leunig, R. K. Jain, Tumor angiogenesis and interstitial hypertension, *Cancer Res.* **1996**, *56*, 4264–4266.

35 G. Helmlinger, P. A. Netti, H. C. Lichtenbeld, R. J. Melder, R. K. Jain, Solid stress inhibits the growth of multicellular tumor spheroids, *Nat. Biotechnol.* **1997**, *15*, 778–783.

36 R. K. Jain, Transport of molecules in the tumor interstitium: a review, *Cancer Res.* **1987**, *47*, 3039–3051.

37 R. K. Jain, Barriers to drug delivery in solid tumors, *Sci. Am.* **1994**, *271*, 58–65.

38 P. E. Thorpe, F. J. Burrows, Antibody-directed targeting of the vasculature of solid tumors, *Breast Cancer Res. Treat.* **1995**, *36*, 237–251.

39 H. Maeda, The enhanced permeability and retention effect in tumor vasculature, the key role of tumor sensitive macromolecular drug targeting, *Adv. Enzyme Regul.* **2001**, *41*, 189–207.

40 H. Maeda, Tumor vascular permeability and the EPR effect in macromolecular therapeutics: a review, *J. Control. Release* **2000**, *65*, 271–284.

41 P. Tartaj, M. Morales, S. Veintemillas Verdaguer, T. Gonzalez-Carreno, C. J. Serna, The preparation of magnetic nanoparticles for applications in biomedicine, *J. Phys. D Appl. Phys.* **2003**, *36*, R182–183

42 R. Gref, Y. Minamitake, M. T. Peracchia, V. Trubetskoy, V. Torchilin, R. Langer, Biodegradable long-circulating polymeric nanospheres, *Science* **1994**, *263*, 1600–1603.

43 S. M. Moghimi, A. C. Hunter, J. C. Murray, Long circulating and target specific nanoparticles: theory to practice, *Pharm. Rev.* **2001**, *53*, 283–218.

44 G. Storm, S. O. Belliot, T. Daemen, D. D. Lasic, Surface modification of nanoparticles to oppose uptake by the mononuclear phagocyte system, *Adv. Drug. Deliv. Rev.* **1995**, *17*, 31–48.

45 J. Kreuter, Drug targeting with nanoparticles, *Eur. J. Drug. Metab. Pharmacokinet.* **1994**, *19*, 253–256.

46 L. Araujo, R. Lobenberg, and J. Kreuter, Influence of the surfactant concentration of the body distribution of nanoparticles, *J. Drug. Target* **1999**, *6*, 373–385.

47 V. Lenaerts, J. F. Nagelkerke, T. J. Van Berkel, P. Couvreur, L. Grislain, M. Roland, P. Speiser, In vivo uptake of polyisobutyl cyanoacrylate nanoparticles by rat liver Kupffer, endothelial, and parenchymal cells, *J. Pharm. Sci.* **1984**, *73*, 980–982.

48 P. Couvreur, B. Kante, V. Lenaerts, V. Scailteur, M. Roland, P. Speiser, Tissue distribution of antitumor drugs associated with polyalkylcyanoacrylate nanoparticles, *J. Pharm. Sci.* **1980**, *69*, 199–202.

49 A. Chonn, S. C. Semple, P. R. Cullis, Separation of large unilamellar liposomes from blood components by a spin column procedure: towards identifying plasma proteins which mediate liposome clearance in vivo, *Biochim. Biophys. Acta* **1991**, *1070*, 215–222.

50 F. Yuan, Transvascular drug delivery in solid tumors, *Semin. Radiat. Oncol.* **1998**, *8*, 164–175, 1998.

51 Y. Noguchi, J. Wu, R. Duncan, J. Strohalm, K. Ulbrich, T. Akaike, H. Maeda, Early phase tumor accumulation of macromolecules: a great difference in clearance rate between tumor and normal tissues, *Jpn. J. Cancer Res.* **1998**, *89*, 307–314.

52 R. Weissleder, H. C. Cheng, A. Bogdanova, A. Bogdanov Jr., Magnetically labeled cells can be detected by MR imaging, *J. Magn. Reson. Imaging* **1997**, *7*, 258–263.

53 T. C. Yeh, W. Zhang, S. T. Ildstad, C. Ho, Intracellular labeling of T–cells with superparamagnetic contrast agents, *Magn. Reson. Med.* **1993**, *30*, 617–625.

54 U. Schoepf, E. M. Marecos, R. J. Melder, R. K. Jain, R. Weissleder, Intracellular magnetic labeling of lymphocytes for in vivo trafficking studies, *Biotechniques* **1998**, *24*, 642–646, 648–651.

55 A. Moore, J. P. Basilion, E. A. Chiocca, R. Weissleder, Measuring transferrin receptor gene expression by NMR imaging, *Biochim. Biophys. Acta* **1998**, *1402*, 239–249.

56 J. Panyam, W. Z. Zhou, S. Prabha, S. K. Sahoo, V. Labhasetwar, Rapid endo-lysosomal escape of poly(DL-lactide-co-glycolide) nanoparticles: implications for drug and gene delivery, *FASEB J* **2002**, *16*, 1217–26.

57 T. T. Shen, A. Bodganov, A. Bogdanov, K. Poss, T. J. Brady, R. Weissleder, Magnetically labeled secretin retains receptor affinity to pancreas acinar cells, *Bioconjug. Chem.* **1996**, *7*, 311–316.

58 D. Portet, B. Denizot, E. Rump, J. J. Lejeune, P. Jallet, Nonpolymeric coatings of iron oxide colloids for biological use as magnetic resonance imaging contrast agents, *J. Coll. Interface Sci.* **2001**, *238*, 37–42.

59 I. Brigger, C. Dubernet, P. Couvreur, Nanoparticles in cancer therapy and diagnosis, *Adv. Drug. Del. Rev.* **2002**, *54*, 631–651.

60 L. M. Lacava, Z. G. Lacava, M. F. Da Silva, O. Silva, S. B. Chaves, R. B. Azevedo, F. Pelegrini, C. Gansau, N. Buske, D. Sabolobic, P. C. Morais, Magnetic resonance of a dextran coated magnetic fluid intravenously administered in mice, *Biophys. J.* **2001**, *80*, 2483–2486.

61 A. E. Hawley, L. Illum, S. S. Davis, Preparation of biodegradable, surface engineered PLGA nanospheres with enhanced lymphatic drainage and lymph node uptake, *Pharm. Res.* **1997**, *14*, 657–661.

62 A. E. Hawley, L. Illum, S. S. Davis, Lymph node localisation of biodegradable nanospheres surface modified with poloxamer and poloxamine block copolymers, *FEBS Lett.* **1997**, *400*, 319–323.

63 G. Schwab, C. Chavney, Antisense oligonucleotides adsorbed to polyalkylcyanoacrylate nanoparticles specifically inhibit mutated HA-RAS-mediated cell-proliferation and tumoricenicity in nude mice, *PNAS* **1994**, *91*, 10460–10464.

64 G. Blume, G. Cevc, Liposomes for the sustained drug release in vivo, *Biochim. Biophys. Acta* **1990**, *1029*, 91–97.

65 R. L. Hong, C. J. Huang, Y. L. Tseng, V. F. Pang, S. T. Chen, J. J. Liu, F. H. Chang, Direct comparison of liposomal doxorubicin with or without polyethylene glycol coating in C-26 tumor bearing mice: Is surface coating with polyethylene glycol beneficial? *Clin. Cancer Res.*, **1999**, *5*, 3645–3652.

66 K. Moribe, K. Maruyama, M. Iwatsuru, Estimation of surface state of poly(ethylene glycol)-coated liposomes using an aqueous two-phase partitioning technique, *Chem. Pharm. Bull. (Tokyo)* **1997**, *45*, 1683–1687.

67 H. Ishiwata, S. B. Sato, S. Kobayashi, M. Oku, A. Vertut-Doi, K. Miyajima, Poly(ethylene glycol) derivative of cholesterol reduces binding step of liposome uptake by murine macrophage-like cell line J774 and human hepatoma cell line HepG2, *Chem. Pharm. Bull. (Tokyo)* **1998**, *46*, 1907–1913.

68 J. Lode, I. Fichtner, J. Kreuter, A. Berndt, J. E. Diederichs, R. Reszka, Influence of surface-modifying surfactants on the pharmacokinetic behavior of 14C-polymethylmethacrylate nanoparticles in experimental tumor models, *Pharm. Res.* **2001**, *18*, 1613–1619.

69 D. Bazile, C. Prud'homme, M. T. Bassoulet, M. Marlard, G. Spenlehauer, M. Veillard, Stealth Me PEG-PLA nanoparticles avoid uptake by mononuclear phagocyte system, *J. Pharm. Sci.* **1995**, *84*, 493–498.

70 M. T. Peracchia, C. Vauthier, Puisieux, P. Couvreur, Development of sterically stabilized poly isobutyl 2 cyano acrylate nanoparticles by chemical coupling of poly ethylene glycol, *J. Biomed. Mater. Res.* **1997**, *34*, 317–326.

71 M. T. Peracchia, C. Vauthier, D. Desmaele, A. Gulik, J. C. Dedieu, M. Demoy, J. d'Angelo, P. Couvreur, Pegylated nanoparticles from a novel methoxypolyethylene glycol cyanoacrylate hexadecyl cyanoacrylate amphiphilic copolymer, *Pharm. Res.* **1998**, *15*, 550–556.

72 D. Sharma, T. P. Chelvi, J. Kaur,
K. Chakravorty, T. K. De, A. Maitra, R. Ralhan,
Novel taxol formulation: polyvinyl-pyrroli-
done nanoparticles encapsulated taxol for
drug delivery in cancer therapy, *Oncol. Res.*
1996, *8*, 281–286.

73 S. Mitra, U. Gaur, P. C. Gosh, A. N. Maitra,
Tumor targeted delivery of encapsulated dex-
tra-doxorubicin conjugate using chitosan
nanoparticles as carrier, *J. Control Release*
2001, *74*, 317–323.

74 C. Verdun, F. Brasseur, H. Vranckx,
P. Couvreur, M. Roland, Tissue distribution
of doxorubicin associated with polyheylcya-
noacrylate nanoparticles, *Cancer Chemother.
Pharmacol.* **1990**, *26*, 13–18.

75 P. Couvreur, B. Kante, V. Lenaerts,
V. Scailteur, M. Roland, P. Speiser, Tissue distribution of antitumor drugs associated with
polyalkylcyanoacrylate nanoparticles,
J. Pharm. Sci. **1980**, *69*, 199–202

76 S. Gibaud, J. P. Andreux, C. Weingarten,
M. Renard, P. Couvreur, Increased bone marrow toxicity of doxorubicin bound to nanoparticles, *Eur. J. Cancer A* **1994**, *30*, 820–826.

77 P. Couvreur, G. Couarraze, J. P. Devissaguet,
F. Puisieux, Nanoparticles: preparation and
characterization, in Benita S (ed) Microencap-
sulation: methods and industrial application,
Marcel Dekker, New York, **1996**, pp. 183–211.

78 D. A. LaVan, T. McGuire, R. Langer, Small
scale systems for in vivo drug delivery, *Nature
Biotech.* **2003**, *21*, 1184–1191.

79 S. M. Moghimi, A. C. Hunter, J. C. Murray,
Long circulating and target specific nanoparticles: theory to practice, *Pharm. Rev.* **2001**,
53, 283–218.

80 C. Damge, C. Mitchel, M. Aprahamian,
P. Couvreur, J. P. Devissaguet, Nanocapsules
as carriers for oral peptide delivery, *J. Control
Release* **1990**, *13*, 233–239.

81 Y. Nishioka, H. Yoshino, Lymphatic targeting
with nanoparticulate system, *Adv. Drug. Delivery Rev.* **2001**, *47*, 55–64.

82 K. Yang, Y. Wen, L. Li, C. Wang, S. Hou, C. Li,
Preparation of cucurbitacinBE polylactic acid
nano-particles for targeting cervical lymph
nodes, *Hua Xi Kou Qiang Yi Xue Za Zhi* **2001**,
19, 347–50.

83 K. Yang, Y. Wen, L. Li, C. Wang, X. Wang,
Acute toxicity and local stimulate test of
cucurbitacinBE polylactic acid nano-particles
of targeting cervical lymph nodes, *Hua Xi
Kou Qiang Yi Xue Za Zhi* **2001**, *19*, 380–382.

84 F. Shikata, H. Tokumitsu, H. Ichikawa,
Y. Fukumori, In vitro cellular accumulation of
gadolinium incorporated into chitosan nanoparticles designed for neutron-capture therapy of cancer, *Eur. J. Pharm. Biopharm.* **2002**,
53, 57–63.

85 H. Tokumitsu, J. Hiratsuka, Y. Sakurai,
T. Kobayashi, H. Ichikawa, Y. Fukumori, Gadolinium neutron-capture therapy using novel
gadopentetic acid–chitosan complex nanoparticles: in vivo growth suppression of experimental melanoma solid tumor, *Cancer Lett.*
2000, *150*, 177–82.

86 T. Watanabe, H. Ichikawa, M. Fukumori,
Tumor accumulation of gadolinium in lipid
nanoparticles intravenously injected for neutron capture therapy in cancer, *Eur. J. Pharm.
Biopharm.* **2002**, *54*, 119–124.

87 N. Nishiyama, S. Okazaki, H. Cabral,
M. Miyamoto, Y. Kato, Y. Sugiyama, K. Nishio,
Y. Matsumura, K. Kataoka, Novel cisplatin-
incorporated polymeric micelles can eradicate
solid tumors in mice, *Cancer Res.* **2003**, *63*,
8977–8983.

88 R. Reszka, P. Beck, I. Fichtner, M. Hentschel,
L. Richter, J. Kreuter, Body distribution of
free, liposomal and nanoparticles associated
mitoxantrone in B16 melanoma bearing
mice, *J. Pharmacol. Exp. Ther.* **1997**, *280*,
232–237.

89 A. E. Gulyaev, S. E. Gelperina, I. N. Skidan,
A. S. Antropov, G. Y. Kivman, J. Kreuter, Significant transport of doxorubicin into the
brain with polysorbate 80-coated nanoparticles, *Pharm. Res.* **1999**, *16*, 1564–1569.

90 J. Kreuter, Nanoparticulate systems for brain
delivery of drugs, *Adv. Drug. Deliv. Rev.* **2000**,
47, 65–81.

91 I. Brigger, J. Morizet, G. Aubert, H. Chacun,
M. J. Terrier-Lacombe, P. Couvreur, G. Vassal,
Poly(ethylene glycol)-coated hexadecylcyanoa-
crylate nanospheres display a combined effect
for brain tumor targeting, *J. Pharmacol. Exp.
Ther.* **2002**, *303*, 928–936.

92 H. Onishi, Y. Machida, Y. Machida, Antitumor properties of irinotecan containing nanoparticles prepared using poly(DL–lactic) acid and poly(ethylene glycol)–block-poly(propylene glycol)-block-poly(ethylene glycol), *Biol. Pharm. Bull.* **2003**, *26*, 116–119.

93 C. Fonseca, S. Simoes, R. Gaspar, Paclitaxel loaded PLGA nanoparticles: preparation, physicochemical characterization and in vitro anti-tumoral activity, *J. Control Release* **2002**, *83*, 273–286.

94 Z. Lu, T. H. Yeh, M. Tsai, J. Au, G. M. Wientjes, Paclitaxel loaded gelatin nanoparticles for intravesical bladder cancer therapy, *Proc. Am. Assoc. Cancer Res.* **2003**, *44*, 3675.

95 E. Tatou, C. Mossiat, V. Maupoil, F. Gabrielle, M. David, L. Rochette, Effects of cyclosporin and cremophor on working rat heart and incidence of myocardial lipid peroxidation, *Pharmacology* **1996**, *52*, 1–7.

96 R. T. Dorr, Pharmacology and toxicology of Cremophor EL diluent, *Ann. Pharmacother.* **1994**, *28*, S11–S14.

97 P. Mankad, J. Spatenka, Z. Slavik, G. Oneil, A. Chester, M. Yacoub, Acute effects of cyclosporine and cremophor EL on endothelial function and vascular smooth muscle in the isolated rat-heart, *Cardiovasc. Drug. Ther.* **1992**, *6*, 77–83.

98 L. Mu, S. S. Feng, A novel controlled release formulation for the anticancer drug paclitaxel (Taxol): PLGA nanoparticles containing vitamin E TPGS, *J. Control Release* **2003**, *86*, 33–48.

99 P. L. Weiden, J. Pratt, G. Brand, Tocosol paclitaxel (vitamin E paclitaxel emulsion): multicenter phase 2A studies with weekly dosing in non small cell lung, bladder, ovarian and colorectal cancers, *Proc. Am. Assoc. Cancer Res.* **2003**, *44*, # R 3665.

100 N. K. Ibrahim, N. Desai, S. Legha, P. Soon-Shiong, R. L. Theriault, E. Rivera, B. Esmaeli, S. E. Ring, A. Bedikian, G. N. Hortobagyi, J. A. Ellerhorst, Phase I and pharmacokinetic study of ABI-007, a cremophor free protein stabilized nanoparticles formulation of paclitaxel, *Clin. Cancer Res.* **2002**, *8*, 1038–1044.

101 N. Desai, T. De, A. Yang, B. Beals, P. Soon-Shiong, Pulmonary delivery of a novel cremophor free proteinbased nanoparticle preparation of paclitaxel, *Proc. Am. Assoc. Cancer Res.* **2003**, *44*, # 3672.

102 N. Desai, T. De, A. Yang, B. Beals, V. Trieu, P. Soon-Shiong, Pulmonary delivery of a novel cremophor free proteinbased nanoparticle preparation of paclitaxel, *Proc. Am. Assoc. Cancer Res.* **2003**, *44*, # 3673.

103 J. S. Chawla, M. M. Amiji, Biodegradable poly(ε-caprolactone) nanoparticles for tumor-targeted delivery of tamoxifen, *Internat. J. Pharmaceutics* **2002**, *249*, 127–138.

104 T. Ameller, V. Marsaud, P. Legrand, R. Gref, J. M. Renoir, In vitro and in vivo biologic evaluation of long-circulating biodegradable drug carriers loaded with the pure antiestrogen RU 58668, *Int. J. Cancer* **2003**, *106*, 446–454.

105 R. K. Gilchrist, R. Medal, W. Shorey, R. C. Hanselman, J. C. Parrott, C. B. Taylor, Selective inductive heating of lymph nodes, *Ann. Surg.* **1957**, *146*, 596–606.

106 A. E. Merbach, E. Toth, The chemistry of contrast agents, in: *Medical magnetic resonance imaging*, Chichester, Wiley, **2001**.

107 A. S. Luebbe, C. Bergeman, H. Riess, F. Schriever, P. Reichardt, K. Possinger, M. Matthia, B. Doerken, F. Herrmann, R. Guertler, P. Hohenberger, N. Haas, R. Sohr, B. Sander, A. Lemke, D. Ohlendorf, W. Huhnt, D. Huhn, Clinical experiences with magnetic drug targeting: A Phase I study with 4-epidoxorubicin in 14 patients with advanced solid tumors, *Cancer Res.* **1996**, *56*, 4686–4693.

108 B. R. Bacon, D. D. Stark, C. H. Park, S. Saini, E. V. Groman, P. F. Hahn, C. C. Compton, J. T. Ferrucci Jr., Ferrite particles: a new magnetic resonance imaging contrast agent. Lack of acute or chronic hepatotoxicity after intravenous administration, *J. Lab. Clin. Med.* **1987**, *110*, 164–171.

109 M. G. Harisinghani, J. Barentsz, P. F. Hahn, W. M. Deserno, S. Tabatabaei, C. H. van de Kaa, J. de la Rosette, R. Weissleder, Noninvasive detection of clinically occult lymph-node metastases in prostate cancer, *N. Engl. J. Med.* **2003**, *348*, 2491–2499.

110 A. Senyei, K. Widder, C. Czerlinski, Magnetic guidance of drug carrying microspheres, *J. Appl. Phys.* **1978**, *49*, 3578–3583.

111 P. A. Voltairas, D. I. Fotiadis, L. K. Michalis, Hydrodynamics of magnetic drug targeting, *J. Biomech.* **2002**, *35*, 813–821.

112 A. S. Luebbe, C. Bergemann, W. Huhnt, T. Fricke, H. Riess, J. W. Brock, D. Huhn, Preclinical experiences with magnetic drug targeting: tolerance and efficacy, *Cancer Res.* **1996**, *56*, 4694–4701.

113 C. Alexiou, W. Arnold, R. J. Klein, F. G. Parak, P. Huhn, C. Bergemann, W. Erhardt, S. Wagenpfeil, A. S. Luebbe, Locoregional cancer treatment with magnetic drug targeting, *Cancer Res.* **2000**, *60*, 6641–6648.

114 C. Alexiou, R. Jurgons, R. J. Schmid, C. Bergemann, J. Henke, W. Erhardt, E. Huenges, F. Parak, Magnetic drug targeting – biodistribution of the magnetic carrier and the chemotherapeutic agent mitoxantrone after locoregional cancer treatment, *J. Drug Target* **2003**, *11*, 139–149.

115 A. Moore, E. Marecos, A. Bogdanow, R. Weissleder, Tumoral distribution of long circulating dextran coated iron oxide nanoparticles in anrodent model, *Radiology* **2000**, *214*, 568–574.

116 S. Goodwin, C. Peterson, C. Hob, C. Bittner, Targeting and retention of magnetic targeted carrier (MTC) enhancing intra arterial chemotherapy, *J. Magn. Magn. Mater.* **1999**, *194*, 132–139.

117 S. Goodwin, C. A. Bittner, C. L. Peterson, G. Wong, Single dose toxicity study of hepatic intra arterial infusion of doxorubicin coupled to a novel magnetically targeted drug carrier, *Toxicol. Sci.* **2001**, *60*, 177–183.

118 A. S. Luebbe, C. Bergemann, J. Brock, and D. G. McClure, Physiological aspects in magnetic drug targeting, *J. Magn. Magn. Mater.* **1999**, *194*, 149–155.

119 S. K. Pulfer, S. L. Ciccotto, J. M. Gallo, Distribution of small magnetic particles in brain tumor bearing rats, *J. Neurol. Oncol.* **1999**, *41*, 99–105.

120 T. Kubo, T. Sugita, S. Shimose, Y. Nitta, Y. Ikuta, T. Murakami, Targeted delivery of anticancer drugs with intravenously administered magnetic liposomes in osteosarcoma bearing hamsters, *Int. J. Oncol.* **2000**, *17*, 309–315.

121 T. Kubo, T. Sugita, S. Shimose, Y. Nitta, Y. Ikuta, T. Murakami, Targeted systemic chemotherapy using magnetic liposomes with incorporated adriamycin for osteosarcoma in hamsters, *Int. J. Oncol.* **2001**, *18*, 121–126.

122 M. O. Oyewumi, R. J. Mumper, Engineering tumor-targeted gadolinium hexanedione nanoparticles for potential application in neutron capture therapy, *Bioconjug. Chem.* **2002**, *13*, 1328–1335.

123 Y. Zhang, N. Kohler, M. Zhang, Surface modification of superparamagnetite nanoparticles and their intracellular uptake, *Biomaterials* **2002**, *23*, 1553–1561.

124 K. Na, T. Bum Lee, K. H. Park, E. K. Shin, Y. B. Lee, H. K. Choi, Self-assembled nanoparticles of hydrophobically-modified polysaccharide bearing vitamin H as a targeted anticancer drug delivery system, *Eur. J. Pharm. Sci.* **2003**, *18*, 165–173.

125 Y. Rabin, Is intracellular hyperthermia superior to extracellular hyperthermia in the thermal sense? *Int. J. Hyperthermia* **2002**, *18*, 194–202.

126 A. M. Granov, O. V. Muratov, V. F. Frolov, Problems in the local hyperthermia of inductively heated embolized tissues, Theoretical Foundations of *Chem. Engin.* **2002**, *36*, 63–66.

127 W. J. Atkinson, I. A. Brezpvich, D. P. Chakraborty, Usable frequencies in hyperthermia with thermal seeds, *IEEEE Trans. Biomed. Eng. BME* **1984**, *31*, 70–75

128 E. Fattal, C. Vauthier, I. Aynie, Y. Nakada, G. Lambert, C. Malvy, P. Couvreur, Biodegradable polyalkylcyanoacrylate nanoparticles for the delivery of oligonucleotides, *J. Control Release* **1998**, *53*, 137–143.

129 I. Aynie, C. Vauthier, H. Chacun, E. Fattal, P. Couvreur, Spongelike alginate nanoparticles as a new system for the delivery of antisense oligonucleotides. Antisense, *Nucleic Acid Drug Dev.* **1999**, *9301–9312.*

130 G. Lambert, E. Fattal, H. Pinto-Alphandary, A. Gulik, P. Couvreur, Polyisobutylcyanoacrylate nanocapsules containing an aqueous core as a novel colloidal carrier for the delivery of oligonucleotides, *Pharm. Res.* **2000**, *17*, 707–714.

131 G. Liu, D. Li, M. K. Pasumarthy, T. H. Kowalczyk, C. R. Gedeon, S. L. Hyatt, J. M. Payne, T. J. Miller, P. Brunovskis, T. L. Fink, O. Muhammad, R. C. Moen, R. W. Hanson, M. J. Cooper, Nanoparticles of compacted DNA transfect postmitotic cells, *J. Biol. Chem.* **2003**, *278*, 32578–32586.

132 D. M. Lynn, D. G. Anderson, D. Putnam, R. Langer, Accelerated discovery of synthetic transfection vectors: parallel synthesis and screening of a degradable polymer library, *J. Am. Chem. Soc.* **2001**, *123*, 8155–8156.

133 J. Panyam, W. Z. Zhou, S. Prabha, S. K. Sahoo, V. Labhasetwar, Rapid endo-lysosomal escape of poly(DL-lactide-co-glycolide) nanoparticles: implications for drug and gene delivery, *FASEB J.* **2002**, *16*, 1217–1226.

134 C. Mah, I. Zolotukhin, T. J. Fraites, J. Dobson, C. Batich, B. J. Byrne, Microsphere mediated delivery of recombinant AAV vectors in vitro and in vivo, *Mol. Ther.* **2000**, *1S*, 239.

135 C. Mah, T. J. Fraites, I. Zolotukhin, S. Song, T. R. Flotte, J. Dobson, C. Batich, B. Byrne, Improved method of recombinant AAV2 delivery for systemic targeted gene therapy, *Mol. Ther.* **2002**, *6*, 106–112.

136 J. D. Hood, M. Bednarski, R. Frausto, S. Guccione, R. A. Reisfeld, R. Xiang, D. A Cheresh, Tumor regression by targeted gene delivery to the neovasculature, *Science* **2002**, *296*, 2404–2407.

137 L. Xu, P. Frederik, K. Pirollo, W. H.Tang, A. Rait, L. M. Xiang, W. Huang, I. Cruz, Y. Yin, E. Chang, Self-assembly of a virus-mimicking nanostructure system for efficient tumor targeted gene delivery, *Human Gene Ther.* **2002**, *13*, 469–481.

138 M. Hiraoka, S. Jo, K. Akuta, Y. Nishimura, M. Takahashi, M. Abe, Radiofrequency capacitive hyperthermia for deep-seated tumors. II. Effects of thermoradiotherapy, *Cancer* **1987**, *60*, 128–135.

139 R. Cavaliere, E. C. Ciocatto, B. C. Giovanella, C. Heidelberger, R. O. Johnson, M. Margottini, B. Mondovi, G. Moricca, A. Rossi-Fanelli, Selective heat sensitivity of cancer cells. Biochemical and clinical studies, *Cancer* **1967**, *20*, 1351–1381.

140 P. Wust, B. Hildebrandt, G. Sreenivasa, B. Rau, J. Gellermann, H. Riess, R. Felix, P. M. Schlag, Hyperthermia in combined treatment of cancer, *Lancet Oncol.* **2002**, *3*, 487–497.

141 A. Jordan, R. Scholz, P. Wust, H. Faehling, R. Felix, Magnetic fluid hyperthermia (MFH): cancer treatment with AC magnetic field induced excitation of biocompatible superparamagnetic nanoparticles, *J. Magn. Magn. Mater.* **1999**, *201*, 413–419.

142 I. Hilger, R. Hergt, W. A. Kaiser, Effects of magnetic thermoablation in muscle tissue using iron oxide particles: an in vitro study, *Invest. Radiol.* **2000**, *35*, 170–179.

143 M. Shinkai, M. Matsui, T. Kobayashi, Heat properties of magnetoliposomes for local hyperthermia, *Jpn. J. Hyperthetm. Oncol.* **1994**, *10*, 168–177

144 R. Hiergeist, W. Andrae, N. Buske, R. Hergt, I. Hilger, U. Richter, W. Kaiser, Application of magnetite ferrofluids for hyperthermia, *J. Magn. Magn. Mater.* **1999**, *201*, 420–422.

145 I. Hilger, K. Fruhauf, W. Andra, R. Hiergeist, R. Hergt, W. A. Kaiser, Heating potential of iron oxides for therapeutic purposes in interventional radiology, *Acad. Radiol.* **2002**, *9*, 198–202.

146 R. E. Rosenzweig, Heating magnetic fluid with alternating magnetic field, *J. Magn. Magn. Mater.* **2002**, *252*, 370–374.

147 A. V. Brusentsov, V. V. Gogosov, T. N. Buntsova, A. V. Sergeev, N. Y. Jurchenko, A. Kuznetsov, O. Kuznetsov, L. I. Shumakov, Evaluation of ferromagnetic fluids and suspensions for the site-specific radiofrequency-induced hyperthermia of MX11 sarcoma cells in vitro, *J. Magn. Magn. Mater.* **2001**, *225*, 113–119.

148 A. Jordan, R. Scholz, P. Wust, H. Fahling, J. Krause, W. Wlodarczyk, B. Sander, T. Vogl, R. Felix, Effects of magnetic fluid hyperthermia (MFH) on C3H mammary carcinoma in vivo, *Int. J. Hypertherm.* **1997**, *13*, 587–605.

149 R. T. Gordon, J. R. Hines, D. Gordon, Intracellular hyperthermia. A biophysical approach to cancer treatment via intracellular temperature and biophysical alterations, *Med. Hypoth.* **1979**, *5*, 83–102.

150 Q. A. Pankhurst, J. Connolly, S. K. Jones, J. Dobson, Applications of magnetic nanoparticles in biomedicine, *J. Phys. D Appl. Phys.* **2003**, *36*, R167–R181.

193 A. K. Larsen, A. E. Escargueil,
A. Skladanowski, Resistance mechanisms
associated with altered intracellular distribution of anticancer agents, *Pharmacol. Ther.*
2000, *88*, 217–229.

194 S. Bennis, C. Chapey, P. Couvreur, J. Robert,
Enhanced cytotoxicity of doxorubicin encapsulated in polyisohexylcyanoacrylate nanospheres against multidrug-resistant tumour
cells in culture, *Eur. J. Cancer A* **1994**, *30*,
89–93.

195 A. Colin de Verdiere, C. Dubernet, F. Nemati,
M. F. Poupon, F. Puisieux, P. Couvreur,
Uptake of doxorubicin from loaded nanoparticles in multidrug-resistant leukemic murine
cells, *Cancer Chemother. Pharmacol.* **1994**, *33*,
504–508.

196 A. Astier, B. Doat, M. J. Ferrer, G. Benoit,
J. Fleury, A. Rolland, R. Leverge, Enhancement of adriamycin antitumor activity by its
binding with an intracellular sustained-
release form, polymethacrylate nanospheres,
in U-937 cells, *Cancer Res.* **1988**, *48*,
1835–1841.

197 A. C. de Verdiere, C. Dubernet, F. Nemati,
E. Soma, M. Appel, J. Ferte, S. Bernard,
F. Puisieux, P. Couvreur, Reversion of multidrug resistance with polyalkylcyanoacrylate
nanoparticles: towards a mechanism of
action, *Br. J. Cancer* **1997**, *76*, 198–205.

198 D. Goren, A. T. Horowitz, D. Tzemach,
M. Tarshish, S. Zalipsky, A. Gabizon, Nuclear
delivery of doxorubicin via folate-targeted
liposomes with bypass of multidrug-resis-
tance efflux pump, *Clin. Cancer Res.* **2000**, *6*,
1949–1957.

199 C. Cuvrie, Doxorubicin loaded nanospheres
bypass tumor cell multi-drug resistance, *Biochem. Pharm.* **1992**, *44*, 509–517.

12
Diagnostic and Therapeutic Applications of Metal Nanoshells

Christopher Loo, Alex Lin, Leon Hirsch, Min-Ho Lee, Jennifer Barton, Naomi Halas, Jennifer West, and Rebekah Drezek

Abstract

Metal nanoshells are a novel type of composite spherical nanoparticle consisting of a dielectric silica core covered by a thin metallic shell, which is typically gold. Nanoshells possess highly favorable optical and chemical properties for biomedical imaging and therapeutic applications. By varying the relative dimensions of the core and the shell, the optical resonance of these nanoparticles can be precisely and systematically varied over a broad wavelength region from the near-UV to the mid-infrared. This range includes the near-infrared (NIR) region where tissue transmissivity peaks. In addition to spectral tunability, nanoshells offer other advantages over conventional organic dye imaging agents, including improved optical properties and reduced susceptibility to chemical/thermal denaturation. Furthermore, the same conjugation protocols used to bind biomolecules to gold colloid are easily modified for nanoshells. In this article, we first review the synthesis of gold nanoshells and illustrate how the core/shell ratio and overall size of a nanoshell influences its scattering and absorption properties. We then describe several examples of nanoshell-based diagnostic and therapeutic approaches including the development of nanoshell bioconjugates for molecular imaging, the use of scattering nanoshells as contrast agents for optical coherence tomography (OCT), and the use of absorbing nanoshells in NIR thermal therapy of tumors.

12.1
Introduction

There is a significant clinical need for novel methods for detection and treatment of cancer which offer improved sensitivity, specificity, and cost-effectiveness. In recent years, a number of groups have demonstrated that photonics-based technologies are valuable in addressing this need [14–17]. Optical technologies promise high-resolution, noninvasive functional imaging of tissue at competitive costs. However, in many cases, these technologies are limited by the inherently weak optical signals of endogenous chromophores and the subtle spectral differences between normal and diseased tissue.

Nanofabrication Towards Biomedical Applications. C. S. S. R. Kumar, J. Hormes, C. Leuschner (Eds.)
Copyright © 2005 WILEY-VCH Verlag GmbH & Co. KGaA, Weinheim
ISBN 3-527-31115-7

Over the past several years, there has been increasing interest in combining emerging optical technologies, with the development of novel exogenous contrast agents designed to probe the molecular specific signatures of cancer, to improve the detection limits and clinical effectiveness of optical imaging. For instance, Sokolov *et al.* [1] recently demonstrated the use of gold colloid conjugated to antibodies to the epidermal growth factor receptor (EGFR) as scattering contrast agents for biomolecular optical imaging of cervical cancer cells and tissue specimens. In addition, optical imaging applications of nanocrystal bioconjugates have been described by multiple groups including Bruchez *et al.* [2], Chan and Nie [3], and Akerman *et al.* [4]. More recently, interest has developed in the creation of nanotechnology-based platform technologies which couple molecular-specific early detection strategies with appropriate therapeutic intervention and monitoring capabilities.

Metal nanoshells are a new type of nanoparticle composed of a dielectric core such as silica coated with an ultrathin metallic layer, typically gold. Gold nanoshells possess physical properties similar to those of gold colloid, in particular, a strong optical absorption due to the collective electronic response of the metal to light. The optical absorption of gold colloid yields a brilliant red color which has been of considerable utility in consumer-related medical products, such as home pregnancy tests. In contrast, the optical response of gold nanoshells depends dramatically on the relative size of the nanoparticle core and the thickness of the gold shell. By varying the relative core and shell thicknesses, the optical resonance of gold nanoshells can be varied across a broad range of the optical spectrum that spans the visible and

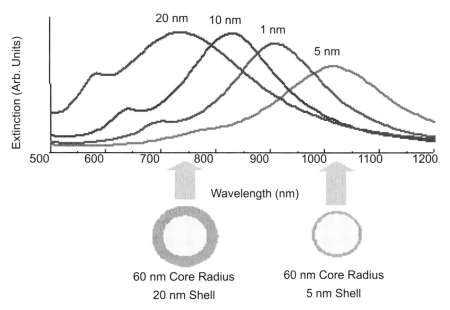

Figure 12.1. Mie scattering plot of optical resonances of gold shell/silica core nanoshells as a function of their core/shell ratio. Respective spectra correspond to the nanoparticles depicted beneath.

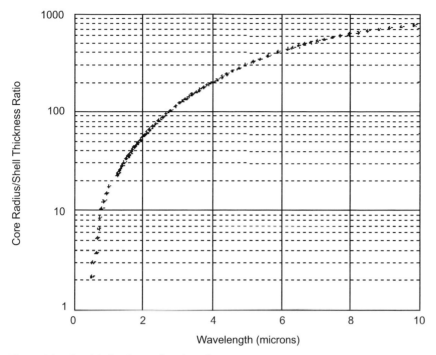

Figure 12.2. Core/shell ratio as a function of resonance wavelength for gold/silica nanoshells.

the NIR spectral regions [5, 6]. Gold nanoshells can be made to either preferentially absorb or scatter light by varying the size of the particle relative to the wavelength of the light at their optical resonance. Figure 12.1 shows a Mie scattering plot of the nanoshell plasmon resonance wavelength shift as a function of nanoshell composition for the case of a 60-nm-core gold/silica nanoshell. In this figure, the core and shell of the nanoparticles are shown to relative scale directly beneath their corresponding optical resonances. Figure 12.2 displays a plot of the core/shell ratio versus resonance wavelength for a silica core/gold shell nanoparticle [6]. The highly agile "tunability" of the optical resonance is a property unique to nanoshells: in no other molecular or nanoparticle structure can the resonance of the optical absorption properties be so systematically "designed."

Halas and colleagues have completed a comprehensive investigation of the optical properties of metal nanoshells [7]. Quantitative agreement between Mie scattering theory and the experimentally observed optical resonant properties has been achieved. Based on this success, it is now possible to design gold nanoshells predictively with the desired optical resonant properties, and then to fabricate the nanoshell with the dimensions and nanoscale tolerances necessary to achieve these properties [6]. The synthetic protocol developed for the fabrication of gold nanoshells is very simple in concept:

1. Grow or obtain silica nanoparticles dispersed in solution.
2. Attach very small (1- to 2-nm) metal "seed" colloid to the surface of the nano-particles via molecular linkages; these seed colloids cover the dielectric nano-particle surfaces with a discontinuous metal colloid layer.
3. Grow additional metal onto the "seed" metal colloid adsorbates via chemical reduction in solution.

This approach has been successfully used to grow both gold and silver metallic shells onto silica nanoparticles. Various stages in the growth of a gold metallic shell onto a functionalized silica nanoparticle are shown in Fig. 12.3. Figure 12.4 shows the optical signature of nanoshell coalescence and growth for two different nano-shell core diameters.

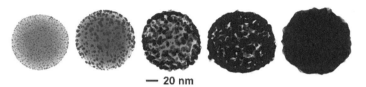

— 20 nm

Figure 12.3. Transmission electron microscope images of gold/silica nanoshells during shell growth.

Based on the core/shell ratios that can be achieved with this protocol, gold nano-shells with optical resonances extending from the visible region to approximately 3 µm in the infrared can currently be fabricated. This spectral region includes the 800–1300 nm "water window" of the NIR, a region of high physiological transmis-sivity which has been demonstrated as the spectral region best suited for optical bioi-maging and biosensing applications [18]. The optical properties of gold nanoshells,

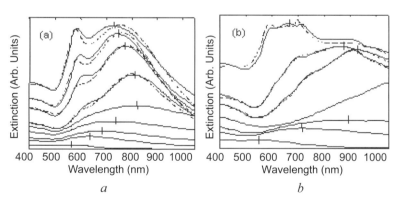

a

b

Figure 12.4. (a) Growth of gold shell on 120-nm-diameter silica nanoparticle. The lower spectral curves follow the evolution of the opti-cal absorption as coalescence of the gold layer progresses. Once the shell is complete, the peak absorbance is shifted to shorter wave-lengths. Corresponding theoretical peaks are plotted with dashed lines. (b) Growth of gold shell on 340-nm-diameter silica nanoparticles. Here the peak shifts are more pronounced with only the shoulder of the middle curve visible in our instrument range.

when coupled with their biocompatibility and their ease of bioconjugation, render these nanoparticles highly suitable for targeted bioimaging and therapeutics applications. By controlling the physical parameters of the nanoshells, it is possible to engineer nanoshells which primarily scatter light, as would be desired for many imaging applications, or, alternatively, to design nanoshells which are strong absorbers, permitting photothermal-based therapy applications. The tailoring of scattering and absorption cross-sections is demonstrated in Fig. 12.5, which shows sample spectra for two nanoshell configurations, one designed to scatter light and the other to preferentially absorb light.

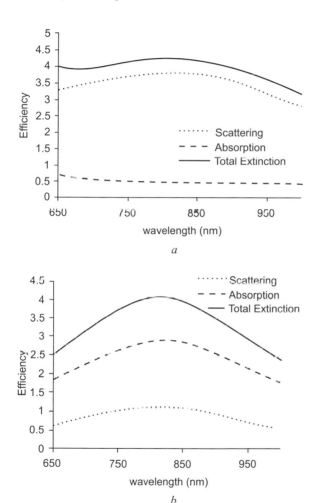

Figure 12.5. Nanoshells may be designed to be predominantly scattering or predominantly absorbing by tailoring the core and shell fabrication materials. To demonstrate this concept, the predicted scattering efficiency, absorption efficiency, and extinction are shown for two nanoshells: (a) a scattering configuration (core radius = 40 nm; shell thickness = 20 nm) and (b) an absorbing configuration (core radius = 50 nm; shell thickness = 10 nm).

Because the metal layer of gold nanoshells is grown using the same chemical reaction as gold colloid synthesis, the surfaces of gold nanoshells are chemically virtually identical to the surfaces of the gold nanoparticles universally used in bioconjugate applications. The use of gold colloid in biological applications began in 1971 when Faulk and Taylor invented the immunogold staining procedure. Since that time, the labeling of targeting molecules, especially proteins, with gold nanoparticles has revolutionized the visualization of cellular or tissue components by electron microscopy. The optical and electron beam contrast qualities of gold colloid have provided excellent detection qualities for such techniques as immunoblotting, flow cytometry, and hybridization assays [8]. Conjugation protocols exist for the labeling of a broad range of biomolecules with gold colloid, such as protein A, avidin, streptavidin, glucose oxidase, horseradish peroxidase, and IgG. Successful gold nanoshell conjugation with enzymes and antibodies has previously been demonstrated [13]. In this article, we present data demonstrating the potential of nanoshells for several biomedical applications including the use of nanoshell bioconjugates as biological labels for optical imaging, the development of nanoshell-based scattering contrast agents for optical coherence tomography, and the use of absorbing nanoshells for photothermal therapy of tumors.

12.2
Methodology

Gold nanoshell fabrication

Cores of silica nanoparticles were fabricated as described by Stober *et al.* [9] in which tetraethylorthosilicate was reduced in NH_4OH in ethanol. Particles were sized with a Philips XL30 scanning electron microscope. Polydispersity of less than 10% was considered acceptable. Next, the silica surface was aminated by reaction with aminopropyltriethoxysilane in ethanol. Gold shells were grown using the method of Duff *et al.* [10]. Briefly, small gold colloid (1–3 nm) was adsorbed onto the aminated silica nanoparticle surface. More gold was then reduced onto these colloid nucleation sites using potassium carbonate and $HAuCl_4$ in the presence of formaldehyde. Gold nanoshell formation and dimensions were assessed with a UV-VIS spectrophotometer and scanning electron microscopy (SEM). The nanoshells used in the darkfield scattering imaging studies described consisted of a 120-nm silica core radius with a 35-nm-thick gold shell. The nanoshells used in the optical coherence tomography (OCT) imaging consisted of a 100-nm core radius and 20 nm thick shell. The nanoshells used in the therapy application described used a 60-nm core radius and a 10-nm-thick shell which absorb light with an absorption peak at ~815 nm. The reader is referred to Ref. [6] for a detailed description of nanoshell synthesis procedures.

Antibody conjugation

Ortho-pyridyl-disulfide-*n*-hydroxysuccinimide polyethylene glycol polymer (OPSS-PEG-NHS, MW=2000) was used to tether antibodies onto the surfaces of gold nanoshells. Using NaHCO$_3$ (100 mM, pH 8.5), OPSS-PEG-NHS was resuspended to a volume equal to that of either HER2 (specific) or IgG (nonspecific) antibodies. At this concentration, the concentration of polymer was in molar excess to the amount of HER2 or IgG antibody used. The reaction was allowed to proceed on ice over-night. Excess, unbound polymer was removed by membrane dialysis (MWCO=10,000). PEGylated antibody (0.67 mg mL^{-1}) was added to nanoshells (~10^9 nanoshells mL^{-1}) for 1 h to facilitate targeting. Unbound antibody was removed by centrifugation at 650 g, supernatant removal, and resuspension in potassium carbonate (2 mM). Following antibody conjugation, nanoshells surfaces were further modified with PEG-thiol (MW=5000, 1 μM) to block nonspecific adsorption sites and to enhance biocompatibility.

Cell culture
HER2-positive SKBR3 human mammary adenocarcinoma cells were cultured in McCoy's 5A modified medium supplemented with 10% FBS and antibiotics. Cells were maintained at 37 °C and 5% CO$_2$.

Molecular imaging, cytotoxicity, and silver staining
SKBR3 cells were exposed to 8 μg mL^{-1} of bioconjugated nanoshells for 1 h, washed with phosphate-buffered saline, and observed under darkfield microscopy, a form of microscopy sensitive only to scattered light. The calcein-AM live stain (Molecular Probes, 1 μM) was used to assess cell viability after nanoshell targeting. A silver enhancement stain (Amersham Pharmacia), a qualitative stain capable of detecting the presence of gold on cell surfaces, was used to assess cellular nanoshell binding. Cells incubated with targeted nanoshells were fixed with 2.5% glutaraldehyde and exposed to silver stain for 15 min. Silver growth was monitored under phase con-trast, with further silver enhancement blocked by immersion in 2.5% sodium thio-sulfate. Darkfield and silver stain images were taken with a Zeiss Axioskop 2 plus microscope equipped with a black–white CCD camera. All images were taken at 40× magnification under the same lighting conditions.

Optical coherence tomography
Optical coherent tomography (OCT) is a state-of-the-art imaging technique which produces high-resolution (typically 10–15 μm), real-time, cross-sectional images through biological tissues. The method is often described as an optical analog to ultrasound. OCT detects the reflections of a low-coherence light source directed into a tissue and determines at what depth the reflections occurred. By employing a het-erodyne optical detection scheme, OCT is able to detect very faint reflections relative to the incident power delivered to the tissue. In OCT imaging, out-of-focus light is strongly rejected due to the coherence gating inherent to the approach. This permits deeper imaging using OCT than is possible using alternative methods such as reflectance confocal microscopy, where the out-of-focus rejection achievable is far lower. The imaging depth of OCT depends on tissue type but is usually up to several millimeters. In the OCT experiments described in this paper, a conventional OCT system with an 830-nm superluminescent diode was used to obtain m-scans of the cuvette (images with time as the x-axis and depth as the y-axis). The axial and lateral

resolution of the OCT system were 16 μm and 12 μm, respectively. Each image required approximately 20 s to acquire. System parameters remained the same throughout the experiment.

In vitro photothermal nanoshell therapy

SKBR3 breast cancer cells were cultured in 24-well plates until fully confluent. Cells were then divided into two treatment groups: nanoshells + NIR-laser and NIR-laser alone. Cells exposed to nanoshells alone or cells receiving neither nanoshells nor laser were used as controls. Nanoshells were prepared in FBS-free medium (2×10^9 nanoshells mL^{-1}). Cells were then irradiated under a laser emitting light at 820 nm at a power density of ~35 W cm^{-2} for 7 min with or without nanoshells. After NIR-light exposure, cells were replenished with FBS-containing media and were incubated for an additional hour at 37 °C. Cells were then exposed to the calcein-AM live stain for 45 min in order to measure cell viability. The calcein dye causes viable cells to fluoresce green. Fluorescence was visualized with a Zeiss Axiovert 135 fluorescence microscope equipped with a filter set specific for excitation and emission wavelengths at 480 and 535 nm, respectively. Membrane damage was assessed using an aldehyde-fixable fluorescein dextran dye. Cells were incubated for 30 min with the fluorescent dextran, rinsed, and immediately fixed with 5% glutaraldehyde. Photothermal destruction of cells was attributed to hyperthermia induced via nanoshell absorption of NIR light.

12.3
Results and Discussion

As an initial demonstration of the potential of nanoshells in cancer imaging and therapy, we designed and fabricated nanoshells suitable for both scattering- and absorption-based photonics applications. For proof-of-principle imaging studies, we fabricated nanoshells with a 120-nm radius and 35-nm shell thickness. It should be noted that nanoshells over a broad range of sizes can be fabricated for scattering-based imaging applications. Figure 12.6 displays the predicted scattering and absorption spectra for these nanoshells obtained using software extensively verified against Mie theory which numerically computes optical spectra for gold nanoshells. As Fig. 12.6 demonstrates, these nanoshells scatter light strongly throughout the visible and NIR regions. This permits the same nanoshells to be used in light-based microscopy studies employing silicon CCDs and in NIR tissue imaging studies using reflectance confocal microscopy and OCT. We also fabricated nanoshells with a 100-nm radius and 20-nm shell thickness for OCT imaging. These nanoshells have very similar scattering and absorption spectra to the larger nanoshells; however, the scattering and absorption cross-sections are smaller, due largely to the smaller particle size. In addition, smaller 60-nm radius nanoshells with a 10-nm shell were fabricated for photothermal therapy applications. Figure 12.7 shows SEM images of the nanoshells fabricated at all three sizes.

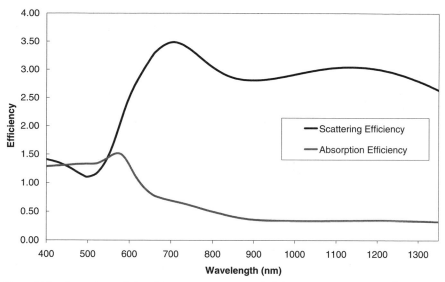

Figure 12.6. Scattering and absorbing properties of nanoshells with a 120-nm silica core radius and a 35-nm-thick gold shell predicted analytically. Scattering maximum (705–710 nm) is 2.5 times greater than absorption maximum (570 nm) and extends into the NIR region. Nanoshell dimensions were assessed using scanning electron microscopy (SEM). Shell thickness was mathematically corroborated by matching experimental measurements to scattering theory and confirmed with SEM.

As an initial demonstration of the molecular imaging potential of nanoshell bioconjugates, we imaged carcinoma cells which overexpress HER2, a clinically significant molecular marker of breast cancer. Under darkfield microscopy, a form of microscopy sensitive only to scattered light, significantly increased optical contrast due to HER2 expression was observed in HER2-positive SKBR3 breast cancer cells targeted with HER2-labeled nanoshells compared to cells targeted by nanoshells nonspecifically labeled with IgG (Fig. 12.8). In addition, greater silver staining intensity was seen in cells exposed to HER2-targeted nanoshells than in cells exposed to IgG-targeted nanoshells, providing additional evidence that the increased contrast seen under darkfield may be specifically attributable to nanoshell targeting of the HER2 receptor. No differences were observed under darkfield or silver stain in HER2 and IgG-targeted nanoshells using the HER2-negative MCF7 breast cancer cell line (data not shown). More extensive descriptions of imaging experiments using nanoshell bioconjugates are described in Ref. [11].

Although darkfield microscopy is suitable for *in vitro* cell level imaging experiments, *in vivo* imaging applications will require the use of appropriate scattering-based imaging technologies such as OCT. To assess the suitability of nanoshells for OCT applications, we computed the scattering efficiencies of gold nanoshells (in saline) over a range of core radii and shell thicknesses at 830 nm as shown in Fig. 12.9. The promising scattering cross-sections (approximately several times the geometric cross-sections) computed for nanoshells based on physical parameters which

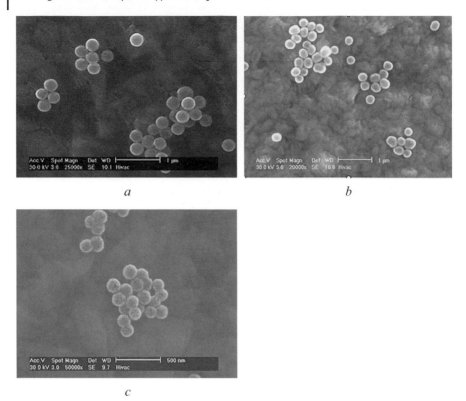

Figure 12.7. SEM images of nanoshells used in the described studies. (a) The larger-diameter nanoshells used in the darkfield imaging experiments. (b) The nanoshells used in the OCT experiments. (c) The smaller-diameter nanoshells used for photothermal therapy applications. The scale bars in (a) and (b) are 1 μm while the scale bar in (c) is 500 nm.

could be readily fabricated encouraged further experimental investigation. To provide a basis for comparison of scattering efficiencies, a 150-nm-diameter polystyrene sphere in saline at 830 nm has a scattering efficiency of 0.009; a 300-nm polystyrene sphere has an efficiency of 0.09. As a visual demonstration of the potential of nanoshells for OCT imaging applications, we imaged a 1-mm-pathlength cuvette containing one of three solutions: saline, a microsphere-based scattering solution, or a solution of scattering nanoshells in water (Fig. 12.10). The microsphere mixture was 0.1% solids by volume of 2-μm polystyrene spheres in saline at a concentration which provided a scattering coefficient, $\mu_{s,} = 16$ cm^{-1}, and an anisotropy factor, $g = 0.96$. The nanoshell (100 nm radius, 20 nm shell) concentration was approximately 10^9 mL^{-1}. Figure 12.10 shows OCT images of the cuvette with saline, microspheres and nanoshells. The images consist of 100 scans in the same lateral location. The average grayscale value inside the cuvette walls was calculated using the National Institutes of Health Image Analysis Program. The OCT intensity is based on a log scale where black (255) corresponds to the noise floor of −100 dB and white (0) to −40 dB. To provide an approximate comparison of measured scattered intensity,

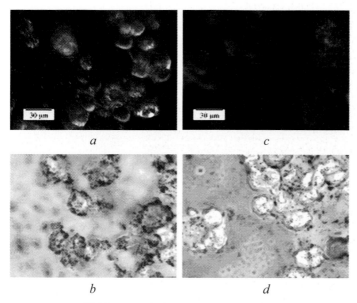

Figure 12.8. Darkfield (a, c) and silver stain (b, d) images of HER2-positive SKBR3 breast cancer cells exposed to nanoshells conjugated with either (a, b) HER2 (specific) or (c, d) IgG (nonspecific) antibodies. As demonstrated here, it is possible to exploit the optical properties of predominantly scattering nanoshells to image overexpressed HER2 in living cells. Similar scattering intensities were observed when comparing cells exposed to IgG-targeted nanoshells and cells not exposed to nanoshell bioconjugates.

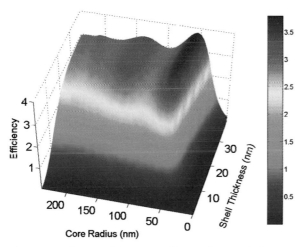

Figure 12.9. Computed scattering efficiency for nanoshells as a function of core radius and shell thickness at 830 nm, a wavelength commonly used in OCT imaging applications.

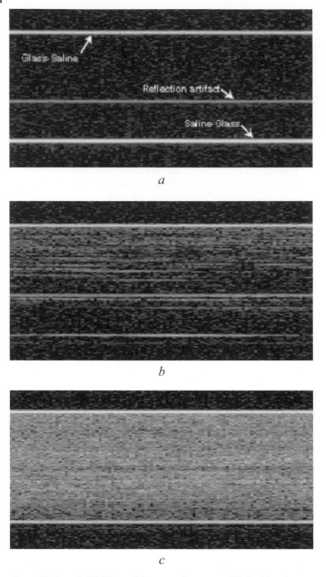

Figure 12.10. OCT (830 nm) images of a cuvette filled with saline (a), a cuvette containing microspheres to approximate a scattering coefficient of 16 cm^{-1} (b), and a cuvette containing nanoshells at a concentration of ~10^9 ml^{-1} (c).

the average grayscale intensity for saline was 247 while the average intensity within the cuvette walls containing nanoshells was 160. Current work is more carefully exploring the potential of nanoshells as contrast agents for OCT through *in vivo* imaging studies of mice after direct injection of scattering nanoshells into the vasculature via a tail vein catheter [12].

Future efforts will be directed towards coupling nanoshell-based molecular imaging technologies to some form of triggerable therapeutic intervention. Recent studies have considered a novel approach to cancer therapy based on the use of metal nanoshells as NIR absorbers [13]. In biological tissue, tissue transmissivity is highest in the NIR spectral range due to low inherent scattering and absorption properties within the region. Figure 12.6 demonstrates that nanoshells can be developed to highly scatter within this spectral region; alternatively, nanoshells may be engineered to function as highly effective NIR absorbers as well. As an example of the intense absorption possible using nanoshells, the conventional NIR dye indocyanine green has an absorption cross-section of $\sim10^{-20}$ m^2 at $\sim$800 nm while the cross-section of the absorbing nanoshells described in this article is $\sim4 \times 10^{-14}$ m^2, an approximately million-fold increase in absorption cross-section [13]. By combining NIR-absorbing nanoshells with an appropriate light source, it is possible to selectively induce photothermal destruction of cells and tumors treated with gold nanoshells. Nanoshell-mediated photothermal destruction of carcinoma cells is demonstrated in Fig. 12.11. After laser exposure at 35 W cm^{-2} for 7 min, all cells within the laser spot underwent photothermal destruction as assessed using calcein AM viability staining, an effect that was not observed in cells exposed to either nanoshells alone or NIR light alone. In addition, evidence of irreversible cell membrane damage was noted in the cells within the laser spot via imaging of fluorescent dextran dye (data not shown). This dye is normally impermeable to healthy cells. The dye was found in the intracellular space of cells exposed to both NIR nanoshells and the laser but was not observed in cells exposed to either the NIR nanoshells or the laser alone. The calcein AM stain and the fluorescent dextran stain can be used to indicate that the cells are not viable and that membrane damage has occurred but do not determine the underlying cause of cell death.

In an animal study described in Ref. [13], absorbing nanoshells (10^9 ml^{-1}, 20–50 μl) were injected interstitially ($\sim$5 mm) into solid tumors ($\sim$1 cm) in female SCID mice. Within 30 min of injection, tumor sites were exposed to NIR light

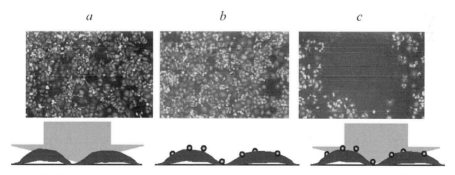

a *b* *c*

Figure 12.11. Calcein AM staining of cells (green fluorescence indicates cellular viability). (a) Cells after exposure to laser only (no nanoshells). (b) Cells incubated with nanoshells but not exposed to laser light. (c) Cell incubated with nanoshells after laser exposure. The dark circle seen in (c) corresponds to the region of cell death caused by exposure to laser light after incubation with nanoshells.

(820 nm, 4 W cm^{-2}, 5 mm spot diameter, <6 min). Temperatures were monitored via phase-sensitive, phase-spoiled gradient-echo MRI. Magnetic resonance temperature imaging (MRTI) demonstrated that tumors reached temperatures which caused irreversible tumor damage ($\Delta T = 37.4 \pm 6.6\,°C$) within 4–6 min. Controls which were exposed to a saline injection rather than nanoshells experienced significantly reduced average temperatures after exposure to the same NIR light levels ($\Delta T = 10\,°C$). These average temperatures were obtained at a depth of ~2.5 mm below the surface of the skin. The MRTI findings demonstrated good agreement with gross pathology indicators of tissue damage. Histological indications of thermal damage including coagulation, cell shrinkage, and loss of nuclear staining were noted in nanoshell-treated tumors; no such changes were found in control tissue. Silver enhancement staining provided further evidence of nanoshells in regions with thermal damage.

The initial work described here established nanoshell and laser dosages which provided effective nanoshell-mediated photothermal therapy. Based on the parameters identified through these initial investigations, survival studies are now underway. Future work will also consider nanoshells conjugated to surface markers overexpressed within tumors.

12.4
Conclusions

Combining advances in biophotonics and nanotechnology offers the opportunity to significantly impact future strategies towards the detection and therapy of cancer. Today, cancer is typically diagnosed many years after it has developed, usually either after the discovery of a palpable mass or based on relatively low-resolution imaging of smaller but still significant masses. In the future, it is likely that contrast agents targeted to molecular markers of disease will routinely provide molecular information that enables characterization of disease susceptibility long before pathologic changes occur at the anatomic level. Currently, our ability to develop molecular contrast agents is at times constrained by limitations in our understanding of the earliest molecular signatures of specific cancers. Although the process of identifying appropriate targets for detection and therapy is ongoing, there is a strong need to develop the technologies which will allow us to image these molecular targets *in vivo* as they are elucidated. In this article, we have described the optical properties and several emerging clinical applications of nanoshells, one class of nanostructures which may provide an attractive candidate for specific *in vivo* imaging and therapy applications. We have reviewed our preliminary work towards the development of nanoshell bioconjugates for molecular imaging applications and described an important new approach to photothermal cancer therapy. More extensive *in vivo* animal studies for both cancer imaging and therapy applications are currently underway in order to investigate more thoroughly both the potential and any limitations of nanoshell technologies. Additional studies are in progress to assess more thoroughly the biodistribution and biocompatibility of nanoshells used in *in vivo* imaging and therapy applications. We believe there is tremendous potential for synergy between the rapidly developing fields of biophotonics and nanotechnology. Combining the tools of both fields – together with the latest advances in understanding the molecular origins of cancer – may provide a fundamentally new approach to detection and treatment of cancer, a disease responsible for over one-quarter of all deaths in the United States today.

Acknowledgments

Funding for this project was provided by the National Science Foundation (BES 022–1544), the National Science Foundation Center for Biological and Environmental Nanotechnology (EEC-0118007), and the Department of Defense Congressionally Directed Medical Research Program (DAMD17–03–1-0384).

References

1 Sokolov, K., Follen, M., Aaron, J., Pavlova, I., Malpica, A., Lotan, R., Richards-Kortum, R., Real-time vital optical imaging of precancer using anti-epidermal growth factor receptor antibodies conjugated to gold nanoparticles. *Cancer Res.* **2003**, *63*, 1999–2004.

2 Bruchez, M., Moronne, M., Gin, P., Weiss, S., Alivisatos, A. P., Semiconductor labels as fluorescent biological labels. *Science* **1998**, *281*, 2013–2016.

3 Chan, W. C. W., Nie, S., Quantum dot bioconjugates for ultrasensitive nonisotopic detection. *Science* **1998**, *281*, 2016–2018.

4 Akerman, M. E., Chan, W., Laakkonen, P., Bhatia, S. N., Ruoslahti, E., Nanocrystal targeting in vivo. *PNAS* **2002**, *99*, 12617–12621.

5 Brongersma, M. L., Nanoshells: gifts in a gold wrapper. *Nat. Mater.* **2003**, *2*, 296–297.

6 Oldenburg, S. J., Averitt, R. D., Westcott, S. L., Halas, N. J., Nanoengineering of optical resonances. *Chem. Phys. Lett.* **1998**, *288*, 243–247.

7 Averitt R. D., Sarkar D., Halas N. J., Plasmon resonance shifts of Au-coated Au2S nanoshells: insights into multicomponent nanoparticles growth. *Phys. Rev. Lett.* **1997**, *78*, 4217–4220.

8 Faulk, W. T., Taylor G., An immunocolloid method for the electron microscope. *Immunochemistry* **1971**, *8*, 1081–1083.

9 Stober, W., Fink, A., Bohn, E., Controlled growth of monodisperse silica spheres in the micron size range. *J. Colloid Interface Sci.* **1968**, *26*, 62–69.

10 Duff, D. G., Baiker, A., Edwards, P. P., A new hydrosol of gold clusters. 1. Formation and particle size variation. *Langmuir* **1993**, *9*, 2301–2309.

11 Loo, C. H., Hirsch, L. R, West, J. L., Halas, N. J., Drezek, R. A. Molecular imaging in living cells using nanoshell bionconjugates. *Optics Letters* **2004** (in review).

12 Barton, J., Romanowski, M., Halas, N., Drezek, R., Nanoshells as an OCT contrast agent. *Proc SPIE* 5316, **2004** (in press).

13 Hirsch, L. R., Stafford, R. J., Bankson, J. A., Sershen, S. R., Rivera, B., Price, R. E., Hazle, J. D., Halas, N. J., West, J. L., Nanoshell-mediated near-infrared thermal therapy of tumors under magnetic resonance guidance. *PNAS* **2003**, *100*, 13549–13554.

14 Cerussi, A. E., Jakubowski, D., Shah, N., et al., Spectroscopy enhances the information content of optical mammography. *J. Biomed. Optics* **2002**, *7*, 60–71.

15 Ntziachristos, V., Chance, B., Probing physiology and molecular function using optical imaging: applications to breast cancer. *Breast Cancer Res.* **2001**, *3*, 41–46.

16 Shah, N., Cerussi, A., Eker, C., et al., Noninvasive functional optical spectroscopy of human breast tissue. *Proc. Natl. Acad. Sci. U. S. A.* **2001**, *98*, 4420–4425.

17 Tromberg, B. J., Shah, N., Lanning, R., et al., Non-invasive in vivo characterization of breast tumors using photon migration spectroscopy. *Neoplasia* **2000**, *2*, 26–40.

18 Richards-Kortum, R., Sevick-Muraca, E.,
Quantitative optical spectroscopy for tissue
diagnosis. *Annu. Rev. Phys. Chem.* **1996**, *47*,
555–606.

13

Decorporation of Biohazards Utilizing Nanoscale Magnetic Carrier Systems

Axel J. Rosengart and Michael D. Kaminski

Disclaimer

This report was prepared as an account of work sponsored by an agency of the United States Government. Neither the United States Government nor any agency thereof, nor The University of Chicago, nor any of their employees or officers, makes any warranty, express or implied, or assumes any legal liability or responsibility for the accuracy, completeness, or usefulness of any information, apparatus, product, or process disclosed, or represents that its use would not infringe privately owned rights. Reference herein to any specific commercial product, process, or service by trade name, trademark, manufacturer, or otherwise, does not necessarily constitute or imply its endorsement, recommendation, or favoring by the United States Government or any agency thereof. The views and opinions of document authors expressed herein do not necessarily state or reflect those of the United States Government or any agency thereof, Argonne National Laboratory, or The University of Chicago.

13.1
introduction

We are developing a novel, integrated system based on superparamagnetic, biocompatible nanospheres for selective and rapid decorporation of biological, chemical, and radioactive biohazards from humans. The system utilizes polymer-based magnetic nanospheres that are injected directly into the blood stream of biohazard-exposed humans. The composition of the injected nanospheres is biodegradable polymers of lactic acid or lactide–glycolide. Since it is well recognized that systemic injection of naked polymer spheres will result in rapid bioclearance by the reticuloendothelial system, the surface must be coupled to long chains of polyethylene glycol. Magnetic iron oxide nanocrystals are encapsulated within the polymer. Receptors are terminally attached to the polyethylene glycol or encapsulated within the polymer sphere, depending on the application. In total, the chemical components of the spheres confer nontoxicity and biocompatibility, avoid rapid bioclearance, and are biostabilized temporarily. Once injected, the spheres circulate freely through the blood stream, selectively capturing blood-borne toxins to specific recep-

Nanofabrication Towards Biomedical Applications. C. S. S. R. Kumar, J. Hormes, C. Leuschner (Eds.)
Copyright © 2005 WILEY-VCH Verlag GmbH & Co. KGaA, Weinheim
ISBN 3-527-31115-7

tors (see Fig. 13.1). After an appropriate time interval (perhaps <1 h), the toxin-bearing nanospheres are removed from the human blood stream using an extracorporeal, closed-loop tubing system attached to a compact magnetic separator device. This hand-held device passes blood through channels designed to avoid clotting and to enhance suitable laminar flow conditions. Permanent magnets contained within the device magnetically separate the spheres from the blood flow. The detoxified blood is returned into the body and the toxin-bound particles are stored inside the device for bioassay, bioforensics, or disposal.

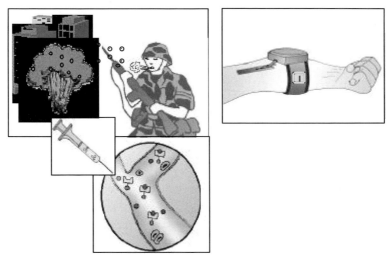

Figure 13.1. Conceptual rendition of proposed biotoxin detoxification technology. An exposed soldier self-administers nanoparticles and a strap-on magnetic filtration unit is attached by vascular access to his arm to remove the toxin.

If developed successfully such a system would provide a combination of strategic advantages.

1. *Uniqueness and therapeutic improvement:* The system is highly innovative and would provide, for the first time, a method of selective and efficient removal of a variety of biotoxins from humans with internal contaminations. Toxin removal, not merely binding of blood-toxin alone, is most important mainly because (i) many biohazards, e.g., chemicals and radioactive substances, are not effectively neutralized by simple *in vivo* antitoxin–toxin binding and (ii) antitoxin–toxin complexes remaining within the body may induce secondary illnesses, e.g., kidney failure from immune complex deposition, and rebound toxicemia from release dissociation and tissue deposition.

2. *Biocompatibility:* Nontoxic, biodegradable nanospheres are composed of magnetite (or other magnetic material) encapsulated within a biocompatible polymer. Therefore, injection of the magnetic nanospheres with subsequent incomplete or no removal will pose no harm to the subject.

3. *Diversity and repeatability:* The therapeutic diversity lies in the vast number of already existing and newly designed antitoxins, antibodies, and ligands that can be attached to the biocompatible magnetic nanoparticles. Chronic exposures or exposures with low blood but high tissue toxin concentrations can be treated with repeated injection and removal of antitoxin-bearing particles.

4. *System compactness and portability:* As the total nanospheres injected for toxin binding can be expected to be less than 2 mg per kilogram body weight, various antitoxin injectants and the actual magnetic removal device can be engineered as a hand-held, single-use, presterilized, self- or helper-applicable unit.

The future biomedical realization of this technology would be an expandable system also applicable to (i) the *diagnosis of diseases*, e.g., rapid bioassay examination of milligram quantities of concentrated toxins bound to the nanospheres, and (ii) the *treatment of diseases in biohazard-exposed humans*, e.g., sequestering a wide variety of blood-borne toxins and biohazards *in vivo* with subsequent removal from the body. Since the nanoparticles will be stabilized against opsonization and macrophage engulfment, they can be circulated within the blood stream for extended periods facilitating in-field use, e.g., by military and first-response personnel, or its application as a mass-screening tool in a triage setting. Importantly, if successful, this detoxification system will provide a unique platform nanotechnology for the treatment of other medical illnesses, such as autoimmune diseases and medication overdoses, in which disease-specific receptors (chelators, ligands) already have been developed but subsequent removal of toxic compound from the blood stream is not yet possible.

This chapter will provide a brief review to familiarize the reader with the pertinent chemical and biomedical engineering background and outline our currently ongoing parallel and sequential efforts to successfully establish this decorporation system. In our discussion, we will focus on the synthesis and *in vivo* testing of biostabilized and biodegradable magnetic nanospheres, the incorporation of toxin-binding ligands to the magnetic nanospheres, and the development of a prototype magnetic sequestration system.

13.2
Technological Need

Several blood detoxification methods are currently available. The clinically more important ones can be summarized as follows.

Hemodialysis and hemofiltration applies an osmotic gradient across a semipermeable membrane to dialyze/filtrate hydrophilic substances out of the blood. The major limitations are long procedure duration, extracorporeal circulation of large blood volumes requiring large-bore arterial access, nonselective substance removal, and effectiveness limited to hydrophilic substances with lower molecular weight. Its use is mostly restricted to patients with kidney failure and in some medication-related intoxications.

Plasmapheresis utilizes extracorporeal, nonspecific exchange of plasma (which is cell-free blood) with albumin or saline solutions. This method removes most of the blood fluid phase and therefore can only be used for a limited period of time and in specific clinical situations where the toxic substance is present in abundant concentration. Its utility is generally restricted to autoimmune diseases.

Extracorporeal immunoabsorption is a variation of hemodialysis in which extracorporeal circulated blood is exposed to a larger exchange surface saturated with immune-absorbent materials (e.g., antibodies). It is a more specific removal method but less effective, requires the circulation of large blood volumes, and is restricted to specific antibody–antigen interactions.

We can perform *direct injection of chelators and antibodies*, in which, for example, injected antibodies neutralize some actions of circulating antigen (e.g., medication or bacterial toxin interactions). However, complete antigen binding can often not be achieved, and also relatively high antibody dosing is required, increasing the risk of allergic (anaphylactic) and systemic (kidney failure, etc.) side effects. Furthermore, the antibody–toxin complex is not removed from the blood and remaining toxin can dissociate, leading to rebound intoxication.

Obviously, there is currently no adequate detoxification system and, for the majority of biohazard exposures, no therapies other than supportive measures. In agreement with the clinically most successful acute and chronic detoxification system, hemodialysis, we are proposing that a novel versatile detoxification system must *remove* the offending agent(s) from the blood stream. Simple blood biohazard sequestration within the blood stream as achieved by toxin–ligand or antibody–antigen binding will insufficiently protect humans from harmful toxin exposure. This is evidently exemplified in (i) treatment approaches solely based on *in vivo* antibody–antigen binding, in which safe antibody treatment is problematic due to reduced antibody affinity, systemic side effects (i.e., antibody–antigen complex-mediated diseases, renal failure, etc.), anti-antibody production – all inherent limitations if higher or repeated antibody injections are required; as well as in (ii) cases of radioactive and chemical toxin exposures, where ligand–toxin binding does not alter the toxic activity and natural disease course induced by the offending agents.

However, in contrast to conventional hemodialysis and other, less commonly used detoxification systems, the nanospheres-based system will have important advantages such as:

- *Specificity.* Only substances binding specifically to the nanoparticle-ligand surface will be removed.
- *Removal.* Several important advantages exist:
  - Dissociation of already formed toxin–antitoxin complexes becomes less important as the compounds are continuously removed from the body.
  - Active toxin blood stores are permanently removed; therefore, additional protein-, tissue-, and membrane-bound toxins are continuously released as free toxin into the blood stream and eventually bound to the nanospheres (decreasing toxin body stores).

- Quantification of removed toxins permits (i) direct estimation of toxin removal efficiency, (ii) a clear estimate of required therapy duration, (iii) further laboratory identification of biohazard origin (bioforensics), (iv) improved antitoxin design, etc.
- *Quantitative binding.* Quantitative toxin binding is facilitated by large nanoparticle-antitoxin binding capacity; this becomes especially useful when only-low affinity antitoxins are available.
- *Nontoxicity.* Nanoparticles remaining within the body are metabolized physiologically without adverse effects.
- *Efficiency of removal.* Strong magnetic field gradients will allow high-yield first-pass removal of antitoxin–toxin compounds.
- *Convenience of usage.* Two-step detoxification, simplified for mass usage by a nonverbal visual guide: injection of nanospheres followed by simple needle insertion of filtration unit; miniaturized portable or large-scale hospital-based designs.;
- *Safety of usage.* No risk of disease transmission (as in antibody treatment, blood transfusion, etc.); no blood loss; closed-loop, preheparinized and pre-sterilized, single-use system avoids blood contamination, allowing self- or helper-applied usage by nonmedical personnel, and preservation of the sample toxin.
- *Repeatability.* Re-exposures to biohazards or reaccumulation of toxin from body tissue stores can conveniently be treated with repeated detoxification sessions.

If successful, the most apparent and dramatic advancement is the introduction of a robust, hand-held biodetection and treatment system that can provide a concentrated multianalyte for high-sensitivity bioassay. As such, it is possible to determine the *in vivo*, presymptomatic presence of pathogens through highly selective sequestration and magnetically assisted separation without harmful side effects to surrounding healthy tissues and cells.

The technology is radically novel, but the components have some precedent as was described above. We seek to integrate known technology with advancements and novel design strategies in our laboratories to engineer a powerful detoxification system. Even in its simplest form, a system of *in vivo* detoxification using biostabilized nanospheres will have a scale of diverse applications making it attractive to many military or civilian applications.

13.3
Technical Basis

Upon first inspection of the technology concept several questions arise as to its feasibility and applicability. Such concerns are addressed directly below.

13.3.1
Difference Between Drug Sequestration and Drug Delivery Using Nanospheres and Microspheres

There are many examples of R&D in nanospheres and microspheres systems for the goal of drug delivery; the reader is pointed to Davis [1] and Douglas et al. [2] and the work of Häfeli [3] and Lübbe et al. [4]. At first inspection, drug delivery and drug removal using nanoparticles might be considered parallel research. However, there are important differences in the development of these two systems (Tab. 13.1). First, drug delivery systems must optimize the encapsulation of drug material within the nanoparticle. As such, the best configuration is to use higher volume-to-surface-area particles. Drug removal requires the antithesis – high surface-area-to-volume ratios – in order to maximize the surface functionality (capacity) of the nanospheres for toxin removal. Also, drug delivery systems often do not require long circulation times in the body and thus may not require robust surface properties to avoid opsonization. In detoxification, drug removal will require circulation of the nanospheres in the blood stream for several cycles in order to maximize capture of toxins and permit removal. Next, because the drug is encapsulated within the microspheres matrix for drug delivery systems, the surface can be freely conjugated. In contrast, nanospheres for toxin removal will require the surface to be conjugated with stabilizing ligands *and* ligands to sequester toxins. The reason for this is to maximize the kinetics for toxin removal – a surface effect. Finally, drug delivery systems are often not magnetic, instead relying on surface ligands to bind to tissue sites. Systems that are magnetic must maximize the magnetic moment of particles in order to minimize the externally applied magnetic gradient and field. For drug removal, the magnetic component is important for filtration but may be adjusted depending on the design of the magnetic filtration system. That is, a high-aspect-ratio filtration chamber can be fabricated to produce blood flow rates that range from < 1 cm s^{-1} to >100 cm s^{-1} depending on the actual capture efficiencies. Because of the versatility in designing the filtration system, we envision the magnetic moment of the particles to be much less than required by magnetic drug delivery systems (<10 emu g^{-1} as opposed to >30 emu g^{-1} in drug delivery systems).

Table 13.1 Contrast between drug delivery and the proposed toxin removal system based on magnetic nanoparticles.

Property	Drug delivery	Toxin removal
Size	High volume-to-surface-area	High surface-area-to-volume
Magnetic moment	>30 emu/g	<10 emu/g
Surface functionality	Not necessary	Stabilized against opsonization, anti-toxins anchored to surface
Circulation time	Minutes	>1 h

emu: electromagnetic unit.

13.3.2
Vascular Survival of Nanospheres

There has been initial work investigating particle pharmacokinetics within our group and by others. Particles without appropriate surface characteristics are immediately removed (within minutes) by the reticuloendothelial system (liver, spleen, etc.) [5], and particles that are too large will cause capillary occlusion and cause serious adverse effects and death after systemic injection into the animal ([6]; unpublished data obtained in our monkey experiments using commercially available cellulose-based microspheres). However, success in liposome and nanoparticle systems identifies the importance of hydrophilic polyethylene glycol (PEG)-derivatives in prolonging intravascular survival [7, 8]. Particles coated with PEG chains offer steric and charge stability (near neutral) and prevent antibody formation, opsonization, and phagocytosis [8, 9]. Investigations found that the blood circulation times of particles increase as the molecular weight of covalently linked PEG increases. Five hours after systemic injection, only one-third of 20-kDa PEG-conjugated poly(lactic-co-glycolic acid) nanospheres (140 nm) had been captured by the liver in comparison to uncoated particles [8]. Similar prolongation was summarized by Allen et al. [10, 11] and described by Li et al. [7] and Dunn et al. [12]. However, exact mechanisms for macrophage avoidance are not known and discrepancies exist as to the best PEG length or derivative. Most recent evidence has demonstrated a 20-h half-life for circulation, exemplifying the great progress scientists have made in this area [13].

13.3.3
Toxicity of Components

Nontoxic polymeric nano- and microspheres have been described in the literature and comprise a category of natural and synthetic biopolymers. Biopolymers include but are not limited to poly(lactic acid), poly(lactic-co-glycolic acid), poly(ethylene glycol), poly(caprolactone), albumin, and dextran. Biopolymers are chemically degraded at rates dependent on the particle size, surface properties, cross-linking density, and the molecular weight of the polymer [14]. The acute toxicity of several biopolymers has been evaluated [15] and suggests no ill effects due to the polymers. Instead, the polymers are degraded and metabolized into harmless fragments [16]. Poly(lactic acid), poly(lactic-co-glycolic acid) and poly(caprolactone) are FDA-approved for injection in several forms. From our own work, histopathological examinations of lung, liver, brain, and spleen of a series of monkeys and rats exposed to systemic injections of magnetic particles identified no early toxicological changes (tissue changes, coagulation, and extravasation) and no capillary obstructions were observed in examined animal organs after adjustment of particle size and injection modus.

The presence of magnetic particles incorporated within the polymeric matrix introduces a second source of potential toxicity. However, studies have shown [17, 18] that, in the long term, the magnetite crystals are in part metabolized, increasing hepatic and splenic ferritin stores, and in part incorporated into red blood cells.

Thus, provided that the injected dose of magnetic iron is below the toxic dose threshold, they are safe. (The toxic dose threshold is 10 mg kg^{-1} body mass or 750 mg in standard man [16]. Also, normal serum levels in blood are 80–180 µg dL^{-1} and action levels are >500 µg dL^{-1}.) We estimate that about 100 mg of nanoparticles will be injected to treat a typical biohazard-exposed subject, a fraction of which will be composed of iron (if 50% loading of nanoparticle with elemental iron, then the patient may be exposed to 50 mg of unbound iron). Not only is this amount of injected magnetite much smaller than the dose leading to toxic iron effects but, more importantly, our detoxification system will extract most of the injected magnetite as part of the toxin removal process. Therefore, body storage will only be a minor deposition method. Superparamagnetic iron oxide is approved by the FDA for injection.

13.3.4
Magnetic Filtration of Nanospheres from Circulation

Magnetic separator units for industrial use are available commercially and have been used for various research related immunoseparations [19–23]. However, a magnetic separator suitable for our application of clearing nanospheres from blood flow has not been designed. Based on the performance of commercial separators and our own preliminary investigations, we will implement technically straightforward engineering for the design of the first prototype biomedical separators. Our proposed "filtration" system utilizes small permanent magnets attached to the body of a specialized closed-loop catheter system. The actual design of the device is part of the proposed research as several options are plausible given predefined conditions. Common to all design options, the blood will be diverted from the body to an array of tiny (few hundred micrometer diameter) flow tubes. The tubes will be immersed in a magnetic field gradient, causing the magnetic nanospheres to deflect towards and collect at the tube wall. The precise geometry of the tubing system (size, material, coating, length, shape, etc.) will be defined in our research to minimize interactions with blood coagulation (i.e., thrombosis) and blood cells (i.e., destruction). In addition, our analyses will identify different design strategies for different modes of operation (e.g., in-field vs. unit-based) and user level of training (e.g., self- vs. helper-applied).

Magnetic separation can be illustrated with our following simple experiment. A suspension of monodisperse nanospheres, 400 nm in diameter (measured moment 50 emu g^{-1}, ρ=1.4 g cm^{-1}) was contained in 0.9% saline in a 0.6-cm diameter vial. A hand-held commercially available magnet (0.4 T at surface) was placed against the outside wall of the vial and all particles rapidly deflected to the inner wall close to the magnet surface within 3 s. Assuming constant particle velocity (here 0.6 cm per 3 s) within the contained fluid, we plot the trajectories of particles under flow conditions (Fig. 13.2). The plot estimates the length of tubing that would be needed to deflect particles flowing at 50, 10, and 1 cm s^{-1} in 1-mm diameter tubes. For rapid flow velocities, a tube >20 cm immersed in the magnetic field would be required to separate the particles from the flow. However, for flow rates of 10 cm s^{-1}, easily achieved with microflow tubing designs, only approximately 5 cm of tubing is necessary. In our design, multiple flow tubes will be used to compensate for the reduction in volumetric blood flow rates per tube and lower the pressure drop across the device. Smaller-diameter tubes will facilitate separation and greatly reduce the length of tube necessary for effective separation.

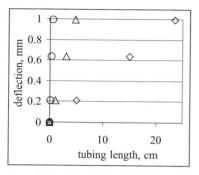

Figure 13.2. Simplified model showing nanoparticle trajectory in various flow velocities ($\Diamond = 50$ cm s^{-1}, $\Delta = 10$ cm s^{-1}, $\bigcirc = 1$ cm s^{-1}). Particles must deflect 1 mm to reach vessel wall.

13.4
Technology Specifications

The main foci are (i) the development of biostabilized magnetic nanospheres that circulate freely in the blood stream, (ii) the design of biodegradable prototype antitoxin-coupled nanospheres, (iii) the demonstration of *in vitro* and *in vivo* toxin sequestration using a model, biological target–receptor pair, and (iv) the design, fabrication, and demonstration of a compact, extracorporeal device to magnetically separate the nanospheres from the blood rapidly.

The state of the art in biocompatible, magnetic nanospheres and magnetic separation devices already provides promising initial data on core technologies needed to realize components of the detoxification system. However, significant scientific challenges must be overcome to determine the functionality and efficacy of our rapid detoxification system. We describe the background and the broad requirements for each of the four necessary technical achievements listed above.

13.4.1
Development of Biostabilized, Magnetic Nanospheres

Polymeric-based microspheres and nanospheres are used primarily as drug carriers for targeted or sustained drug delivery. Several articles have been devoted to a summary of the application of drug-loaded spheres [24–27], magnetic carrier technology [28], and technology status with magnetic targeting of magnetic carriers [29]. In fact, an international conference is held regularly to discuss advances in the medical applications of magnetic carriers [http://www.magneticmicrosphere.com] and attendance has grown rapidly.

For the detoxification technology, we need to synthesize nanospheres with (i) optimal size to avoid obstructing capillary blood flow and immediate vascular clearance, (ii) adequate size uniformity to facilitate modeling and quality control and assurance, and (iii) optimal surface properties to prolong vascular circulation. Specifically, the nanospheres must be single populations between 100 and 5000 nm. If bimodal or multimodal populations are synthesized then the diameters of the individual populations must be sufficiently different to allow for some physical separation method (e.g., filtration). The surface charge must be neutral or slightly negative (<10 mV, measured by zeta-potential) due to surface PEGylation in order to increase blood circulation time and minimize bioclearance by opsonization and phagocytosis.

13.4.1.1 Nanosphere Size

The literature presents many articles describing the synthesis of nanospheres and microspheres composed of natural and synthetic polymers. Many types of polymers have been investigated for potential *in vivo* applications. By far the most discussed polymers are poly(lactic acid) and poly(lactic-co-glycolic acid) macromonomers. There is a plethora of literature too numerous to cite that describes synthesis methods, drug encapsulation, release kinetics, biostability, and *in vivo* properties such as macrophage engulfment, biodegradation, and organ disposition. The *Journal of Controlled Release* contains many articles every year on the subject matter.

The method of nanospheres formation is straightforward. The polymer is dissolved in a volatile organic solvent and added to an aqueous solution and stirred vigorously. An emulsion of oily droplets forms. Surfactants in the solution promote the stabilization of the emulsion until the volatile organic solvent dissolves in the solution and evaporates, leaving a solidified sphere of polymer. Magnetic spheres contain additional magnetic powders in the organic solution. The size of the spheres depends on process parameters. Some process parameters are outlined in Tab. 13.2.

A popular type of polymer composition is polystyrene. Polystyrene spheres are formed by a completely different method called emulsion polymerization [30] and will not be mentioned further except to say that they are uniform in size with a well-known surface chemistry and functionalization. They are not appropriate for *in vivo* use due to the toxicity of polystyrene, but their uniform properties make them attractive model systems during feasibility studies.

The limits set by physical removal of particles flowing in the blood are rather broad. To maximize surface receptor density we seek particles with sufficiently high surface-area-to-volume ratios. To be discussed in the next section, high surface-area-to-volume ratio nanospheres would imply using the smallest particle possible or those <100 nm. However, as one reduces the nanosphere size, it becomes more difficult to sustain the magnetite content or magnetic moment. In other words, the magnetic moment drops at a rate faster than would be expected based on the reduction in particle volume as the diameter is reduced. We have had success in separating 400-nm polystyrene magnetic nanoparticles from blood (specific magnetization=50 emu g^{-1}). Thus, we hypothesize that a nanoparticle 100–400 nm in size will

Table 13.2 Some process parameters in the synthesis of magnetic nano-/microspheres.

Polymers	Poly(D-lactide), poly(L-lactide), poly(D,L-lactide), poly(lactic–co-glycolic acid), polymers and copolymers linked to polyethylene glycol, poly(caprolactone)
Surfactants	Poly(vinyl alcohol), poloxamer, polyethylene glycol, tocopherol polyethylene glycol succinate, sodium dodecyl sulfate
Volatile organics	Dichloromethane, chloroform, acetone, ethyl acetate
Mixing speed	1,000–20,000 rpm, 15–95 W insonation
Mixing modality	High-speed homogenization, ultrasonic insonation, paddle mixing, magnetic stir bar
Polymer/organic solvent ratio	1–0%
Oil/water ratio	1–10%
Mixing time	Several seconds to minutes depending on modality to form initial emulsion, several hours to harden
Polymer MW	2–300 kDa
Surfactant MW	PVA 10–150 kDa
Surfactant concentration	PVA 0.1–10%
Magnetic phase	Magnetite Fe_3O_4, maghemite γ-Fe_2O_3, passivated Fe, passivated Co
Temperature	Unknown, may reduce coalescence

be optimal in this program. There is a parallel requirement that the size of the particles not be broadly distributed (high polydispersity). The reason for this is to facilitate modeling of the system and ensure predictability in particle properties such as magnetic moment, surface area, receptor density per batch size, surface charge, etc.

13.4.1.2 Surface Properties

Covalent attachment of biologically active compounds to polymers and polymeric spheres became one of the methods for alteration and control of biodistribution, pharmacokinetics, and, often, toxicity of these compounds [31]. One of the most popular polymeric materials used for this purpose is polyethylene glycol (PEG). It possesses an ideal array of properties: excellent solubility in aqueous solutions [32], extremely low immunogenicity and antigenicity [33], and has the advantage of being nontoxic and was approved by the FDA for internal use in humans [34]. Gref [27] provides an excellent discussion on PEG biochemistry and protein interaction.

Jeon and Andrade [35] proposed a mathematical model taking into account the four types of interactions between a protein and hydrophobic substrate. They stated that the best conditions for protein repulsions were found to be long PEG chain length and high surface density. If D is the distance from the anchorage to the substrate of the two terminally attached PEG chains, in the case of small proteins (approximately 4 nm in diameter), D should be around 1 nm, whereas for larger proteins (6–8 nm), D should be around 1.5 nm [35].

The challenge to our program is to determine the proper PEG chain length and surface coverage to increase circulation half-lives to permit *in vivo* binding of the toxin and removal of the toxin-loaded nanospheres. PEGs come in various forms. They can be made linear or branched, and with different molecular weight (chain length), and partial substitution (e.g., polyethylene glycol–polypropylene glycol coblock polymers). The literature contains discrepancies as to the best choice of chain length, and computer models [35–38] suggest potentially subtle but important differences to explain PEG behavior towards protein adsorption. One study [8] concludes that polystyrene nanospheres with longer PEG chains, those greater than 10,000 Da, survive longest in the rat, but does not show direct evidence of surface coverage. Another study [12] concentrates on showing the importance of surface coverage density but does not compare these results with those as a function of PEG length. Importantly, long-chain PEGs may sterically interfere with each other during the surface bonding procedure or during copolymerization. Thus, the surface may not be able to accommodate the long-chain PEGs to the 100% surface coverage needed to avoid opsonization. Another study [39] suggests that vascular survival is not enhanced by conjugating PEG chains longer than 5000 Da, while Yamoaka et al. [13] refute this suggested lack of dependency. These discrepancies necessitate confirmatory and supplemental study.

13.4.1.3 **Biodegradability**

We are adapting the parameters found to prolong the vascular survival of the model polystyrene-based nanoparticles (PEG chain length, PEG surface density, receptor site density, biokinetics) to biodegradable magnetic nanospheres. There are several biopolymers to choose from in synthesizing the nanoparticles. There is a plethora of data on poly(lactic acid) (PLA) and poly(lactic-co-glycolic acid) (PLGA) polymers. Gref et al. [40] describe how to synthesize PLGA–PEG nanoparticles <150 nm using an oil-in-water solvent evaporation technique. Li et al. [7] describe a similar method. Häfeli et al. [3] describe an oil-in-water method for synthesizing PLA microspheres >3 μm, and our work has shown that this method can yield less polydisperse nanospheres from 900 nm to 3000 nm (Fig. 13.3). Mosqueira et al. [41] describe the preparation of PLA–PEG nanoparticles with variable PEG content and chain length. Only Häfeli et al. [3] describe incorporating magnetite into the PLA or PLGA particles so we must understand how magnetite affects the size and chemistry of the biodegradable spheres, especially the submicron spheres.

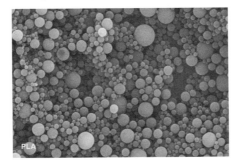

Figure 13.3. Poly(lactic acid) microspheres. Note the variance in sphere diameter.

13.4.1.4 **Surface Receptors**

The nanospheres must have (i) large surface area and surface functional sites for attachment of surface receptors (e.g., antibodies, chelating agents for radionuclides, or ligands for chemical toxins) in sufficient number to reduce or eliminate completely the concentration of toxins in the blood, and (ii) the placement of such surface receptors where they can interact readily with the blood but not facilitate protein adsorption onto the particle surface. The first criteria are interrelated with the design of the nanosphere size since particle diameter is intrinsically related to surface area. To estimate the target receptor site densities we provide an example. Simulating a class A agent *Bacillus anthracis* exposure, we inject 0.6 µg of lethal factor (LF) into a rat (300 g weight, 25 mL blood volume) which leads to a LF blood concentration of 24 ng mL^{-1} or 7.5 pmol per rat (MW 80–90 kDa). Nanospheres we used in our laboratory have a surface receptor capacity of at least 1 µequiv mg^{-1} or 1 μmol mg^{-1} for univalent ligands. Thus, an anticipated injection of 10 mg of nanoparticles into the rat would have a capacity of 10 μmol barring steric hindrance. This value is on the order of 10^6 times greater than the necessary theoretical capacity needed to bind quantitatively LF toxin in the blood. Assuming a steric hindrance offered by an 80- to 90-kDa LF protein (15 nm diameter projected image) on a 400-nm magnetic nanoparticle (10 mg injection), we can expect a binding capacity of 0.7 nmol LF protein – still a factor of 100 greater than necessary for quantitative LF removal from the blood. Therefore, a goal is to maintain this order of receptor site density (1 µequiv mg^{-1}) during the synthesis of biodegradable nanospheres from PEG based copolymers.

There are two options to attaching the antibodies or chelating ligands to the surface: (i) attach them directly to the particle surface using short chain functional groups such as carboxyl and epoxy bridges, or (ii) attach them to the PEG or PEG-derivative chains extending from the particle surface. The shortcomings of the first procedure are that the receptor groups may be sterically hindered from encountering the toxins present in solution due to the long-PEG-chain neighbors. Also, since the receptors would be attached directly to the surface they would be competing with the PEGs for surface coverage. This condition inherently limits the surface density of PEGs and facilitates opsonization and macrophage removal of the nanospheres *in vivo*. Therefore, option (ii) appears to be the most appropriate choice. It has been demonstrated by our research partners and others that the terminal groups of PEG chains can be activated and covalently bonded to other functional groups. Our own research has shown that we can attach streptavidin onto the terminal groups of PEG (300 and 2000 Da). However, we believe it may be difficult to cap the activated terminus of the PEG completely during the receptor attachment step. This may result in a charged surface and increased opsonization. Instead, we are pursuing a method by which the coblock polymer used to make the nanospheres (e.g., PLA–PEG) contains a suitable end group for direct attachment of receptor. For instance, we have synthesized PLA–PEG–biotin copolymer (Fig. 13.4) based on the technique of Salem et al. [42]. Streptavidin was attached after incubation in buffer solution [43]. Similarly, biotinylated antibodies could be directly attached to the streptavidin [44] or attached during the coblock polymer synthesis by substituting the biotinylated antibody for biotin. Although preliminary studies are promising, there are still limitations that need to be addressed, including efficiency of antibody conjugation and effect of conjugation of the PEGs on circulation of the nanospheres *in vivo.*

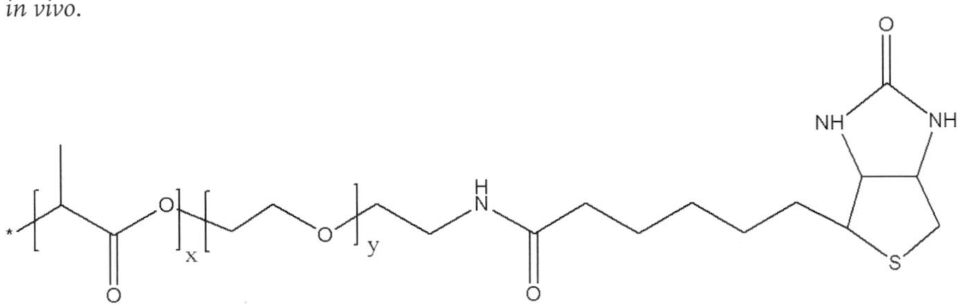

Figure 13.4. Biodegradable coblock polymer with biotinylated end group.

The model system that we are demonstrating in initial experiments is the streptavidin–biotin pair. However, this system provides a lower limit in determining the necessary residence times for nanosphere flow (i.e., the amount of time or blood volumes to which the nanospheres must be exposed to ensure removal of toxins). The follow-up system in subsequent research will concentrate on more realistic systems that display weaker binding constants. These antigen–antibody systems will present a realistic expectation of this system based on the accomplished receptor site densities and vascular survival. Therefore, a goal is to determine if it is possible to functionalize identified candidate magnetic nanoparticles with their respective candidate antibodies and that the antigen–antibody capture efficiency remains stable when applied to the living animal conditions. This would ensure that it is possible to detoxify the antigen from the animal blood.

Inherent with the research challenge of receptor site density is the determination of the suitability of current antibodies to maximize surface density. Antibodies are quite large molecules (tens to hundreds of kilodaltons), but their active components can be quite small. Thus, it is important to quantitatively evaluate whether antibody fragments can enhance the receptor site density achievable (due to steric effects) while maintaining its specificity and stability. In other words, the development of a reliable technology for developing receptors for future implementation of the nanoparticle technology is critical – and a obligatory prerequisite for optimization of receptor density (i.e., maximizing receptor capacity while minimizing opsonization/ engulfment risk). Importantly, it is possible that trying to optimize receptor capacity with conventional antibodies is not useful, since the system cannot be manufactured with conventional antibodies. It is plausible that a reduction in the size of the antibody or fragment needed to achieve selective bonding to the toxin would enhance the manufacturing output of the antibody.

An important aspect of the proposed technology is the possibility for generic attachment of a variety of antitoxins to the nanoparticle substrate. For instance, by coupling streptavidin to the terminal groups of the PEGs one has a generic method of attaching biotinylated antibodies to the nanoparticle's surface. We seek to describe methods of generic attachment for chemical toxin ligands and radiological toxins. We initially focus on the streptavidin receptor for biotinylated targets. Successful designs will be moved from the streptavidin–biotinylated protein system of sequestration to the biotinylated antibody–antigen system to more realistically determine binding kinetics *in vitro* and *in vivo*. If encouraging results are obtained, we will test the ability of the magnetic nanospheres to deplete the antigens in complex mixtures, such as rat blood. Positive results would serve as the prelude to the development of animal models as part of subsequent testing. We note the success of Sakhalkar et al. [44] in selective binding of *in vivo* targets with similar nanosphere composition.

13.3.2
Magnetic Filtration of Toxin-Bound Magnetic Nanospheres

We are designing an extracorporeal magnetic filtration unit, a small external catheter system, that (i) allows dialysis-like blood circulation through a well-defined tubing or microchannel system while generating no blood clots or coagulation events, (ii) will permit quantitative removal of the freely circulating, toxin-bound nanospheres via application of hand-held permanent magnets, (iii) is easily portable and in-field applicable (i.e., the entire device is hand-held), and (iv) is designed for administration by nonmedical staff or trained medics (both designs will be developed).

The current knowledge of perfusion devices and blood flow is sufficient for the development of such a filtration device without additional scientific discovery. Our own research provides evidence for particle filtration in high-velocity blood flow using small, permanent magnets (Fig. 13.5). Thus, a carefully orchestrated series of experiments utilizing existing technology should be successful for production of the device. The device is predicated on:

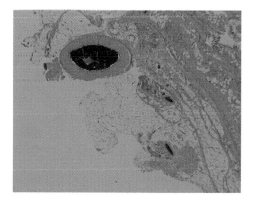

Figure 13.5. Pathology of magnetic particles (black, peak vel.=105 cm s^{-1}) using a 20-mm permanent magnet placed 8–10 mm from artery.

1   The ability to successfully separate the particles from the blood. This will require a sufficient magnetic field to draw the particles to the wall and hold them there under shear flow.

2   A flow rate of at least 100–200 ml min^{-1} will be required for total body cleansing. Lower flow rates can easily be obtained from a large-bore venous puncture at mid-arm level. A simple siphon hand-pump will ensure proper flow rate. Higher flow rates are achieved from a commonly performed femoral arterial puncture with a double-bore needle [double-lumen (inflow and outflow) catheterization avoids a second vascular puncture]. A trained technical person should be capable of performing such a procedure.

3   Anticoagulation locally in the perfusion chamber by dissolution of heparin from the walls of the tubes into the flowing fluid, which should permit adequate anticoagulation locally without inducing it systemically. Shear rates and stresses will be kept at levels compatible with minimizing both clot formation and thrombosis in the design of the magnetic field so that particle removal or "filtration" is quantitative and highly efficient, minimizing the length of the device and its size and weight. The design will include a clot-filtering device at the blood return port.

Further design aspects depend on future investigations using the prototype magnetic field chamber. Certainly, the inherent physical properties will vary for each magnetic filtration device variation necessary to accommodate different biomedical and user applications and these design characteristics cannot be predicted *a priori* purely on theoretical grounds. However, realistic modeling of various future design strategies will become possible if the first prototype separation device is characterized and its performance tested in living animals and healthy volunteers. Important design strategy questions can then be adequately answered; examples are:

1   What are the most feasible vascular access modes (venous, bivenous, arterial?) and site (antecubital, brachial, femoral?) for easy and fast (self- vs. helper?) use?

2   What are the optimal catheter (single vs. dual lumen, cannulation size?) and tubing system (coating, flexibility, filter system, branching?) for unobstructed blood circulation and to avoid blood clotting?

3   What is the most favorable magnetic device design (one magnet/catheter unit or a "snap-on" tubing design?), allowing easy fixation and portability?

Our first prototype detoxification system will be designed to utilize safe and practical venous access at the inner mid-arm regions (antecubital) of an exposed human via 14-gauge, dual-lumen needle insertion (self- or helper-applied) which can easily be performed by trained nonmedical personnel. Using this approach, the careful estimation of practical low blood flow rates at about 40 ml min^{-1} will allow total body blood (estimated 6 L) turnover with removal of toxin in about 150 min. Variations of this approach will permit faster detoxification; for example, helper-applied femoral artery needle access at the groin site can reduce the clearance time by a factor of 8–10, establishing blood clearance within approximately 15 min. More complex vascular access and device design strategies can be instituted when exposed humans are treated in medical units or hospital settings.

13.4
Technical Progress

This program is in its infancy and we are pursuing proof-of-principle experiments. *In vitro* sequestration of a biotinylated enzyme from simple fluids and whole rat blood was performed under static and dynamic flow conditions. Particles were composed of nanocrystalline maghemite (γ-Fe$_2$O$_3$) encapsulated in polystyrene nanospheres. Several variations were tested including various PEG lengths (MW 330–6000) and particle sizes (250–3000 nm). Streptavidin, the model receptor, was either bonded to the carboxylated terminal group of the PEG or attached directly to the nanoparticle surface. Biotinylated horseradish peroxidase (HRP) was used as the model "toxin." The results (Fig. 13.6) indicate a reduction of the free enzyme to about 50% maximum levels in the blood in all tests. Equilibrium was reached within 20–30 min. We have more recently achieved 72% separation of biotinylated HRP from saline in heparinized whole rat blood [43].

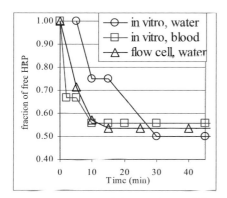

Figure 13.6. Horseradish peroxidase (HRP) "toxin" levels in water and blood after HRP and nanosphere injection.

We synthesized biodegradable PLA–PEG–biotin–streptavidin nanospheres and microspheres and achieved up to 42% separation of the biotinylated HRP from normal saline (0.25 mg nanospheres per milliliter of heparinized rat blood, 0.375 µg HRP mL^{-1} blood [43]). The surface charge is neutral from pH 4 to pH 9 making them, to a first order, suitable for *in vivo* trials. We have encapsulated rhodamine-B and measured detection limits in whole rat blood. The signal-to-noise levels are suitable for quantification.

In vivo experiments, performed on retired breeder rats, included (i) the design of a closed-loop, adjustable-flow blood recirculation unit permitting blood turnover and sampling over several hours in the live animal; (ii) kinetic studies of several candidate magnetic nanospheres and toxins; (iii) first magnetic filtration experiments. In the latter investigations, continuous extracorporeal blood circulation was achieved via carotid–jugular cannulation and external pump support with filtration of magnetic nanospheres using 1-mm-diameter closed-loop tubing and a single NdFeB magnet (0.4 T at surface, 18 mm diameter) (Fig. 13.7). Procedural blank experiments monitored the circulation half-life ($T_{1/2}$) of biotinylated HRP in the rat model (Fig. 13.8), showing that the HRP levels decrease immediately ($T_{1/2}$=15–20 min). A longer circulating marker would be useful. Experimental results on toxin sequestration and rat pathology are pending.

Toward the development of the magnetic filter, we employed computational fluid dynamics and computational magnetic field models developed by The Department of Energy to predict the capture efficiency of magnetic microparticles passing through a simple magnetic field profile. This design provides the simplest case of a

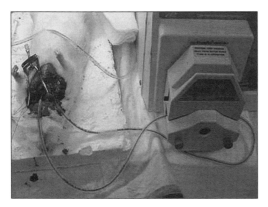

Figure 13.7. Continuous extracorporeal blood circulation in the rat.

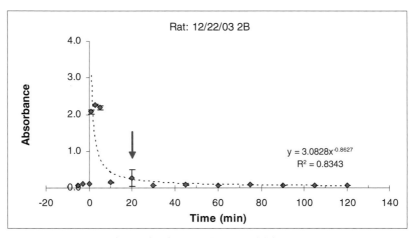

Figure 13.8. Removal of biotinylated HRP in the rat model during normal circulation [34].

filter design (Fig. 13.9) and shows that microspheres with a mean diameter of $7\,\mu$m (specific magnetization≈20 emu/g) can be separated with 50% efficiency in a single pass in a 2×2 mm channel. The model also predicts that a 1-μm change in particle diameter yields a 10% change in the capture efficiency. If we make the assumption that we cannot achieve a significant increase in the ambient magnetic field and field gradients across a small section of tube, then we must design the filter with three considerations in order to achieve >99% capture of circulating nanospheres (100–200 nm diameter) in a single pass. First, the tube diameter must be reduced to decrease the radial path length (assume round tubes) the spheres must traverse to reach the tube wall. The decreased tube diameter produces a concomitant increase in flow velocity. To reduce the flow velocity, then, we must secondly increase the number of tubes in the filter to create a bundle of tubules. Finally, we can design the filter such that the tubules pass multiple times through the magnetic field. These design features will be modeled and compared to experiments.

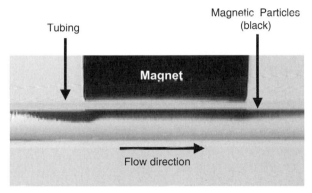

Figure 13.9. Magnetic capture of magnetic microparticles in a tube under laminar flow (NdFeB magnet, 0.4 T at surface, saline fluid velocity ~ 10 cm s^{-1}).

References

1 S. S. Davis, Biomedical applications of nano-technology – implications for drug targeting and gene therapy, *Trends in Biotechnology* **1997**, 15, 217–224.

2 S. J. Douglas, S. S. Davis, L. Illum, Nanoparti-cles in drug delivery, *CRC Critical Reviews in Therapeutic Drug Carrier Systems*, **1987**, 3, 233–261.

3 U. O. Häfeli, S. M. Sweeney, B. A. Beresford, E. H. Sim, R. M. Macklis, Magnetically direct-ed poly(lactic acid) ^{90}Y-microcapsules: novel agents for targeted intracavitary radiotherapy, *Journal of Biomedical Materials Research*, **1994**, 28, 901–908.

4 A.S. Lübbe, C. Bergemann, W. Huhnt, T. Fricke, H. Riess, J. W. Brock, D. Huhn, Clinical experiences with magnetic drug tar-geting: a phase I study with 4'-epidoxorubicin in 14 patients with advanced solid tumors, *Cancer Research*, **1996**, 56, 4694–4701.

5 H. Yoshioka, Surface modification of haemo-globin-containing liposomes with polyethyl-ene glycol prevents liposome aggregation in blood plasma, *Biomaterials*, **1991**, 12, 861–864.

6 J.D. Slack, M. Kanke, G. H. Simmons, P. P. DeLuca, Acute hemodynamic effects and blood pool kinetics of polystyrene microspheres following intravenous administration, *Journal of Pharmaceutical Sciences*, **1981**, 70, 660–664.

7 Y.-P. Li, Y. Y. Pei, X. Y. Zhang, Z. H. Gu, Z. H. Zhou, W. F. Yuan, J. J. Zhou, J. H. Zhu, X. J. Gao, PEGylated PLGA nanoparticles as protein carriers: synthesis, preparation and biodistribution in rats, *Journal of Controlled Release*, **2001**, 71, 203–211.

8 R. Gref, Y. Minamitake, T. M. Peracchia, V. Trubetskoy, V. Torchilin, R. Langer, Biodegradable long-circulating polymeric nanospheres, *Science*, **1994**, 263, 1600–1603.

9 S.K. Huang, E. Mayhew, S. Gilani, D. D. Lasic, F. J. Martin, D. Papahadjopoulos, Pharmacokinetics and therapeutics of sterically stabilized liposomes in mice bearing C-26 colon carcinoima, *Cancer Research*, **1992**, 52, 6774–6781.

10 T. M. Allen, C. B. Hansen, D. E. L. Demenezes, Pharmacokinetics of long-circulating liposomes, *Advanced Drug Delivery Reviews*, **1995**, 16, 267–284.

11 T. M. Allen, E. H. Moase, Therapeutic opportunities for targeted liposomal drug delivery, *Advanced Drug Delivery Reviews*, **1996**, 21, 117–133.

12 S. E. Dunn, A. Brindley, S. S. Davis, M. C. Davies, L. Illum, Polystyrene-poly(ethylene glycol) (PS-PEG2000) particles as model systems for site specific drug delivery. 2. The effect of PEG surface density on the in vitro cell interaction and in vivo biodistribution, *Pharmaceutical Research*, **1994**, 11, 1016–1022.

13 T. Yamaoka, Y. Tabata, Y. Ikada, Comparison of body distribution of poly(vinyl alcohol) with other water-soluble polymers after intravenous administration, *The Journal of Pharmacy and Pharmacology*, **1995** 47, 479.

14 R. Arshady and M. Monshipouri, Targeted delivery of microparticulate carriers, in R. Arshady (ed.) Microspheres microcapsules and liposomes, vol. 2, London, Citus Books, **1999**, pp. 403–432.

15 U. O. Häfeli and G. J. Pauer, In vitro and in vivo toxicity of magnetic microspheres, *Journal of Magnetism and Magnetic Materials*, **1999**, 194, 76–82.

16 B. Erbas, M. T. Ercan, B. Caner, Biodistribution and localization of radiolabeled microparticles, in R. Arshady (ed.) Microspheres Microcapsules and Liposomes Series, vol. 3, London, Citus Books, **2001**, pp. 249–280.

17 E. Okon, D. Pouliquen, P. Okon, Z. V. Kovaleva, T. P. Stepanova, S. G. Lavit, B. N. Kudryavtsev, P. Jallet, Biodegradation of magnetite dextran nanoparticles in the rat – a histologic and biophysical study, *Laboratory Investigation*, **1994**, 71, 895–903.

18 D. Pouliquen, J. J. Le Jeune, R. Perdrisot, A. Ermias, P. Jallet, Iron oxide nanoparticles for use as an MRI contrast agent: pharmacokinetics and metabolism, *Magnetic Resonance Imaging*, **1991**, 9, 275–283.

19 Z. Tang, H. T. Karnes, Heterogeneous postcolumn immunoreaction detection using magnetized beads and a laboratory-constructed electromagnetic separator, *Biomedical Chromatography*, **2003**, 17, 118–125.

20 G. A. Martin-Henao, M. Picon, B. Amill, S. Querol, J. R. Gonzalez, C. Martinez, R. Martino, C. Ferra, S. Brunet, A. Granena, J. Sierra, J. Garcia, Isolation of CD34+ progenitor cells from peripheral blood by use of an automated immunomagnetic selection system: factors affecting the results, *Transfusion*, **2000**, 40, 35–43.

21 M. Berger, J. Castelino, R. Huang, M. Shah, R. H. Austin, Design of a microfabricated magnetic cell separator, *Electrophoresis*, **2001**, 22, 3883–3892.

22 L. Sun, M. Zborowski, L. R. Moore, J. J. Chalmers, Continuous, flow-through immunomagnetic cell sorting in a quadrupole field, *Cytometry*, **1998**, 33, 469–475.

23 A. J. Richards, O. S. Roath, R. J. Smith, J. H. Watson, High purity, recovery, and selection of human blood cells with a novel high gradient magnetic separator, *Journal of Hematotherapy*, **1996**, 5, 415–426.

24 M. Downbrow (ed.), Microcapsules and nanoparticles in medicine and pharmacy, Boca Raton, CRC Press, **1992**.

25 R. Arshady (ed.), Microspheres, microcapsules, and liposomes, Vol. 1–3, London, Citus Books, **1999**.

26 U. Edlund, A. C. Albertsson, Degradable polymer microsphere for controlled drug delivery, *Advances in Polymer Science*, **2002**, 157, 67–112.

27 R. Gref, Surface-engineered nanoparticles as drug carriers, in M.-I. Baraton (ed.) Synthesis, functionalization and surface treatment of nanoparticles, Stevenson Ranch, CA, American Scientific Publishers, **2003**, 234–257.

28 U. O. Hafeli, W. Schutt, J. Teller, and M. Zborowski, Scientific and clinical applications of magnetic carriers, New York, Plenum Press, **1997**.

29 M. D. Kaminski, A. Ghebremeskel, L. Nunez, K. Kasza, F. Chang, T. Chien, P. Fischer, J. Eastman, A. J. Rosengart, R. L. McDonald, Y. Xie, L. M. Johns, P. Pytel, U. O. Hafeli, Magnetically responsive microparticles for targeted drug and radionuclide delivery: a review of recent progress and future challenges, Argonne National Laboratory Report ANL-03/28, **2003**.

30 A. Brindley, M. C. Davies, R. A. P. Lynn, S. S. Davis, J. Hearn, J. F. Watts, The surface characterization of model charged and sterically stabilized polymer colloids by SSIMS and XPS, *Polymer*, **1992**, 33, 1112–1115.

31 R. Duncan and J. Kopecek, Soluble synthetic polymers as potential drug carriers, *Advances in Polymer Science*, **1984**, 57, 51–101.

32 S. N. J. Pang, Final report on the safety assessment of polyethylene glycols (PEG3)-6, -8, -32, -75, -150, 14M, -20M, *Journal of the American College of Toxicology*, **1993**, 12, 429–457.

33 S. Dreborg, and E. B. Akerblom, Immunotherapy with monomethoxypolyethylene glycol modified allergens, *Critical Reviews in Therapeutic Drug Carrier Systems*, **1990**, 6, 315–365.

34 J. Harris, Laboratory synthesis of polyethylene glycol derivatives, *Journal of Macromolecular Science, Part C – Reviews in Macromolecular Chemistry and Physics*, **1985**, C25, 325–373.

35 S. I. Jeon and J. D. Andrade, Protein-surface interactions in the presence of polyethylene oxide: II. Effect of protein size, *Journal of Colloid and Interface Science*, **1991**, 142, 159–166.

36 V. P. Torchilin and V. S. Trubetskoy, Which polymers can make nanoparticulate drug carriers long-circulating?, *Advanced Drug Delivery Reviews*, **1995**, 16, 141–155.

37 V. P. Torchilin, V. G. Omelyanenko, M. I. Papisov, A. A. Bogdanov Jr., V. S. Trubetskoy, J. N. Herron, C. A. Gentry, Poly(ethylene glycol) on the liposome surface: on the mechanism of polymer-coated liposome longevity, *Biochimica et Biophysica Acta*, **1994**, 1195, 11–20.

38 I. Szleifer, Protein adsorption on surfaces with grafted polymers: a theoretical approach, *Biophysical Journal*, **1997**, 72, 595–612.

39 T. M. Allen, C. Hansen, F. Martin, C. Redemann, A. Yau-Young, Liposomes containing synthetic lipid derivatives of poly(ethylene glycol) show prolonged circulation half-lives in vivo, *Biochimica et Biophysica Acta*, **1991**, 1066, 29–36.

40 R. Gref, A. Domb, P. Quellec, T. Blunk, R. H. Muller, J. M. Verbavatz, R. Langer, The controlled intravenous delivery of drugs using PEG-coated sterically stabilized nanospheres, *Advanced Drug Delivery Reviews*, **1995**, 16, 215–233.

41 V. C. F. Mosqueira, P. Legrand, J. L. Morgat, M. Vert, E. Mysiakine, R. Gref, J. P. Devissaguet, G. Barratt, Biodistribution of long-circulating PEG-grafted nanocapsules in mice: effects of PEG chain length and density, *Pharmaceutical Research*, **2001**, 18, 1411–1419.

42 A. K. Salem, S. M. Cannizzaro, M. C. Davies, S. J. B. Tendler, C. J. Roberts, P. M. Williams, K. M. Shakesheff, Synthesis and characterization of a degradable poly(lactic acid)-poly(ethylene glycol) copolymer with biotinylated end groups, *Biomacromolecules*, **2001**, 2, 575–580.

43 C. J. Mertz, M. D. Kaminski, Y. Xie, M. R. Finck, S. G. Guy, A.J. Rosengart, In vitro studies of functionalized magnetic nanospheres for selective removal of a simulant biotoxin, *Journal of Magnetism and Magnetic Materials*, accepted August 2004.

44 H. S. Sakhalkar, M. K. Dalal, A. K. Salem, R. Ansari, J. Fu, M. F. Kiani, D. T. Kurjiaka, J. Hanes, K. M. Shakesheff, D. J. Goetz, Leukocyte-inspired biodegradable particles that selectively and avidly adhere to inflamed endothelium in vitro and in vivo, *Proceedings of the National Academy of Scinces of the United States of America*, **2003**, 100, 15895–15900.

14

Nanotechnology in Biological Agent Decontamination

Peter K. Stoimenov and Kenneth J. Klabunde

14.1
Introduction

Decontamination from dangerous biological organisms is of a considerable interest not only for eliminating the hazard of potential biological warfare agents on a battle-field, but also in cases of industrial accidents, terrorist attacks, etc. Biological agents could be of several different types, like bacteria, fungi, viruses and toxins. It is very unlikely that a universal decontaminant could be developed since the biologically dangerous agents are very diverse in their structure, which leads to different survival ability upon treatment with a certain agent. For example, spore cells are much more difficult to kill than regular gram-positive and gram-negative bacteria due to their much stronger and thicker cell wall. Some bacteria generate biofilms in which they are embedded, that prevent them from contact with the disinfecting agent. In some cases the disinfection of the biological threat is not enough as some of the microorganism cells contain a considerable amount of toxins which remains active even if the generating cells are not alive.

The conventional approach for disinfection is to use disinfectant liquids, solutions, or gases. The application of solid materials as decontaminants for biological agents is very limited, mostly because of the incomplete interaction of the solid materials with the biological agents. Generally, conventional solid materials cannot be used for extensive biological decontamination even if they are very efficient, mostly because these solids are composed of very large particles. There is a high possibility that a large particle or aggregate would not get into contact with the much smaller bacterial or viral particles. The presence of irregularities in the surfaces, cracks, openings, etc., reduces the contact probability, making the applicability of solid materials for disinfection even more questionable.

However, with the development of solid materials in nanoscale size, all of these disadvantages can be significantly alleviated. Nanoparticles are much smaller than bacterial cells and hence smaller than virus particles. They can penetrate into irregular surfaces, cracks, etc. Nanoparticles can be easily dispersed in a gas stream or a liquid, taking advantage of the penetrating capabilities of these media while at the same time provide the benefits of solid materials, such as easier cleanup and less damage to the contacted surfaces.

Nanofabrication Towards Biomedical Applications. C. S. S. R. Kumar, J. Hormes, C. Leuschner (Eds.)
Copyright © 2005 WILEY-VCH Verlag GmbH & Co. KGaA, Weinheim
ISBN 3-527-31115-7

14.2
Standard Methods for Chemical Decontamination of Biological Agents

There are several approaches generally used for chemical decontamination of biological agents. The most widely used one is when the disinfectant agent is applied as a liquid or solution. Examples of such disinfectants are bleach solutions, chlorine solutions, chloramine T, glutaraldehyde solutions, phenolic solutions, and ethanol/water mixtures. This approach has the advantage of being relatively quick and very efficient for many bacteria and viruses. Some of the disinfecting compositions, such as bleach, are able to detoxify certain chemical warfare agents as well (e.g., mustard gas). A major disadvantage of this approach is that the solutions deteriorate most surfaces rather rapidly. Although bleach is well suited for situations which require frequent disinfection, like hospitals, public areas, and others, it is not very useful for certain cases, such as industrial spills or terrorist attacks, where many types of surfaces as well as air and water are simultaneously exposed to the biological agent. The application of solutions or liquids is not desirable in the case of sensitive surfaces like those of electronic gear, paper documents, etc. Residues from the solutions are usually corrosive to many surfaces and difficult to remove once the disinfection process is over. Another significant problem of this approach is the need to use relatively concentrated solutions for disinfection of spore cells. Aging of these solutions is a serious issue which is often overlooked – the activity of most disinfecting solutions diminishes drastically with time [1, 2]. Eventually this could lead to overestimation of the efficiency of the disinfection.

An advantage in the liquid-based technology is the development of the so-called nanoemulsions developed by Baker Jr. et al. [3–5]. The nanoemulsions are based on small oil droplets in water stabilized by surfactant (diameter of the order of hundreds of nanometers). They have an advantage over standard liquid treatments in their ability to induce spore proliferation and subsequent death of the generated vegetative cells, which makes these nanoemulsions more effective against spore-forming microorganisms [4]. Although this approach improves the biocidal efficiency and is based on harmless chemicals, it is still limited to the type of surfaces to which it can be applied. It could not be used for air cleaning and would be difficult to remove once the disinfection is over.

Disinfection can also be achieved with gases such as chlorine or chlorine dioxide. Gases are applicable to closed environments only, and they are also very corrosive to practically all surfaces. Gases usually damage sensitive materials and surfaces like those of electronics, art artifacts, paper documents, and many others. Their application also requires specially trained personnel and a well-controlled environment (temperature, humidity, etc.) for the disinfection to be successful.

In general the "traditional" methods of disinfection are not suitable for decontamination from both biological and chemical agents. In these cases separate treatment has to be used for each potential threat, bringing up chemical incompatibility issues.

14.3
Nanomaterials for Decontamination

The most serious disadvantage of common solid materials as biological decontaminants is lack of good contact with the biological agent, whether virus particles or bacterial cells. Nanomaterials can compensate this disadvantage since they can be fabricated and dispersed as very small particles or aggregates. Additionally, solid nanomaterials can be dispersed either in a liquid or as a relatively stable aerosol, which increases their contact with potential biological hazards. Besides alleviating the contact problem, the nanomaterials introduce much higher chemical and higher biological activity.

The surface of solid materials is a very important factor concerning their overall chemical activity. When a material is broken down to smaller and smaller particles, more and more building elements are exposed on the surface. These exposed molecules have significantly higher energy than those inside the crystal, which translates into higher chemical reactivity. The activity of specific crystal morphologies such as edges, corners, defects, and high index planes is even higher. This is evident in the case of nanomaterials, where the concentration of these highly active sites is much higher compared to the corresponding bulk material. The higher surface reactivity translates into the capability of nanomaterials to adsorb many more molecules per unit surface area and bind them more strongly than a regular solid material can. Besides much higher chemical reactivity, the nanomaterials can be dispersed in much finer form [as airborne particles (aerosol) or dispersed in a liquid (sol)] than other solid materials, which significantly enhances their application and efficiency.

The high surface area and higher chemical reactivity which are inherent in nanoparticles make them suitable for a dual purpose. Besides playing a role as a biological decontamination agent, they could serve as excellent adsorbents for toxic chemicals and chemical warfare agents and convert them to benign compounds by breaking the labile bond responsible for their toxicity [6–12]. Other toxins are retained by physical adsorption only [13].

An important requirement for the solid material is that it should be practically nontoxic. Because of their small particle size, the nanoparticles and their aggregates can bypass the aerosol-gathering systems of the organism and penetrate to the lungs if they are applied as aerosol. Examples of harmless materials are asbestos and soot, which become dangerous only in the form of tiny aerosol particles. This restricts the choice of nanomaterials as disinfectants to a very small number of compounds. This review concentrates on the two which have the greatest potential and have been most researched in their nanoparticulate form: magnesium oxide and titanium dioxide.

14.4
Magnesium Oxide

Magnesium oxide is a highly ionic compound with high lattice energy. This means that magnesium oxide has a strong driving force to crystallize and a probable amorphous state would be unstable. However once "fixed" in nanocrystalline form, the MgO nanoparticles are stable and can endure heating to temperatures as high as 550 °C without considerable sintering. Magnesium oxide has the advantage of being benign to organisms as well as practically harmless to most surfaces. One of its major advantages is that when left in contact with the atmosphere it converts to harmless magnesium carbonate.

Magnesium oxide nanocrystals can be prepared in sizes of the order of 4 nm and very high surface area (exceeding 500 m^2 g^{-1}) [14]. The particles arrange themselves in the form of a voluminous network made of separate nanoparticles (Fig. 14.1). For comparison, the commercially available magnesium oxide (comprised of large, micrometer-sized crystallites) has surface area in the range of 5–80 m^2 g^{-1}. The small size of the particles and the porous aggregates is of importance for the stability of the aerosol. Such an aerosol of magnesium oxide nanoparticles is stable for as long as 1 h [11, 15]. This allows the application of nanoparticle materials as aerosols for air volume decontamination combined with surface decontamination.

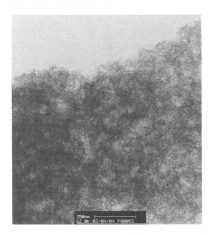

Figure 14.1. TEM image of magnesium oxide nanoparticles. The nanoparticles are interconnected in a very porous network.

Magnesium oxide in the form of nanoparticles is a good bactericidal agent [15, 16] capable of killing more than 90% of gram-positive bacteria (*B. cereus*), gram-negative bacteria (*E. coli*), and spores (*B. globigii*) in minutes. This is in contrast with the fact that "normal" magnesium oxide comprised of micrometer-sized particles does not exhibit appreciable bactericidal activity. The mechanism of the bactericidal activity is described later.

The increase of the surface area has been used to adsorb strong disinfectants such as chlorine, bromine, or interhalogen compounds. The resulting solid materials (MgO/X$_2$, where X= Cl, Br, I) have excellent bactericidal and sporicidal properties [17], while at the same time they are not as harsh to the surfaces as most disinfectants. These halogen-loaded nanoparticle compositions contain large amounts of very active halogen (up to 43 wt% in the case of IBr), while at the same time the halogen is "activated" compared to its free state, enhancing its biocidal activity [17].

Nanoparticulate magnesium oxide successfully removed traces of toxins from solution such as aflatoxins at parts per million levels [15]. This capability was observed earlier for both micro- and nanoparticles and is expected to be due to physical adsorption. However, the capacity to retain toxin molecules is expected to be higher for the nanoparticles due to their higher surface area and higher activity surface.

14.5
Mechanism of Action

It is of fundamental as well as of practical interest to understand the mechanism, of the bactericidal activity of nanoparticles.

Perhaps the most important disadvantage, as emphasized earlier, of using solid materials as disinfectants is the contact issue. Some of the cells are not touched by particles and survive the treatment. Electrostatic interaction is an important parameter of how colloidal particles such as nanoparticles could interact with other colloidal particles (bacteria cells or viruses). In aqueous solution it is described by the ζ-potential. In water the magnesium oxide nanoparticles are slowly converted to hydroxide. Upon dissociation of hydroxide anions the nanoparticles become positively charged. As determined by ζ-potential measurement, all magnesium nanoparticles composites (MgO/X_2) are positively charged. This is favorable for an interaction with bacterial cells as they are negatively charged at biological pH values [18]. Confocal and optical microscopy studies show that mixing of magnesium oxide nanoparticles in suspension and bacterial cells causes spontaneous coagulation into large aggregates composed of nanoparticle aggregates and bacterial cells. This phenomenon was observed with all types microorganisms tested [16] (Fig. 14.2). The sticking effect was observed with atomic force microscopy, showing that upon contact the bacterial cells suffered considerable cell wall damage [13]. Upon contact with bacterial cells the aggregates of nanoparticles can fall apart into smaller aggregates [16].

Several other effects are of importance for the bactericidal activity of the nanoparticles in the dry state or as a suspension: their charge attraction with bacteria; their enhanced abrasive action; the oxidation capability (in cases where halogen or interhalogens are present); and their basicity.

The bactericidal activity of the nanoparticles is not limited to suspension; they are also bactericidal in the dry state. The important factors for the activity in dry states are explained by the abrasive properties of the particles, which can mechanically damage the cells. Since the nanoparticles adsorb relatively large amounts of water, they are desiccants (in the particles' vicinity) and generate a high pH. Both factors are considered responsible for the bactericidal activity of the nanoparticles against vegetative bacterial cells.

This is supported by the fact that spores, which have mechanically stronger cell walls and are not sensitive to the amount of water or high pH, are not influenced significantly by the magnesium oxide nanoparticles. However, it is observed that halogen-loaded nanoparticles (nanoparticle compositions prepared by adsorbing free halogens on MgO nanoparticles, MgO/X_2, X=Cl, Br, I) have excellent sporicidal

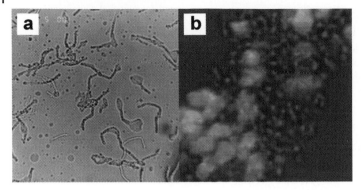

Figure 14.2. (a) Optical microscopy image (magnification 1000×) of *B. megaterium* bacterial culture mixed with MgO nanoparticles in suspension (5 min contact time). The aggregates of particles stick to the bacteria and some of them burst into pieces due to cell wall damage. (b) Confocal fluorescence microscopy image (magnification 630×) of *E. coli* bacterial culture mixed with MgO/fluorescein nanoparticle suspension (5 min contact time). The fluorescent nanoparticle aggregates (green) coagulate with the bacterial cells (red).

activity. The explanation of this fact lies in the synergistic combination of two factors: the basicity of the nanoparticles and the halogen presence. Halogens are generally poor sporicides because the outer spore coat is resistant to halogens [19]. On the other hand, the outermost spore coat is base-sensitive, which means that the nanoparticles, which generate a high local pH environment, can partially or completely dissolve this shell [20, 21]. This is confirmed by observation in the literature that spores pretreated with sodium hydroxide are much more sensitive to chlorine than is the control [20, 21]. In the case of halogen-loaded nanoparticles the two factors are present together – highly basic environment and high local concentration of active halogen. Another factor found to be of importance is the higher activity of the halogen

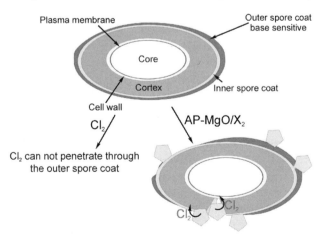

Figure 14.3. Mechanism of the sporicidal activity of MgO/Cl$_2$. The MgO nanoparticles sensitize the spore cell and allow chlorine to damage the cell further.

compared to gas-phase free halogen [17], combined with the high local halogen concentration. The sporicidal activity of the halogen-loaded nanoparticles is schematically shown in Fig. 14.3.

The advantage of a solid disinfectant is apparent in this example: the nanoparticle brings to the spore cell a high concentration of adsorbed active halogen, and its oxidizing activity is promoted by the nanoparticle carrier, which removes the most resistant layer of the spore protection. At the same time, the overall concentration of the halogen in the environment is low and the halogen is released primarily upon contact with a cell and secondarily due to desorption.

14.6
Titanium Dioxide

Another nontoxic metal oxide of interest is titanium dioxide (titania). It becomes a very potent oxidation photocatalyst when illuminated with UV light. Nanoparticulate titania was found to be an active bactericide in water suspensions. Compared to magnesium oxide, its biocidal activity is optimal only in the presence of UV light [22, 23]. Films of titanium dioxide were found to have detoxification abilities by oxidizing biological toxins such as the *E. coli* endotoxin under UV light in aqueous solutions [24]. The conclusion from the literature reports is that as a very powerful photocatalyst the titania nanoparticles catalyze the catalytic oxidation and destruction of the cell wall by generating high concentrations of highly active inorganic radicals [22]. It was observed that smaller particles and higher anatase content perform better than other titania nanoparticulate composites [23]. An advantage of the nanoparticles versus bulk titania is again easier dispersibility and higher biocidal activity.

14.7
Summary

Nanomaterials have properties different from those of the corresponding bulk material they are made of. They have different chemical and physical properties and sometimes exhibit properties never observed in bulk matter. One direction which still hides many unknowns is how the change in size and properties influences their biological activity. Some materials, as discussed in this chapter, have been found to interact with biological materials in a manner different from the bulk. These new properties and behavior in biosystems can be put to good use as a new line of defense against biological threats.

References

1 Sagripanti, J., Bonifacino, A. Bacterial spores survive treatment with commercial sterilants and disinfectants, *Appl. Environ. Microbiol.* **1999**, *65*, 4255–4260.

2 Sagripanti, J., Bonifacino, A. Comparative sporicidal effects of liquid chemical agents, *Appl. Env. Microbiol.* **1996**, *62*, 545–551.

3 Hamouda, T., Myc, A., Donovan, B., Shih, A., Reuter, J., Baker Jr, J. A novel surfactant nanoemulsion with a unique non-irritant topical antimicrobial activity against bacteria, enveloped viruses and fungi, *Microbiol. Res.* **2001**, *156*, 1–7.

4 Hamouda, T., Hayes, M., Cao, Z., Tonda, R., Johnson, K., Craig, W., Brisker, J. A novel surfactant nanoemulsion with broad band sporicidal activity against *Bacillus* species, *J. Infect. Dis.* **1999**, *180*, 1939–1949.

5 Hamouda, T., Baker Jr., J. Antimicrobial mechanism of action of surfactant lipd preparations in enteric Gram-negative bacilli, *J. Appl. Microbiol.* **2000**, *89*, 397–403.

6 Rajagopalan, S., Koper, O., Decker, S., Klabunde, K. J. Nanocrystalline metal oxides as destructive adsorbents for organophosphorous compounds at ambient temperatures, *Eur. J. Chem.* **2002**, *8*, 2602–2607.

7 Wagner, G. W., Koper, O., Lucas, E., Decker, S., Klabunde, K. J. Reactions of VX, GD, and HD with nanosize CaO: autocatalytic dehydrohalogentaion of HD, *J. Phys. Chem. B* **2000**, *104*, 5118–5123.

8 Wagner, G., Procel, L. R., O'Connor, R. J., Munavalli, S., Carnes, C. L., Kapoor, P. N., Klabunde, K. J. Reactions of VX, GB, GD, and HD with nanosize Al_2O_3. Formation of aluminophosphonates, *J. Am. Chem. Soc.* **2001**, *123*, 1636–1644.

9 Wagner G., Bartam, P. W., Koper, O., Klabunde, K. J. Reactions of VX, GD, and HD with nanosize MgO, *J. Phys. Chem. B* **1999**, *103*, 3225–3228.

10 Koper, O., Klabunde, K. J. Nanoparticles for the destructive sorption of biological and chemical contaminants, US patent 6,057,488.

11 Koper, O., Klabunde, K. J. Reactive nanoparticles as destructive adsorbents for biological and chemical contamination, US patent 6,417,421 B1.

12 Narske, R. M., Klabunde, K. J., Fultz, S. Solvent effects of the heterogenous adsorption and reactions of (2-chloroethyl)ethyl sulfide on nanocrystalline magnesium oxide, *Langmuir* **2002**, *18*, 4819–4825.

13 Ecker, E. E., Weed, L. A. Studies of the adsorption of diphtheria toxin to nad elution from magnesium hydroxide, *J. Immunol.* **1932**, *22*, 61–66.

14 Utampanya, S., Klabunde, K., Schlup, J. Nanoscale metal oxide particles/clusters as chemical reagents. Synthesis and properties of ultrahigh surface area magnesium hydroxide and magnesium oxide, *Chem. Mater.* **1991**, *3*, 175–181.

15 Koper, O., Klabunde, K. J., Marchin, G. L., Klabunde, K. J., Stoimenov, P., Bohra, L. Nanoscale powder and formulations with biocidal activity toward spores and vegetative cells of Bacillus species, viruses, and toxins, *Curr. Microbiol.* **2002**, *44*, 49–55.

16 Stoimenov, P. K., Klinger, R. L., Marchin, G. L., Klabunde, K. J. Metal oxide nanoparticles as bactericidal agents, *Langmuir* **2002**, *18*, 6679–6686.

17 Stoimenov, P. K., Zaikovski, V., Klabunde, K. J. Novel halogen and interhalogen adducts of nanoscale magnesium oxide, *J. Am. Chem. Soc.* **2003**, *125*, 12907–12913.

18 Busscher, H. J. B. R., van der Mei, H. C., Handley, P. S., eds., *Physical Chemistry of Biological Interfaces*, Marcel Dekker, New York, **2000**.

19 Block, S. S., ed. *Disinfection, Sterilization, and Preservation*, Lea & Febiger, Philadelphia, London, **1991**.

20 Bloomfield, S. F., Arthur, M. Interaction of *Bacillus subtilis* spores with sodium hypochlorite, sodium dichloroisocyanurate and chloramines T, *J. Appl. Bacteriol.* **1992**, *72*, 166–172.

21 Bloomfield, S. F., Arthur, M. Effect of chlorine-releasing agents on *Bacillus subtilis* vegetative cells and spores, *Lett. Appl. Microbiol.* **1989**, *8*, 101–104.

22 Lu, Z.X. et al., Cell damage induced by photocatalysis of TiO_2 thin films, *Langmuir* **2003**, *19*, 8765–8768.

23 Jang, H. D., Kim, S. K., Kim, S. J. Effect of particle size and phase composition of titanium dioxide nanoparticles on the photocatalytic properties, *J. Nanopart. Res.* **2001**, *3*, 141–147.

24 Sunada, K., Kikuchi, Y., Hashimoto, K., Fujishima, A. Bactericidal and detoxification effects of TiO_2 thin film photocatalysts, *Env. Sci. Technol.* **1998**, *32*, 726–728.

IV
Impact of Biomedical Nanotechnology on Industry, Society, and Education

15

Too Small to See: Educating the Next Generation in Nanoscale Science and Engineering

Anna M. Waldron, Keith Sheppard, Douglas Spencer, and Carl A. Batt

15.1
Introduction

The challenges of educating the next generation in nanoscale science and engineering are considerable as the foundation for advancing these concepts in the minds of students cannot be easily established without an understanding of the world that is too small to see. The field of nanoscale science and engineering, however, presents opportunities to motivate students in the areas of physics, chemistry, and the life sciences. The basis for understanding the world that is too small to see is complex but is a foundation upon which educators need to build. The nano world is one that can excite young students and this excitement can be carried over into their graduate student years. To ground young students in the world of nanoscale science and engineering initially requires an understanding of size and scale and the relative sizes of macroscopic, microscopic, and nanoscopic objects. But this understanding must begin in the visible world (i.e., concrete) and then progress down to smaller and smaller (i.e., abstract) dimensions. Concepts in nanoscale science and engineering coalesce with much broader ideas of how students learn about the physical world and the submicroscopic parts of matter. Addressing nanoscale science and engineering education is important, as the technology impacts the general public and the future of the technology is often feared.

The following notes the opportunities of using nanotechnology as a motivator to students and the challenges associated with the world that is too small to see. It begins with the notion that engaging students requires that they have a firm grasp on size and scale, a prerequisite for any further inquiry into nanoscale science and engineering. Much of this is based upon the authors' observations over the past few years and reflects attempts to introduce these concepts into formal and informal science education venues.

15.2
Nanotechnology as a Motivator for Engaging Students

The impact that nanotechnology is currently having on new and existing industries is significant, but the potential for the future is enormous. It is estimated that nano-

Nanofabrication Towards Biomedical Applications. C. S. S. R. Kumar, J. Hormes, C. Leuschner (Eds.)
Copyright © 2005 WILEY-VCH Verlag GmbH & Co. KGaA, Weinheim
ISBN 3-527-31115-7

technology will have a one-trillion-dollar impact on the global economy in the next decade. The challenges are to insure that the general public is informed and cognizant of the potential. Mistakes made by failing to promote public understanding can lead to a wholesale rejection of the technology [1]. Existing industries including those not typically characterized as "high-tech" will see their product lines as well as the way they manufacture them influenced by our growing knowledge in nanotechnology. Moreover, aspects of nanotechnology will help to drive small companies whose products are developed for niche market areas such as sensors, bio and chemical analytical devices, and boutique chemicals and ingredients. These technologies are not likely to require the multi-billion-dollar investments that chip manufacturers must face. Therefore progress will be even more rapid as the relative risk from investing in nanotechnology will be lower. Nevertheless, significant investment in research and development is needed, especially in the academic sector.

Nanotechnology will lead a renaissance in manufacturing in more rural areas abandoned by traditional manufacturing over the past 50 years. It has the potential for reviving communities that used to be the home of skilled laborers who contributed to the last industrial revolution. While "traditional" chip-based manufacturing has contributed to economic growth in a select number of regional areas, nanotechnology and especially its applications to the interface with biology will have a more widespread geographic impact.

"The impact of nanotechnology on health, wealth, and lives of people will be at least the equivalent of the combined influences of microelectronics, medical imaging, computer-aided engineering, and man-made polymers developed in this century". [2]

There are three compelling technical reasons to predict that nanobiotechnology will have an impact in the future:

- *The development of more portable, more robust devices that can be deployed in the field.* Sensors can be developed and deployed that will be small enough to be distributed and collect data from a wide area. Given the state of the art in micro- and nanofabrication, sensors as small as a particle of dust could be created. The challenges are in powering these devices and the effective distance that they can transmit their signal.
- *The creation of novel analytical devices capable of interrogating single molecules.* These devices will have unprecedented sensitivity and specificity by virtue of their ability to isolate single molecules in an exceedingly small volume. Novel approaches to optical or electromagnetic interrogation schemes will be a key factor. Efforts in this area will entail the use of highly sophisticated techniques to understand such basic phenomena as how proteins fold. Progress in this area will translate the vast information reservoir of genomics into vital insights that illuminate structure:function relationships in nature. Other efforts seek to understand how biological systems interact and communicate.
- *The fabrication of separation modules that force molecules into confined environments.* Unique separation effects can be realized that afford a more rapid and

in some cases a more specific separation based upon the behavior of molecules in a microfluidic environment.

Nanotechnology is a powerful motivator for students and there is a potential to use this enthusiasm to engage them in more classical fields such as physics, chemistry, and the life sciences. The challenge is to maintain rigor in the fundamental concepts that are important to these fields and not be swayed too much by the excitement. Pictures of "nanobots" are a compelling image to engage students, but they also need to understand that they will simply not work. There are fundamental barrier to the movement of nanobots which are a function of their tiny size [3].

15.3
The Nanometer Scale

15.3.1
Too Small to See

Our ability to explain the world that is too small to see is complex, and extrapolating beyond the size of the smallest thing that we can see is difficult even for a trained scientist. Alice in Lewis Carroll's *Alice's Adventures in Wonderland* described her experience in nanotechnology after consuming a mysterious beverage that caused her to shrink. When faced with the prospect of shrinking to a height even smaller than 10 inches:

First, however, she waited for a few minutes to see if she was going to shrink any further: she felt a little nervous about this; "for it might end, you know," said Alice to herself; "in my going out altogether, like a candle. I wonder what I should be like then? [4]

The notion is that once you are too small to see, you go "out" and cease to exist. Things too small to see are not easy to comprehend simply because "seeing" is a fundamental part of our belief in most things outside of the spiritual world.

The world too small to see has been studied using more or less the same observational tools that date back to cave man. We observe, we see, and while we are perplexed by many of these observations, we are more confused by things that we cannot see. Size and scale are a complex set of concepts for children and the smallest thing that they can think of is typically the smallest thing that they can see. Without a basis for understanding, this microscopic world cannot be easily distinguished from any other world including the fictional world.

15.3.2
How Do We See Things Too Small to See?

The microscopic and nanoscopic world is defined in part by the wavelength of light. With our naked eye (and it varies with different people and age) we can see approximately 100 µm or the width of a hair. The optical microscope can resolve features down to approximately 0.7 µm (700 nm) or about the wavelength of visible light. To see the nanoscopic world, you need more powerful microscopes, ones that use electrons to illuminate the surface. One of the only ways to "see" things on the atomic scale is through the use of atomic force microscopy.

The first simple microscopes revealed a world that previously was too small to see [5]. The resolution of these simple single-lens microscopes was rather remarkable and objects of a few microns could be seen. The first images of the microscopic world were of objects that could be seen with the naked eye but whose details were revealed by microscopy. One image in Robert Hooke's *Micrographia* which was published in 1665 was his rendering of the microscopic image of a "dot," a period, a common printed mark on a piece of paper. The details of this "dot" on a microscopic scale are quite spectacular. These first descriptions of the microscopic world were greeted with some derision "a Sot, that has spent 2000£ in Microscopes, to find out the nature of Eels in Vinegar, Mites in Cheese, and the Blue of Plums which he has subtly found out to be living creatures."

For most young students the world that is too small to see is first revealed using a simple magnifying glass. While operating a magnifying glass is relatively simple, a survey of 600 children ages 5–8 showed that most know or could figure out how to use a magnifying glass properly. However, almost 60% of kindergarten and first-grade children confused a microscope with a telescope. A smaller but significant percentage of second- and third-graders knew that a microscope was a magnification tool, but did not know its name or how to operate it. Yet the transition from the visible world to the microscopic world is best accomplished through a continuous process where the child can barely see the object and then the object is better revealed through the use of a magnifying glass. The optical microscope (depending upon the school and its resources) is not introduced until grades 2–4 or later. In New York State, the use of a compound microscope is not required until the eighth-grade assessment; the requisite skills are as follows [6]: "The student will be able to:

1. Manipulate a compound microscope to view microscopic objects
2. Determine the size of a microscopic object, using a compound microscope
3. Prepare a wet mount slide
4. Use appropriate staining techniques."

Without a foundation in the world that is too small to see, the opportunities to effectively integrate learning activities in nanoscale science and engineering into the curriculum before grade 8 would be difficult. While students can appreciate the world that is too small to see in an abstract sense, separation of reality and science fiction is not facile without a firm grounding in the micro- and nanoscopic worlds. Optical microscopes help students grasp that connection, giving them the link between something they can barely see to the microscopic world.

Below approximately 500–700 nm, objects cannot be resolved with an optical microscope. Electron microscopes and atomic force microscopes reveal that nanoscopic world. Both of these instruments present a more difficult challenge in student learning. First, they are expensive and beyond the reach of most schools and even universities for instructional use. Second, and again for younger students, because of the operating nature of electron and atomic force microscopes there is a disconnect between the sample and the image viewed by the student. Most modern electron microscopes and all atomic force microscopes interface with a computer and the image is viewed on a monitor. Without an understanding of the operation of an electron or an atomic force microscope, the student is then faced with an image on a computer screen. Confounding the viewer's ability to grasp the nature of images in the nanoworld is the lack of color. Individual atoms and molecules have no color, as color is a macroscopic property that depends upon the collective action of atoms [7]. All color is lost around 400 nm; the last colors are blues and purples. Scientists use color enhancement of electron and atomic force micrographs to highlight certain features or simply to make the images more artistically attractive. While illuminating, these colorized versions can confuse the viewer's perception of the scale of the object under study.

Atomic force microscopy is used to see the smallest objects, and subnanometer resolutions have been achieved. The historical origins of atomic force microscopy date back to the 1920s, when stylus profilers were first developed. In the 1980s scientists at IBM created some of the first scanning probe microscopes, which could see by physically interacting with the surface; these then gave rise to the current generation of atomic force microscopes. At this resolution the double helix of DNA, only 2 nm across, can be visualized [8]. The view into the atomic world provided by atomic force microscopy is useful as it provides students with images that show the particulate nature of atoms and the relationship between the physical interactions and the chemical formula. Most high-resolution microscopes are beyond the reach of undergraduate and almost all high-school classroom laboratories except perhaps as a demonstration. A number of efforts to give world-wide virtual access to atomic force microscopes exist, and these provide students with a graphical interface to the instrumentation [9]. While these efforts have been made to put atomic force microscopes "on-line" with frequent sample changes and the opportunity to control the instrument, these demonstrations are limited and not optimally connected. The translation of the sensing component from the visualization in both electron and atomic force microscopes is a difficult concept for students to comprehend. Nevertheless, learning activities which are based upon student visits to a research laboratory appear to introduce critical concepts [10].

15.3.3
How Do We Make Things Too Small to See?

Making things small is the answer to many of the challenges that we face. With the exception of some automobiles and ourselves, we have attempted to make most things smaller. One approachable challenge came in 1959 in the form of the following question: "Why cannot we write the entire 24 volumes of the *Encyclopedia Britannica* on the head of a pin?" [11]. Feynman predicted much of the current events in nanotechnology. Curiously, in his seminal talk "There's plenty of room at the bottom," Feynman never used the word "nanotechnology."

The focused beam of light or electrons in a microscope can be directed to carry out chemical reactions with a high degree of resolution. In addition, atomic force microscopes can be used to physically move objects with atomic precision, including moving individual atoms. In other words, microscopes can be used not only to "see" but also to "make" things too small to see. The iconic images produced by scientists at IBM have proven to be striking and illustrate the power of the technology. From the first logos spelled out with individual xenon atoms, to the stick figures created out of individual carbon monoxide molecules, these efforts have demonstrated the promise of nanotechnology. We should be reminded however, that although the challenge of moving around single atoms has been met, the issue of making it practical remains [12].

Most of the very small integrated electronic circuits are fabricated using a process called photolithography. Photolithography combines optics and chemistry to produce three-dimensional structures. Alois Senefelder first invented lithography in 1798, taking advantage of the observation that oily substances could not be wetted with water. The first demonstration of lithography involved drawing an object on a surface using a greasy crayon and then applying ink to produce a printing template. Curiously, the first surfaces that he tried were limestone. Lithography is still practiced today more as an art form than a technology to reproduce images. In photolithography, literally meaning *light–stone–writing* in Greek, an image can be produced on a surface by drawing with light or electrons in much the same way that you might scratch away the crayon on a scratch board.

Modern photolithography is a process that involves photosensitive chemicals called photoresists, light (typically deep ultraviolet $\approx$180 nm) and optics. Lenses are used to shrink the pattern, and for students there is an opportunity to reinforce concepts in optics including focal length and relatively simple calculations to determine the fold reduction in the image. The pattern produced via photolithography is essentially two-dimensional and is first written on a mask which is similar to a stencil (Fig. 15.1). Photolithography involves using energy (e.g., light or electrons) to change the solubility of the photoresist. The photoresist protects the underlying material from being etched when a caustic chemical is applied. Therefore, the unprotected material is eroded or etched, a process which can be controlled to achieve a certain depth of etching. Through a series of steps, layer upon layer of materials can be deposited, patterned, and etched to generate a multilayered structure. The current generation of Pentium chips contain approximately 40 layers and millions of individual transistors.

Today, the average computer chip carries a series of electrical circuits that are so small thousands could fit onto the head of a pin. The latest microprocessors in Apple's G-5 computers are pushing the optical limits of photolithography. Advances in nanotechnology now allow wires to be built that are literally just a few atoms wide. Eventually practical circuits will be created using series of individual atoms strung together like beads serving as switches and information storage devices.

Making things too small to see is a significant technical challenge, and for students the hurdles to comprehending how to make these things, let alone see them, are many. The fundamental technical issues are not easily translatable to the macroscale. Models are used to help communicate concepts. We can, for example, demon-

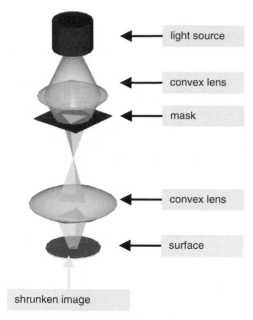

light source

convex lens

mask

convex lens

surface

shrunken image

Figure 15.1. Modern photolithography.

strate with ball and stick models how molecules are assembled, but these balls and sticks do not need to conform to the basic rules of chemistry. Bonds can be formed by a student using a model that are strictly prohibited by virtue of the reactants. Individual blocks in a set of Legos can be assembled into larger structures but these do not present the limitations that truly exist in the nanoscale world. Constraints can be introduced: for example, a ball representing a carbon atom can have four holes, limiting the student to attaching four other atoms [5]. While valence can be properly represented to a limited extent, other chemical and physical phenomena cannot. So electrostatic attraction and repulsion, which play an important role in nanoscale behavior, are not easily translated into the macroscale. Finally, the material properties of models do not always reflect the material properties of the actual materials. In fact, the material properties are a function of the size scale, with behavior as simple as the flow of a liquid radically different as the dimensions approach the nanoscale.

Models do serve a critical role by making the world that is too small to see tangible in terms of the student's perception. Physical models are able to represent space-filling more accurately than models on a computer screen. On the other hand computer-generated models can be more complex and obviously more easily shared than physical models. Regardless, since neither model is fully obliged to obey the laws of chemical bonding and other limits in their assembly, they are potentially misleading. A growing number of computer-generated pictures purport to show molecular assembled devices whose structures are simply the result of computer graphics. The challenges of mechanosynthesis (a term coined by Eric Drexler) are still formidable, without a clear path to success.

15.4
Understanding Things Too Small to See

15.4.1
What They Know

Introduction of nanoscale science and engineering to any student must be within the context of what they know. Otherwise the concepts are simply abstract and the ability of a student to discriminate between fact and science fiction is restricted. So an understanding of what students know is a critical element and there are no generalities that can be easily drawn. More broadly, in a cognitive sense, how we come to "know" is a multifaceted issue and is different for different people [13]. Students are not exposed to nanotechnology in grades K–12, and in fact, in a recent survey of approximately 50 children that we conducted in science museums in New York and Arizona, 75% responded "No" when asked, "Have you heard of nanotechnology?" [14]. A 2003 national study conducted by Edu. Inc. of 1000 youth age 6–18 showed 80% had not heard of nanotechnology. Eighteen percent who had heard of nanotechnology showed serious misconceptions. Only 2% could give an accurate example of nanotechnology. In a related survey over 90% of 250 adults surveyed were not familiar with nanotechnology [15]. In 2004 Edu. Inc. interviewed 100 youth age 12–18. All of the youth showed interest, sometimes exuberant interest, in learning more about nanotechnology when it was explained to them. Clearly there is a need to introduce nanoscale science and engineering into the classroom and this will impact public understanding in a direct and indirect fashion.

The paucity of understanding about nanoscale science and engineering among students is not restricted to grades K–12. Over the past 4 years, the authors have surveyed 184 undergraduate and graduate students enrolled in a course entitled "Nanobiotechnology" at Cornell University. The course is fully interactive and a collective effort. Lectures are given by faculty at the respective institutions and "teams" are assembled at these institutions and between institutions. The teams work on a design project to build a device to explore a biological question. One of the major outcomes for the students is to experience working as part of an interdisciplinary team. Course presurveys are collected to establish the background of the students, and teams are then assembled to achieve a balance in the background and experience of the students [16]. At the beginning of the semester when asked, "What do you think of when you hear the word nanobiotechnology?", many students have shown clear misconceptions in describing tiny machines, nanobots, or similar fictional assessments of the technology. Responses have included, "small robots in the bloodstream" and "little guys that go into people and fix stuff." An overwhelming 92% of these graduate and undergraduate students have responded with answers that are neither right nor wrong but that do not demonstrate knowledge of the field. Only 8% of students give a satisfactory definition of nanobiotechnology. In postsurvey data collected at the end of the course, 78% of students have achieved a level of knowledge at which they can describe the field accurately. Clearly, this course effectively grounds students in the true nature of nanobiotechnology and corrects the initial misperceptions held by many students.

15.4.2
Particle Theory

The first fundamental challenge for young students is understanding that there exists a microscopic and nanoscopic world that they cannot see, and that events in this world have a profound impact in their world. For example, dust particles that are too small to see are transported halfway around the world by global wind currents. Particles too small to see are responsible for allergies. The notion that there are things that we cannot see may be simple to scientists and engineers, and at some point we came to understand this connection. Surveys carried out by the authors as part of formative evaluation for an informal science education effort (It's a Nano World, www.itsananoworld.org) revealed a significant lack of awareness of and appreciation for the world too small to see [17]. Up to approximately 8 years of age, children do not understand that there is a world too small to see and the "smallest thing that they can *think* of" is typically something visible to the naked eye. This mirrors much of the educational research that has been conducted about students' view of matter [15]. When encouraged to use their imagination to think of the smallest thing possible, high-achieving first-grade children progressed down within the macro world: "a lady bug, a spot on a lady bug, the lady bug's tongue, a bump on the lady bug's tongue." During interviews several second- and third-graders who understood the existence of and could explain the function of cells still drew macro objects such as dust, dirt, and bugs as the smallest thing they knew. Seeing microscopic objects allowed students to consider the reality of a microscopic world. However, appreciation and trust of this reality was limited by cognitive development.

Key to the learning process are explorations into how students in grades 7–16 learn about things that are too small to see. Such learning is embedded in broader ideas of how students learn about the physical world in general. Much prior educational research has shown that students bring to their formal education ideas and prior knowledge about how the world works [18]. Students' prior knowledge can and often does interfere with their subsequent learning. Instruction that proceeds without recognizing such knowledge is often viewed as being nonmeaningful – students do not build understanding based on what they already know, but instead simply memorize what they are told. These ideas about student learning and the need for more inquiry-based learning are clearly articulated in the National Science Education Standards and the NYS Math, Science, and Technology standards.

Contained within this body of educational research are studies that begin to address how students view matter and how it is structured on a submicroscopic or molecular level. Many students across a wide range of grade and ability groups hold the view that matter is not particulate [19]. For these students, materials at the atomic or molecular level are simply shrunken versions of their real-world manifestations and the atoms/molecules retain the materials' macroscopic properties [7].

The impact of holding such views on subsequent learning is profound: understanding much introductory chemistry, for instance, requires a robust understanding of the particulate nature of matter and an ability to relate the macroscopic and submicroscopic worlds [20]. Scientists (experts) freely move between these two realms, while students (novices) are often confined only to what they can see, and subsequent learning in more advanced science classes becomes shallow and algorithmic, rather than conceptual and deep. While science textbooks have recently begun to address this issue by introducing submicroscopic representations into their text, much still needs to be done.

To effectively incorporate nanotechnology into classrooms, we need to address some of these issues, by investigating the ages and developmental levels that are appropriate to introduce ideas and concepts about atoms and molecules and their interactions, and to further address issues of scale and size and how nanoscale events impact the physical and life sciences. As Vogel noted, effective designs for large things often work poorly for small things and vice versa, i.e. "size matters" [21]. For example, the phenomenon of molecular diffusion is important over small distances, but unimportant over large distances, and consequently larger organisms need transport systems to supply themselves with food and oxygen [22]. Questions that should be addressed include: What activities (e.g., computer simulations, use of models, etc.) are appropriate for developing understanding of things that cannot be seen? At what age can these most effectively be introduced? Is there a developmentally appropriate conceptual sequence for learning about nanoscale phenomena, and how can this be coordinated across the different science disciplines as they are presently configured? For example, molecular kinetic theory is often taught in chemistry classes after students have encountered such concepts as diffusion and osmosis in previously taken biology classes.

15.5
Creating Hands-On Science Learning Activities to Engage the Mind

Minds-on (hands-on) learning activities help students gain a greater insight into the basic concepts of science [23, 24]. These learning activities also serve to engage students, motivating them through an approach that has come to be known as "activity before content". To help children and even adults understand nanotechnology, we must create a bridge between their knowledge of "what is small" and the tools used to see and make things too small to see. They must first learn about the microscopic world and the scale at which nanotechnology can interact with that world. Once they have established the relative scale of microscale and nanoscale, then they can begin to understand the technologies used to see and make tiny objects. The use of activities that not only inform but engage is critical, and minds-on activities are a critical part of the learning experience. The challenge is to develop activities that involve objects that are too small to see. Seeing the known at a microscopic level helps children create a mental bridge from macro to micro. Spencer notes that during research investigating grade K–3 children's perception of microscopic objects, children who named a mosquito as the smallest thing they knew, looked at prepared slides of a mosquito through a microscope. By viewing the slides at slowly increasing levels of magnification children were able to see direct evidence of the progression from a highly magnified, but recognizable mosquito to the cellular level. Based on affect and comments during the activity and interviews after the activity, children appeared to experience a paradigm shift of sorts. A third-grade girl remarked, "I never knew that's what mosquitoes looked like close up." An excited 8-year-old boy said "Look there's the cells. Right there. You can see them!"

Microscopy and seeing the world too small to see is an important part of our repertoire of minds-on science activities. Microscopes present an opportunity to engage young students, and the parts of a microscope, namely the lenses, provide a segway to photolithography. The concept of magnification is the same as the optical reduction that is used to make things too small to see. Lessons on measuring the focal length of a lens provides the basis for compound lenses and shrinking of a pattern. On the opposite end of the spectrum, we use activities to measure the focal length of a lens as the basis for lessons on telescopes.

Perhaps one of the best examples of efforts of the authors in developing unique minds-on science activities about nanotechnology is the design and fabrication of a portable 5X optical reduction system that can be used by students to perform photolithography (Fig. 15.2). The instrument, built by Alliance for Nanomedical Technologies (Batt is the project leader), allows a student to fabricate a microelectronics circuit with feature sizes down to sub-100 µm. The design team consisted of graduate students, research associates, and undergraduates affiliated with Alliance for Nanomedical Technologies. The instrument has an optical system to achieve a five-fold

Figure 15.2. Portable 5X optical reduction system that can be used by students to perform photolithography.

pattern reduction, and ancillary efforts have resulted in the development of a photo-resist and etching process that is safe for use in middle- or high-school classrooms. Students design a circuit, print it to a transparency (that serves as the mask), and then transfer the pattern to aluminum that has been coated onto a standard microscope slide. They can then test the circuit using batteries and LEDs. We have tested this activity with students as young as 11–12 years old with success.

15.6
Things That Scare Us

15.6.1
The Societal Concerns of Nanotechnology

New technology will always raise the concerns of the public especially when we, the scientific community, do not take the time and have the patience to articulate the field. Surveys document that the general public view nanotechnology as holding great promise and that there is currently not a widespread fear of the technology [1]. What falls into the void that we create by remaining cloistered in our laboratories, are pundits and pseudoscientists whose mission is to, at best, tantalize, and at worse, to strike fear. It makes for great novels, and it makes for even better movies; but the threshold into science fiction is murky. Nanotechnology will not as a technology spawn a new threat to society. History shows that most of the dangers to society that result from the misuse of technology arises not from state-of-the-art technologies but from more mundane technologies in the hands of opportunists. We have in the last twenty years alone seen horrific acts carried out by individuals and groups with some fairly unsophisticated technology.

In the Nevada desert, an experiment has gone horribly wrong. A cloud of nanoparticles – micro-robots – has escaped from the laboratory. This cloud is self-sustaining and self-reproducing. It is intelligent and learns from experience. For all practical purposes, it is alive. It has been programmed as a predator. It is evolving swiftly, becoming more deadly with each passing hour. Every attempt to destroy it has failed. And we are the prey [25].

There are certainly ethical concerns with any new technology that must be considered. The prospects of a run away technology as described in Michael Crichton's book *Prey* would be a sad outcome, but the current state of the art in nanotechnology in no way enables that outcome. Nano-Rome will not be built in a day:

There are real dangers in the world, and those that concern us now are 50-year-old technologies, lethal in the hands of individuals and organizations that would choose to use them [26].

The technology as described in this fictional account is not even close to reality. No enabling technology exists or is on the horizon that could account for the fanciful creatures described there. Yet it received lots of press coverage and through a variety of media, especially the Internet, fanciful predictions of doom proliferated. Unfortunately, the barrier between science and science fiction is only as high as the imagination of a talented novelist or cartoonist. Pictures abound on the Internet of nanobots and other imaginary things, and those of us who choose to be engaged with the general public spend a good deal of time offering reality checks for students and the general public. Even some professional colleagues lean over the line at times, seduced by the publicity and the potential that this notoriety brings in terms of funding and other opportunities. Yet the practical reality of constructing self-assembling, autonomous machines smaller than a single bacterium that can scurry about like little fleas is still only the product of an artist's imagination. What we should be concerned with is the more mundane and the root causes for the growing desire to use them. Education is the key.

15.6.2
The Next Generation

We have elected to focus our attention on the next generation of potential scientists and engineers. We are engaging young people who often do not view science as an educational opportunity let alone a career for them. *They do not see themselves as scientists*, and that is a significant barrier that we seek to overcome. We work in concert with their teachers, recognizing that this partnership will only work if we understand their world. With more and more mandated curricula, we must meet the needs of the schools rather than continue to offer content that has no relevance to the rest of the educational experience. Presently, we operate three middle-school science clubs for girls in an attempt to address the challenge of encouraging young women to consider careers in science. In a recent survey of 38 girls who signed up to join the club, 74% did not see themselves as a scientist [27]. We also host three after-school science clubs for underrepresented minorities at Beverly J. Martin School (Ithaca, NY), the Onondaga Nation School and Shea Middle School (Syracuse, NY), exercising our belief that these young students have all the potential in the world. Furthermore, we offer events for the general public, and every summer we engage more than 3000 people at the Great New York State Fair with the wonders of the nanoworld. Finally, in April, 2003, a traveling museum exhibit, *It's a Nano World*, debuted that was developed with our collaborators from the Sciencenter and Painted Universe in Ithaca. We estimate that the exhibit will be visited by more than one million people during its tour around the United States.

In helping educate and, perhaps more importantly, inspire these young students, we hope to raise the general awareness of the public at large as to what nanotechnology is all about. This technology, and in general most technology, will have a very positive impact on our lives. We have coined this outreach effort "Main Street Science," and over the next 5 years hope to capture the exciting scientific discoveries of our center and others and translate them into practical and approachable concepts for students and the general public. At Main Street Science we will harness the energy of our undergraduates and graduate students to develop hands-on activities, giving them a practical experience in community science.

What scientific discoveries do we hope to share with young students? For example, few students even through high school understand what the term "nano" means in its fullest context. They understand that a nanosecond is pretty fast, but they do not comprehend that a hummingbird beats its wings about 100–200 times per second. That is virtually imperceptible to the human eye and is faster than the flickering of a fluorescent light bulb. Nevertheless, computers operate about a million times faster. They also know that a nanometer is pretty small, but they do not realize that the distance between atoms is on the order of a nanometer. So, for younger children, we consider it a challenge for them to comprehend simply what a billion is, and initially try to expose them to the concept by using thousands of little plastic Lego blocks. In summary, we make an honest effort to engage kids and have them hopefully begin to believe that science is a good thing and learning about science can be exciting. Regardless of whether these kids go on to get their Ph.D. in nanotechnology, it is important to have them believe that they can do it.

A more scientifically literate public is one sure route to ameliorate the fears that seem to accompany many scientific revolutions. History is replete with examples of where new scientific discoveries were initially met with public challenges, and only after a significant backlash did we the scientific community leave our laboratories and begin to actively engage the public. Thus it is not surprising that the National Science Foundation, much to their credit, has now put forth the challenge of public engagement as one of the important criteria to consider for their supported research programs.

15.7
The Road Ahead

The road ahead is one of great promise and potential challenges. It is not simply enough to know how students come to understand ideas and concepts, it is important to know how they can be most effectively taught such things. Many science teachers themselves learned their own science passively as received wisdom. If changing the learning of students is to be effective, then teachers need to know not only more about how students learn, but also about what are the effective pedagogical techniques and strategies for addressing students' prior knowledge. Teachers and textbooks often proceed without such knowledge. In short, for high-quality learning to occur, high-quality teaching is needed.

Nanotechnology as a field evolved from engineering, but its extended roots will be found in fields including physics, chemistry, and materials science. Its major new impact will clearly be in the life sciences, presenting a challenge of organizing interdisciplinary groups that can communicate and function effectively. This is no simple task as considerable boundaries exist in language and culture. But, moreover, advances in the field will be obscured if there is a failure to engage the general public and lay the foundation for articulating how advances in the field will have a positive impact on their lives.

Acknowledgements

This material is based upon the work supported by the STC program of the National Science Foundation under Agreement No. ECS-9876771 through the Nanobiotechnology Center of Cornell.

References

1 Bainbridge, W. S., Public attitudes toward nanotechnology. *J. Nanoparticle Res.* **2002**, *4*, 561–570.

2 Smalley, R., *Prepared Statements for House Science Committee, Subcommittee on Basic Research, June 22, 1999*. **1999**.

3 Purcell, E. M., Life at low Reynolds number. *Am. J. Phys.* **1977**, *45*, 3–11.

4 Carroll, L., *Alice's Adventures in Wonderland*. **1865**.

5 Hitt, A., Townsend, J. S., Models that matter. *Sci. Teacher* **2004**, 29–31.

6 New York State, *New York State Standards*, **2004**.

7 Ben-Zvi, R., Eylon, B., Silberstein, J., Is an atom of copper malleable? *J. Chem. Edu.* **1986**, *63*, 64–66.

8 Binnig, G., Quate C. F., Geber, Ch., Atomic force microscope. *Phys. Rev. Lett.* **1986**, *56*, 930.

9 Bade, N., Amresh, A., Ramakrishna, B., Ong, E., Sun, J., Razdan, A., Remote control and visualization of scanning probe microscopes via the Web. WebNet Journal: *Internet Technologies, Applications & Issues* **2001**, *3*, 20–26.

10 Margel, H., E. B-S, Z. Scherz, We actually saw atoms with our own eyes. *J. Chem. Edu.* **2004**, *81*, 558–566.

11 Feynman, R. P., *There's Plenty of Room at the Bottom*. Caltech Engineering and Science, **1960**, *23*, 22–26.

12 Eigler, D. M., Schweizer, E. K., Positioning single atoms with a scanning tunneling microscope. *Nature* **1990**, *344*, 524–526.

13 Kuhn, D., How Do People Know? *Psychol. Sci.* **2001**, *12*, 18.

14 Waldron, A. M., Batt, C. A., Trautmann, C., *Nanotechnology Awareness at Three Science Museums*. **2004**, unpublished data.

15 Spencer, D., *Evaluating Public Readiness for and Interest in Learning New Science*. **2003**, unpublished data.

16 Waldron, A. M., Batt, C. A., *Nanobiotechnology Graduate Course*. **2004**, unpublished data.

17 Waldron, A. M., Batt, C. A., Spencer, D., *It's a NanoWorld formative evaluation*. **2001**, unpublished data.

18 Gabel, D. L., *Handbook of Research on Science Teaching and Learning*. **1994**, New York, Macmillan.

19 Driver, R., Squires, A., Rushworth, P., Wood-Robinson, V., *Making Sense of Secondary Science: Research into Children's Ideas*. **1994**, London, Routledge.

20 Nakhleh, M. B., Why some students don't learn chemistry. chemical misconceptions. *J. Chem. Edu.* **1992**, *69*, 191–196.

21 Vogel, S., *Cats' Paws and Catapults*. **1998**, New York, Norton.

22 Barnes, G., Physics and size in biological systems. *Phys. Teacher* **1989**, *27*, 234–253.

23 Flick, L. B., The meanings of hands-on science. *J. Sci. Teacher Edu.* **1993**, *4*, 1–8.

24 Welch, W. W., Twenty years of science curriculum development: a look back. *Rev. Res. Edu.* **1979**, *7*, 282–306.

25 Crichton, M., *Prey*. **2002**, New York, Harper Collins.

26 Batt, C.A., *Prepared Testimony for House Science Committee HR766, March 19, 2003*. **2003**.

27 Waldron, A. M., Batt, C. A., Kong, E., *Science Club for Middle School Girls*. **2004**, unpublished data.

16

Nanobiomedical Technology: Financial, Legal, Clinical, Political, Ethical, and Societal Challenges to Implementation

Steven A. Edwards

16.1
Introduction

What are the obstacles to the implementation of "nanobiomedicine?" Will certain roadblocks prove fatal to the development of what many see as a "transforming" technology – one that will forever change the way we do things, perhaps even the way we perceive ourselves? Or is it the case that nanobiomedicine will mature along the conventional paths pioneered by more established industries, like the biotechnology and the implantable medical device industry?

After first introducing the concept of nanobiomedicine as it is currently constituted and describing why nanotechnology has become fascinating to many, this article will describe in turn the financial, legal and regulatory, operational, and clinical challenges to the implementation of nanobiomedicine respectively, followed by political, ethical, and societal challenges considered as a group, since these cannot be easily separated.

First, let us stipulate that certain small steps have already been taken. Drugs formulated as nanocrystals, e.g., Rapamune and Emend [1], have already been approved and are on the shelves. A nanoparticle drug called Estrasorb, from NovaVax, has been approved that delivers estrogen through the skin. Fluorescent quantum dots, a type of nanoparticle, have already been injected into living patients as an experimental aid to surgeons to recognize lymph nodes [2]. Diagnostic tests that employ nanoscale devices and nanofluidics have already been introduced into the drug discovery process [3]. There have been no political protests or social upheavals that occurred with respect to these advances. In fact, they have taken place pretty much unnoticed. It is actually difficult to get people to believe that nanotechnology is not just a technology for the future, but a technology that is already being employed.

What do mean when we say "nanobiomedicine?" We will accept the general definition of nanotechnology as the engineering and fabrication of objects in the scale from 1 to 100 nm, which is a scale that includes molecules and supramolecular structures, like ribosomes or viruses. Nanobiomedicine would be anything that uses nanotechnology in traditional biomedical pursuits.

Table 16.1 makes the point that currently fabricated objects are within the scale of biomolecules and biomolecular structures. For instance, buckminsterfullerene, an

Nanofabrication Towards Biomedical Applications. C. S. S. R. Kumar, J. Hormes, C. Leuschner (Eds.)
Copyright © 2005 WILEY-VCH Verlag GmbH & Co. KGaA, Weinheim
ISBN 3-527-31115-7

iconic nanotech object, is 1 nm in diameter. DNA, the founding molecule of biotechnology, is 2 nm in diameter. Quantum dots, which are already being used in a variety of bioassays, are manufactured as particles 2–10 nm in diameter. DNA is 2 nm in diameter; a single walled nanotube is slightly larger than that. Proteins are generally 5–50 nm in diameter. Dendrimers, sometimes called "artificial proteins," are in the same size range, although they may self-assemble into much larger objects. Viruses are similar in size to manmade nanoparticles.

Table 16.1 Sizes of nanoscale objects – nature vs. fabrication.

Object	Diameter
Hydrogen atom	0.1 nm
Buckminsterfullerene (C_{60})	1 nm
Six carbon atoms aligned	1 nm
DNA	2 nm
Nanotube	3–30 nm
Proteins	5–50 nm
CdSe quantum dot	2–10 nm
Dip-pen nanolithography features	10–15 nm
Dendrimer	10–20 nm
Microtubule	25 nm
Ribosome	25 nm
Virus	75–100 nm
Nanoparticles	2 to 100 nm
Semiconductor chip features	90 nm or above
Secretory vesicle	100–1000 nm
Mitochondria	500–1000 nm
Bacteria	1000–10 000 nm
Capillary (diameter)	8 000 nm
White blood cell	10 000 nm

What are the applications within the biomedical enterprise that we believe will be affected by nanotechnology? There have been various attempts to define this. Richard Freitas, in his forward-looking, three-volume work *Nanomedicine* [4], has concentrated largely on "molecular manufacturing," by which he means something similar to Eric Drexler's original vision of nanotech as described in his *Engines of Creation*. Freitas' work is filled with references to artificial cells and complex nanomachines that do not yet exist.

A somewhat more prosaic attempt to define a "nanomedicine taxonomy" was made by Neil Gordon and Uri Sagman [5] in a briefing document prepared for the Canadian Nanobusiness Alliance, which focuses on applications that are currently under development. S. A. Edwards [6], for the BCC business report *Biomedical Applications of Nanoscale Devices*, independently developed his own outline of medical uses for nanotechnology that is similar to Gordon's and Sagman's although it differs in the naming of categories and the divisions thereof (Tab. 16.2). Gordon and Sagman's categories were: *biopharmaceutics*, which includes drugs, drug delivery, and drug discovery; *implantable materials*, which includes tissue repair and replacement, implant coatings, tissue regeneration scaffolds, and structural implant materials; *implantable devices*, including sensors, retina implants, and cochlear implants; *surgical aids*; including smart instruments and surgical robots; *diagnostic tools*, including genetic testing and imaging; and *understanding basic life processes*, which includes nanotechnology as applied to basic research. Edwards included most of the categories described by Gordon and Sagman, but divided them up somewhat differently. Biopharmaceutics is called *drugs and drug delivery*, but drug discovery is folded into the general topic *diagnostics and analytics*. Imaging, which Gordon and Sagman lumped with diagnostic tools, was given a separate heading – *imaging and nanotools*. Edwards also included the topic *molecular modeling*, which might be thought of as software nanotools. Edwards substituted the term "*artificial organs*" for implantable devices.

Table 16.2 What is nanobiomedicine?

Gordon and Sagman [5]	Edwards [6]
Biopharmaceutics	Drugs and drug delivery
Implantable devices	Artificial organs
Diagnostic tools	Diagnostics and analytics
Implantable materials	Biomaterials
Surgical aids	General nanobiotechnology
Understanding basic life processes	Imaging and nanotools
	Molecular modeling

Perusing the lists in Tab. 16.2, the applications of nanotechnology in biomedicine would appear to be similar in purpose with other technologies applied to biomedicine. The aims of medicine are not generally transformed by the application of new technology. There are, however, ethical debates, discussed later in this paper, that nanotech might render more acute – for instance, the difference between "normative" medicine and medicine in the pursuit of greater than normal performance.

If the aims of medicine are not greatly altered by nanotechnology, however, why then do we worry so much about political, social, legal, and ethical challenges to the pursuit of nanomedicine? In part, this is due to the particular provenance of nanotechnology with respect to public awareness.

16.2
Drexler and the Dreaded Universal Assembler

The term "nanotechnology" came to widespread attention after K. Eric Drexler's book, *The Engines of Creation* [7], was published in 1986. A principal goal of nanotechnology, in Drexler's view at the time, was the creation of an assembler, a device built on a molecular scale that could assemble other devices, molecule by molecule. With appropriate programming, an assembler could presumably also replicate itself.

The advantages of such a device are self-evident – the problem that vexed Drexler was how to control it. He examined the so-called "gray goo" scenario in which assemblers replicate uncontrollably like a new life form. That prospect is still well into the future, but is nevertheless responsible for some of the fascination with nanotechnology by the press and by the public. Such a runaway replicator is also the premise of one of Michael Crichton's recent science fiction novels, ominously titled *Prey* [8].

Although we will generally avoid discussion of the more unlikely scenarios, we would like to discuss the assembler – a hypothetical device that enables molecular manufacturing, as nanotech pioneer Ralph Merkle would have it, by snapping together molecules in the manner of Lego blocks. While such a device is not necessarily devoted to biomedical applications, it could certainly be useful in manufacturing nanoscale biomedical machines, and might lead to the manufacture of devices that would otherwise be prohibitively expensive and impractical.

Mathematician John von Neumann, more famous for his contributions to early ideas in computing and game theory, also was responsible for early conceptions of the assembler. His assembler consisted of two central elements: a universal computer and a universal constructor. The universal computer directed the behavior of the universal constructor. The universal constructor, in turn, is used to manufacture both another universal computer and another universal constructor. The program code contained in the original universal computer is copied to the new universal computer and executed.

"Self-assembly and replication, the paradigms of molecular and cell biology, are being increasingly seen as desirable goals for engineering," said Phillip Ball (quoted in [6], a journalist for *Nature*. "The alternative – laborious fabrication of individual structures 'by hand' – is still the way that electronic and micromechanical devices are made today, by a sequence of deposition, patterning, etching or mechanical manipulation that becomes ever harder as the scales shrink and the device areal density increases."

Self-assembly is a defining property, perhaps *the* defining property of life. A continual increase in the complexity of self-assembly likewise is characteristic of evolution. Surely, the baroque method of reproduction used by human beings could not have been envisioned at the time the first primitive cells began replicating in the primordial ooze, about 4 billion years ago. As any molecular biologist will tell you, any bacteria, any eukaryotic cell, any organism, even a human being, is essentially a marvelously contrived machine. Even the ability to read these words is a consequence of tiny moving parts, protein molecules, electric currents, and nanofluidics.

Eric Drexler's assembler is a specific case of the general von Neumann architecture, but is specialized for dealing with systems made of atoms. The emphasis here (in contrast to von Neumann's proposal) is on small size. The computer and constructor are shrunk to the molecular scale. The constructor must also be able to manipulate molecular structures with atomic precision. The molecular constructor has both a positional capability and some sort of dynamic chemistry at the tip of its arm able to manipulate individual atoms. In other words, an arm and a hand. Another necessary attribute is some means of programming the assembler. It would also be desirable to have a macroscopic interface to a human operator whereby the assembler could respond to commands or communicate its needs.

"How soon will we see the nanoscale robots envisioned by K. Eric Drexler and other molecular nanotechnologists?" asked Nobel laureate Richard Smalley rhetorically [9] in a now famous *Scientific American* article. The answer, he maintains is never. He argues that there are two fundamental problems, which he refers to as "fat fingers" and "sticky fingers." Smalley says that robot will not be able to manipulate molecules within molecular spaces, because grippers cannot be built of the dimensions required, given that they must also be constructed of molecules. He also argues that nanobots capable of attaching its fingers to a particular molecule, presumably through some sort of chemistry, would not necessarily be able to let go, making assembly impossible. What's needed in order to make a universal assembler is really "magic fingers" according to Smalley.

Whether or not a universal molecular assembler is ever built, nanotechnology is already here to stay: a number of applications for nanomaterials, nanotools, and even bioassays are already in the marketplace. However, advanced nanotechnology applied to biomedicine could be delayed by a number of factors – a lack of venture capital, unexpected technical problems, lack of skilled labor, resistance by the public, or over-regulation by a panicked government. Below are listed some factors we believe to be relevant to the continued development of nanotechnology.

16.3
Financial

"Nanotechnology is the design of very tiny platforms upon which to raise enormous amounts of money," according to a definition favored by Lita Nelson, head of MIT's Technology Licensing Office.

Nanotechnology and nanobiotechnology companies are, for the most part, small entrepreneurial firms that cannot support themselves as yet on current revenues. Their financing comes from both government grants and from venture capital. DARPA, the research arm of the defense department, has been particularly active in financing some of the more speculative uses of nanotechnology. But, in general, DARPA uses fixed-term grants that cannot be renewed. Most of the billions coming out of the National Nanotech Initiative and the subsequent $3.7 billion bill passed in 2004 will go to the construction of infrastructure or the support of academic research. Relatively little will pass into the hands of the tiny firms currently struggling to commercialize nanotechnology.

Despite being tiny, nanotechnology does not come cheap. "Even when you're dealing with a very early stage company in nanotechnology, we find that the budget for intellectual property work, the annual budget, may range between $250,000 to $2 million a year, which is a lot for a fledgling company," points out Charles Harris, CEO of venture capital company Harris and Harris [10].

Some venture capitalists, burned by the dot.com meltdown and corporate scandals, have been turned off by some of the excessive hype used by nanotech promoters. The hyping of nanotechnology has been shameless by any standard. For instance, the following from a promotion sponsored by well-respected business publication:

"There's a MAMMOTH technological shocker afoot...If you want to REAP FUTURE PROFITS write this name down now and REMEMBER that you read it here first: NANOTECHNOLOGY."

The ad went on to say that nanotechnology was about to render present day manufacturing techniques obsolete.

Some of the hype, oddly enough, is coming from government sources, for instance the oft-quoted projection by the National Science Foundation that nanotech would account for goods valued at a trillion dollars within a decade. At the request of the National Science and Technology Council, the NSF and the Department of Commerce recently released a report called *Converging Technologies for Improving Human Performance* [11], that, for the most part, belongs in the sci-fi and fantasy category, despite the fact that a number of respected researchers were represented in the report. The extravagant claim made for nanotech may represent a kind of lobbying effort on the part of agency officials designed to appeal to the militaristic aspects of the Bush administration. Indeed, nanotech is often sold in the same nationalistic way that the space program was sold in the early 1960s. "If we don't do it, the Chinese will, or the Japanese, or the Europeans, or the Israelis..." leading, of course, to a loss of American technological dominance. And of course, the same shrill nationalistic rhetoric can also be heard in Europe on behalf of pan-European nanotech or in Japan on behalf of Japanese nanotech. Europe, the US, and Japan contributed about $600–800 million each in nanotech research funds in 2003.

On the US side of the Atlantic, the lobbying effort by NSF and others, like the Nanobusiness Alliance, has proved quite successful, given the passage of the nanotech bill in 2004. The down side is that the cyborgian fantasies emanating from the NSF may scare away private money from anything labeled nanotech, which already suffers from a kind of science-fiction aura.

Entrepreneurial companies, themselves, have been schizophrenic with respect to embracing or denying hyped claims. Some companies have redefined their ongoing projects as nanotechnology in order to profit from the public interest they see developing in this particular buzzword. And nano has become a sexy prefix for many companies to attach to company names: Nanogen, Nanophase, Nanosphere, NanoInk, Nanomet, Nanobio, Bioforce NanoSciences, Nanospectra Biosciences, NanoProprietary, Altair Nanomaterials, etc. Other companies run, run, run away... the word nanotechnology is erased from their vocabulary despite the nanoscale projects that they work on.

Although many venture capital companies have been scared off, a small number of far-seeing firms concentrate on almost nothing but nanotech, including Lux Capital and Harris and Harris, and others like Ardesta, Polaris, and Draper, Fisher, and Jurvetson have a strong concentration on nanotech within their portfolios.

As of March 2004, the American and world economies seem to be in a recovery phase. The stock markets recovered somewhat in 2003 and seem to be at least stable in 2004. The initial public offering (IPO) market, however, so necessary to maintain and enhance the pool of venture capital available for speculative enterprises, has not yet fully recovered from the dot.com debacle at the turn of the millennium and the subsequent corporate scandals. In particular, there has yet to be an IPO from a nanotechnology-focused firm. Until the IPO window opens, venture capitalists have no easy exit strategy, and will continue to be conservative in their financing of nanotech firms.

Financial constraints may delay the implementation of nanobiomedicine at the level of private financing, but are not likely to prevent it for long. Assuming other obstacles do not abort the nanotech revolution, continued government support and the potential financial rewards seem enticing enough to assure deal flow down the line. Charles Harris estimates that the number of "tiny technology" companies (which includes some microscale companies involved in microelectromechanical systems) is about 650 worldwide. Harris and Harris fielded 50 inquiries from those firms in the first 3 months of 2004 alone.

16.4
Legal and Regulatory

"What one finds as a venture capitalist is when you're thinking about Nanopharma, after you get through with your due diligence you're not really thinking about nanotechnology. You're thinking about a biotechnology model company. The economics of it and the business decisions are standard for dealing with a biotech company," said Harris in a Merrill Lynch interview. The same might be said for the legal and regulatory aspects of dealing with nanobiomedicine. So, below we describe the general regulations for dealing with drugs, diagnostics, and medical devices, which represent probably *the* major near-term challenge to the implementation of nanobiomedicine. We also describe some initial attempts to regulate the nanotech industry generally.

In the US, the Food and Drug Administration (FDA) classifies medical products that achieve their intended purpose without being metabolized or performing a chemical action on the body as "devices." Other products are considered as either "drugs" or "biologicals." The latter categories include gene therapy products and cells that have been manipulated *ex vivo* in addition to protein pharmaceuticals.

Many nanoscale devices fall under the category of medical devices. However, devices that are exclusively research-oriented in nature, such as assays used for drug discovery, are not FDA-regulated. The FDA must preapprove before marketing of devices that are intended for diagnosis or for implantation into the body. Drug delivery devices, such as nanoparticles that are ingested or injected, would be regulated as part of the drug formulation, therefore as drugs. Some macromolecular fullerene nanoparticles, such as those being developed by the nanobiotech company C-60, are in fact metabolically active and are therefore considered drugs. Designer proteins used for therapeutic purposes, such as antibodies, even if they are substantially modified beyond what can be seen in nature, are considered biologicals.

There are two tracks of FDA review procedures for medical devices. A new device requires either a Premarket Notification Section 510(k) or a Premarket Approval application (PMA). A "510k" requires either that the marketer of the device prove that it is substantially equivalent to a device marketed prior to May 28, 1976, or that it is equivalent to a device marketed subsequent to that date which has already been classified class I or class II. Class III devices are defined as those which are life-sustaining or life-supporting, those which may prevent substantial impairment of health, or may pose risks of illness or serious injury. For these a PMA application is required. Most microelectronic implants would fall under this category.

PMA approval will usually require proof of clinical effectiveness and safety, and satisfactory demonstration of current Good Manufacturing Processes (cGMP) and quality control. Clinical studies on the safety and effectiveness of a new class III device first require the filing of an Investigational Device Exemption (IDE) application. An IDE application must include the results of preclinical tests, clinical protocols, informed consent forms, and information on biocompatibility testing, manufacturing, and quality control. Further clinical testing may be required even after the submission of a PMA as the result of unforeseen clinical findings or FDA advisory panel requests. The FDA will grant market approval when it has been satisfied of the clinical effectiveness and safety of the product, and that performance and manufacturing standards have been met. Any design change in the product requires a supplement to the PMA application that must be separately approved. Nanoscale devices involved in drug delivery are likely to be considered as part of the drug formulation by the FDA, rather than as a separate device that can be applied to any drug in a mix-and-match fashion.

The approval process for drugs and biologicals is tortuous and complex; on average it takes 12 years for the average drug to go from preclinical research to production, although new drug discovery technologies are likely to shorten this interval. For every five products that enter clinical trials, only one is approved.

Even before the clinic, there are hoops that must be jumped through. Preclinical animal trials must demonstrate probable effectiveness of the product without undue hazards to humans. An Investigation New Drug (IND) application must be filed. Such an application includes information on the investigators, information on the chemistry, composition, pharmacology, and toxicology of the product, and on any previous human tests performed with the product. A clinical design and protocol must be submitted along with agreements among the parties involved, and an approval letter from the Institutional Review boards involved in the studies.

Three phases of clinical testing are usually required. Phase I is primarily to test safety and human tolerance of the product on a small number of volunteers, although information on efficacy and dosage may be collected. Phase II trials test for efficacy of the product, using a larger number of subjects, and studies are designed to determine optimum dosage. Phase III clinical trials consist of additional testing for efficacy, usually at multiple centers. A separate IND application is required for each separate clinical study. Not infrequently, phase III trials raise new concerns for which the FDA may require additional testing.

After completion of phase III trials, either a New Drug Application must be filed or, for biologicals, a Product License Application and Establishment License Application. For some biologicals, these may be combined in a Biological License Application (BLA). All application types require data on safety, efficacy, and human pharmokinetics of the product: patient information, labeling information, and information on manufacturing, chemistry, quality control, as well as samples of the product in question. Environmental impact statements may be required concerning the manufacturing or transport of the product. After the FDA conducts its review, a panel of independent medical experts will usually be called to review all relevant data. The decisions of the advisory panels are not binding on the FDA. When and if the applicant can successfully resolve all questions regarding safety and efficacy to the satisfaction of the agency, the product will be approved for marketing.

Even after approval, companies must submit periodic reports to the FDA, logging side effects or adverse reactions. Quality control records must be maintained. In some cases, the FDA will require phase IV studies after approval to test long-term effects of the drug or biological.

16.4.1
Diagnostics

In principle, in order to market a test that claims to diagnose human disease, FDA approval is required. As with medical devices, clinical trials must be performed and a PMA or 501(k) approval sought. Since 1997, however, the FDA has allowed the sale of "analyte specific reagents (ASRs)," for instance, monoclonal antibodies or specific oligonucleotides, which are components of diagnostic kits. These ASRs must be made with Good Manufacturing Practices but otherwise can generally be marketed without FDA approval. The result has been that many manufacturers, especially with regard to DNA diagnostics, have by-passed the FDA. They sell the ASR components, and let the clinical lab or physician put them together. Another option is to release the kit first through ASRs and then apply for FDA approval for a diagnostic kit at some time in the future, effectively field testing the kit in the meantime. One might expect that FDA approval would be a good selling point, but price is also a consideration. Some insurance companies would rather pay a lower price for tests conducted with a kit concocted from ASRs, rather than pay a premium for FDA-approved diagnostics.

16.4.2
European and Canadian Regulation

In general, other countries have less stringent approval processes than does the United States, particularly for devices. Though these markets are smaller and less lucrative, they provide an opportunity to gain extra clinical experience with a product prior to FDA review. Typically, new implant products, for instance, will be marketed in Europe and/or Canada for 2–3 years before they receive market approval in the US.

The European Union (EU) has established the Committee for Proprietary Medical Products that regulates most medical devices. Medical devices are classified under its directives as class I, class IIa, class IIb, class III, or active implantable medical devices, in order of increasing risk. "Notifying Bodies" within each country may pass judgment on a given device.

The major considerations for the Notifying Body are whether the item is manufactured according to applicable standards (usually ISO standards) and whether the product is efficacious and safe. The Committee theoretically has a 210-day maximum review period within which to bring forth any questions regarding a product's safety or efficacy. The company then has the right to affix a "CE mark" on its product, which allows marketing throughout the Union (some non-EU countries, such as Switzerland, also honor the CE mark). After a product has its CE mark, "off-label" uses are less strenuously regulated there than in the US. In Europe, in general, a product must be proven safe and to perform as advertised. The European regulators are loath to tell physicians how to practice their art, and emphasize safety over efficacy.

Like the FDA, the European Commission has set forth guidelines requiring the reporting of adverse events related to a medical device. Reportable incidents include any malfunction or deterioration of the device, as well as inadequacies in labeling or instructions that lead to death or serious damage to the health of the user, or might have led to death or serious injury. Unlike the FDA, the European authorities do not allow retrospective evaluations to support an existing application.

16.4.3
General Regulation of Nanotechnology

Nanotechnology has benefited perversely from some of its critics in the perception that it is a highly advanced, sci-fi sort of technology still way beyond the horizon. This is not necessarily the case. By dollar volume, for instance, the biggest use of nanoparticles is in sun cream. The cosmetics maker L'Oreal is among the largest holders of nanotechnology patents.

There has already been a reaction among certain environmentalists, bioethicists, and futurologists concerning the destructive potential of nanotechnology. While we feel that the fear of runaway replicators is premature and overblown, other potential problems have a greater basis in reality. Fullerene-based objects, for instance, are diamond-hard, stronger than steel, chemically inert, and nearly indestructible. While fullerene nanotubes and diamondoid nanoparticles, it turns out, are natural products found in soot and crude oil, the environmental effects of producing large quantities of these materials have yet to be worked out. Nanotubes are already being used in a small number of commercial devices and will probably have general application in electronics. So the disposal of these materials may turn out to be a serious problem.

Government regulatory agencies have begun now began to focus on nanotechnology. The Better Regulations Taskforce, which advises the government of the United Kingdom, has already made some general recommendations. According to the Task Force the Government should:

- enable, through an informed debate, the public to consider the risks for themselves, and help them to make their own decisions by providing suitable information;
- be open about how it makes decisions, and acknowledge where there are uncertainties;
- communicate with, and involve as far as possible, the public in the decision-making process;
- ensure it develops two-way communication channels;
- and take a strong lead over the handling of any risk issues, particularly information provision and policy implementation.

While the policy recommendations are mild enough, the consideration of nanotechnology by the task force puts it in the same category as stem cell science and genetic engineering of food organisms, areas that have already suffered heavily at the hands of policy makers, politicians, and public interest groups.

Glenn Harlan Reynolds, a Tennessee law professor, has written a review for the Pacific Research Institute [12], called *Forward to the Future: Nanotechnology and Regulatory Policy*, perhaps the first serious, noninflammatory look at the issue in the US. In the review, Reynolds puts forth three potential scenarios for the regulation of nanotechnology in the U.S: (i) prohibition; (ii) restriction to the military, (iii) moderate regulation of public use. He sees prohibition is unworkable, not least because the seeds of the technology are already widely distributed and available. Prohibition would also be wasteful in the bencfits that to society that would have to be foregone. A military monopoly Reynolds sees as particularly dangerous, in that military versions of nanotechnology would likely involve robust weapon systems and be under the control of Pentagon bureaucrats who are a power unto themselves. He sees as most beneficial a regime of modest regulation emphasizing civilian research and professional responsibility.

"With all new technologies that come into the manufacturing fold, one must assess and attempt to forecast their impact on the environment and humankind and determine ways to prevent brown fields or water and air pollution," says Kelly Kirkpatrick [10], a venture capital consultant, and one of the authors of the National Nanotech Initiative (NNI). "So with the NNI, there have been efforts from very early on to look at the ethical, legal and social aspects of nanotechnology and what the implications could be. The goal would be to mitigate some of the potential problems in the future, to highlight some of the approaches that social scientists and environmental and ecological scientists might be concerned about, and to provide material scientists and physicists and chemists an opportunity to develop devices or processes with benign materials that have little deleterious impact to the earth."

Robin Fretwell Wilson, of the University of South Carolina, School of Law, has pointed out, however, that the Environmental Protection Agency (EPA) is set up as a reactive rather than a proactive agency. The EPA exists to carry out environmental legislation passed by Congress. Such laws are passed usually only after serious problems are recognized in the environment, such as the DDT crisis, or the health problems in the wake of Love Canal pollution. The Environmental Protection Agency has only recently focused on nanotechnology as an industry separate from chemistry. They held meetings in 2003 with nanotech industry leaders to get a feel for the sorts of regulation that might be required in the future.

My own opinion is congruent with that recently expressed by K. Eric Drexler (quoted in [6]): "The tools required to develop nanotechnologies are typically small and unobtrusive. The pace of research is accelerating worldwide. Some suggest stopping it, but it is hard to imagine how. Thus, it seems that this technology, with all its challenges and opportunities, is an unavoidable part of our future."

16.5
Operational

Governments both here and abroad are supporting nanotechnology generously, and academic research in this area is very healthy. However, the sort of interdisciplinary skills required to go from technology to practical biomedical development are in short supply, at present. According to a recent report of the European Commission (quoted in [6]), "There is a dearth of experts able to manage and integrate different disciplines involved at the interface, from concept and development of medical biotechnology products, to clinical practice. This is true for both the more scientific aspects, dealing for instance with labile molecules (e.g. proteins) and their delivery or with complex tissues and organs, as well as for the technological and managerial components involved in development, testing and regulation." "Ideally, you need to have a lot of multi-lingual people, people that speak physics, chemistry and biology and also people who speak science and business and law in order to fully realize the potential of these inventions," says venture capital consultant Laurie Pressman [10].

The situation is particularly problematic in the case of nanoscale medical devices that require approval from regulatory authorities like the US FDA. Despite the best of intentions, the FDA is undermanned and underfunded with respect to its mission. It has enough trouble keeping up with the pharmaceutical industry and recent innovations, like stem cell therapies. Nanotechnology is an area in which the agency has not had time to develop any expertise, and delays in the approval process are likely, as the FDA will err on the side of caution.

A lack of experts is particularly problematic in the US Patent Office, a critical agency for the development of nanobiomedicine. That agency has refused to create a single unit to handle nanotechnology, claiming that the term is so broad that it effectively cuts across all technologies, although they do have units that deal specifically with biotechnology patents, for example.

Classification is a tricky business. Already, over 1000 patents have been issued dealing with dendrimers, an iconic nanotech molecule. The older patents in this field, however, go back to the 1980s, before nanotech was even dimly perceived as a field in its own right.

Nanotech intellectual property (IP) claims will have to rest on new properties or new problems solved with the use of old materials. Just making the architecture of a material smaller is not patentable, on the basis of "obviousness." Carbon nanotubes are a case in point. The chemical formula for a carbon nanotube is just $C_{(n)}$, the same as graphite. So to claim it as a new material, one has to come up with new properties of this new form of carbon. Of course, nanotubes have many unusual properties in terms of structure, electrical conduction, strength, and hardness, etc., that one could not find in graphite. But other cases may not be so clear-cut.

Timothy Hsieh [13], a partner in Min, Hiseh, and Hack, LLP, has pointed out that nanotech is now about where biotech was in the early 1980s with respect to the patent office. As with biotech, the early patents that are given may be overly broad and not stand up to later challenge. "The inherent nature of nanotech," he has said, "will raise some new technical and legal issues that inventors need to be aware of and which the patent system will eventually have to resolve." This is problematic for small nanotech firms who spend a large proportion of their capital to acquire intellectual property, the value and defensibility of which has yet to be established.

16.6
Clinical

Nanomaterials are not small tech as far as medicine is concerned. Molecules as small as single ions are already used medically, as well as biomolecules as large as immunoglobulin, at about 150,000 daltons. On the size scale, nanoscale devices and materials inhabit the territory between immunoglobulin and submicron liposomes and micellar structures that are already used as drug delivery devices. As discussed above in Section 16.4, procedures already in place at the FDA will apply to new nanoscale medicines or devices, and it is not necessary to develop new procedures to handle nanomedicine *per se*. Already, certain types of dendrimers and quantum dots are phase I approved, and as mentioned, nanoparticles are already on the market as drug delivery devices.

The most important qualification with regard to nanomaterials in relation to nanomedicine is whether these new materials are biocompatible or not. Nanoscale devices are in the range where one might expect immunological reaction. The body may see these devices as foreign invaders and react accordingly. Failing to eliminate the materials with immunological attack and phagocytic cells, the body may encyst the device and prevent its work or cause further medical problems. Aggregates of foreign material tend to collect in fine meshwork within the body, like the alveoli in the lungs or the glomeruli of the kidneys, where they can become life-threatening.

There is, of course, a long history of materials research involved with establishing various materials as biocompatible. Medical implants, like pacemakers or artificial joints, tend to be made of materials like titanium or Teflon that are metabolically inactive. Preclinical toxicology work will have to be done on new nanomaterials, as it would for any new medicine. Fullerene devices may prove to be problematic, as one has to distinguish between various buckyballs (C_{60}, C_{70}) and carbon nanotubes. On the basis of very preliminary work, the former seem to be metabolically inert, invisible to the immune system, and innocuous in the body, whereas the latter may collect in the body, particularly the lungs, and cause serious damage.

Commercially available quantum dots are currently made of cadmium selenide, a material that is doubly disadvantageous in that it is highly toxic and insoluble in aqueous media. The manufacturers counter this problem by coating the dot material in zinc sulfate to prevent the cadmium selenide crystal from dissolving, and then coating that with a hydrophilic polymer. The final coating has the advantage that functional biomolecules can be linked to it.

Another sort of nanoparticle has been developed by Kereos, Inc. – a perfluorocarbon/lipid micelle substituted with chelated gadolinium and targeting ligands for use as an MRI contrast agent. The perfluorocarbon micelle was originally developed by Allied Pharmaceuticals in a blood substitute product. Chelated gadolinium is currently approved used for use in MRI contrast agents. Targeting ligands can be monoclonal antibodies or hormone-type ligands. The nanoparticle aggregates a number of elements already used in medicine, for which the toxicology is already well known. The nanoparticle breaks down rapidly. The chelated gadolinium is excreted in the urine, the perfluorocarbons leave the body through the lungs, and the lipids are recycled as cellular lipids. Though this is a piece of nanotechnology, it employs elements already well-known to medicine.

In summary, there is no reason as yet to regard nanobiomedicines as different in kind from other medicines. Until we start talking about autonomously intelligent or self-replicating nanodevices (which this author is not prepared to do), nanobiomedicines do not present unique clinical challenges.

16.7
Political, Ethical And Social Challenges

Eric Drexler [7] was the first one to warn of the possibility of nanotechnology running amok, even though he is an ardent booster. In *Engines of Creation*, he pointed out some clearly dystopian possibilities:

"Plants" with "leaves" no more efficient than today's solar cells could outcompete real plants, crowding the biosphere with an inedible foliage. Tough omnivorous "bacteria" could out-compete real bacteria: they could spread like blowing pollen, replicate swiftly, and reduce the biosphere to dust in a matter of days.

Bill Joy, Chief Technology for Sun Computers, is another one who worries. In a *Wired* magazine article titled "Why the Future Doesn't Need Us" [14], he wrote:

The 21st-century technologies – genetics, nanotechnology, and robotics (GNR) – are so powerful that they can spawn whole new classes of accidents and abuses. Most dangerously, for the first time, these accidents and abuses are widely within the reach of individuals or small groups. They will not require large facilities or rare raw materials. Knowledge alone will enable the use of them. Thus we have the possibility not just of weapons of mass destruction but of knowledge-enabled mass destruction (KMD), this destructiveness hugely amplified by the power of self-replication."

[...]

" ...Rereading Drexler's work after more than 10 years, I was dismayed to realize how little I had remembered of its lengthy section called "Dangers and Hopes," including a discussion of how nanotechnologies can become "engines of destruction." Indeed, in my rereading of this cautionary material today, I am struck by how naive some of Drexler's safeguard proposals seem, and how much greater I judge the dangers to be now than even he seemed to then.

It is important to remember that Drexler and Joy are not burnt-out hippies spreading luddite propaganda. Quite the contrary, they are among the leaders in major areas of technology as well as quite intelligent people. Thus, the dangers that nanotechnology presents, especially when combined with biosciences, should be taken seriously.

Ray Kurzweil, author of *The Age of Spiritual Machines* [15] and generally a nano-tech supporter, points out the potential of deliberate misuse of nanotechnology:

"...the bigger danger is the intentional hostile use of nanotechnology. Once the basic technology is available, it would not be difficult to adapt it as an instrument of war or terrorism.... Nuclear weapons, for all their destructive potential, are at least relatively local in their effects. The self-replicating nature of nanotechnology makes it a far greater danger."

No less an authority than *Jane's International Security News* has echoed Kurzweill's concerns.

Another more immediate concern that has emerged is the possible environmental effects of large scale production of nanomaterials. Barbara Karn, who is in charge of directing nanotechnology research at EPA, recently asked researchers about the potential for nanoparticles causing harm to the environment. According to Mark Wiesner, professor of Civil and Environmental Engineering at Rice University, tests have shown that nanoparticles penetrate living cells and accumulate in the liver of experimental animals. He is especially worried about fullerene derivatives, like carbon nanotubes, which are extremely stable and therefore could be expected to accumulate in the environment over time as companies like Frontier Carbon Corp. and Carbon Nanotech Research Institute gear up to make tons of nanotubes.

Vicki L. Colvin, who is the executive director of the Center for Biological and Environmental Nanotechnology at Rice University (a nanotech powerhouse) has repeatedly pointed out that there has been almost no research into the potential toxicology of nanoparticles. Colvin, however, is a definitely a supporter of the industry. In testimony before the House of Representatives, she discussed recent concerns about nanotechnology, mentioning Michael Crichton's novel *Prey* [8], which she says illustrates "a reaction that could bring the growing nanotechnology industry to its knees: fear. The perception that nanotechnology will cause environmental devastation or human disease could itself turn the dream of a trillion-dollar industry into a nightmare of public backlash."

A report from Abdallah Daar and Peter Singer, from the University of Toronto Joint Centre for Bioethics, has called for a general moratorium on nanomaterial deployment, on the grounds that the ethical, environmental, economic, legal, and social implications of nanotechnology have not yet been taken seriously and pursued on a large enough scale.

Etc (pronounced *et cetera*), a technology watchdog type of organization, prominent in the fight against genetically engineered organisms, has now put its sight on nanotech. They have published an 84-page report [16] titled *The Big Down – Atomtech: Technologies Converging at the Nanoscale.* Not surprisingly, they find nanotech alarming. An excerpt:

> The hype surrounding nano-scale technologies today is eerily reminiscent of the early promises of biotech. This time we're told that nano will eradicate poverty by providing material goods (pollution free!) to all the world's people, cure disease, reverse global warming, extend life spans and solve the energy crisis. Atomtech's [Etc-speak for nanotech] present and future applications are potentially beneficial and socially appealing. But even Atomtech's biggest boosters warn that small wonders can mean colossal woes. Atomtech's unknowns – ranging from the health and environmental risks of nanoparticle contamination to Gray Goo and cyborgs, to the amplification of weapons of mass destruction – pose incalculable risks.

Etc offers as a guiding concept its Precautionary Principle, to wit: "The Precautionary Principle says that governments have a responsibility to take preventive action to avoid harm to human health or the environment, even before scientific certainty of the harm has been established. Under the Precautionary Principle it is the proponent of a new technology, rather than the public, that bears the burden of proof." Etc wants an immediate moratorium on commercial production of new nanomaterials. Molecular manufacturing, says the group, poses "enormous environmental and social risks and must not proceed – even in the laboratory – in the absence of broad societal understanding and assessment."

Etc and groups like them cannot be ignored, if only because they have demonstrated the power to sway public opinion. Under its previous incarnation as RAFI, Etc group members were given credit for putting a halt to Monsanto's so-called "terminator technology," a genetic engineering method to protect agritech patent rights by making second-generation seeds sterile. Etc has recently won a powerful convert in Prince Charles, of England, a living anachronism who has previously inveighed against the horrors of modern architecture as well as "Frankenfoods." Public and political opinion can matter a great deal, as agritech and stem cell companies have learned by bitter experience.

The nightmare scenario for the nanotech entrepreneur is that, as with genetically engineered foods, the public will belatedly come to appreciate real or imagined threats and demand regulation, moratoriums, or outright banning of certain nanotech-related technologies. So, after 10 years and tens or hundreds of millions of developmental costs, a company may find itself unable to manufacture or market its product.

Table 16.3 lists some of the commonly expressed concerns about the development of nanotechnology. The concerns are somewhat overstated for the purposes of discussion; nonetheless, Tab. 16.3 accurately reflects the tenor of discussion among the small percentage of the general population who are aware of nanotechnology. Most of these relate somewhat to the development of nanobiomedicine.

Table 16.3 Political, social or ethical concerns related to nanotechnology development.

1.	Grey goo scenario – environmental disaster due to self-replication.
2.	Green goo scenario – GMO organism takes over the world.
3.	Environmental disaster due to inhalable or ingestible nanoparticles.
4.	End of shortage-based economics.
5.	"People will live forever, leading to overpopulation."
6.	"Only rich people will live forever": nanotech benefits accrue only to those in charge.
7.	"Nanotech will turn us into cyborgs."
8.	"Nanotech can be used to create incredible weapons of mass destruction."

16.7.1
The Gray Goo Scenario

The *gray goo scenario*, already discussed, is the proposition that an intelligently de-signed nanotech agent with the properties of self-replication might escape into the general environment and wipe out the biosphere by sequestering to itself certain elements or materials necessary for life. This has been much discussed by others, both by proponents of nanotechnology and by those against. The best reason for believing this scenario is unlikely is that the biosphere is already full of self-replicat-ing agents, perfected over a billion years of evolution, and these organisms are very jealously acquisitive with respect to life-sustaining materials. There are arguments related to intelligent design that might counter this hopeful outlook. However, I am going to finesse further discussion of this possibility by pointing out that the gray goo scenario has nothing particular to do with nanobiomedicine, as a subcategory of nanotechnology.

16.7.2
The Green Goo Scenario

The *green goo scenario*, somewhat more likely, if only because it is more easily in reach, suggests that a DNA-based artificial organism might escape from the lab and cause enormous environmental damage. Genetically modified organisms have, of course, existed since the 1970s. From early on, biologists recognized the danger and restrictions have been in place to keep organisms that were likely to be dangerous – modified pathogens, for instance – from escaping. More recently, biotechnology has been redefined by some as "wet nanotechnology," which, on the basis of size class, it surely is, because molecular biology has always been about events and structures on the nanoscale.

In the last decade, number of people have been interested in the "synthetic gen-ome." The idea is to create an organism that has the minimal number of standard genes necessary for life. This could then be modified at will to create an organism for specific purposes. Craig Venter, the genius behind the successful private effort to sequence the human genome, is interested in creating such an organism to lessen our dependence on hydrocarbon energy sources. Other uses of such an organism could be to manufacture biomedicines more efficiently than modified bacteria or yeast do, at present.

An ethical dilemma might arise if such a synthetic organism was injected into a person so that the biomedicines it manufactures might be more conveniently delivered where they are needed. Such an organism, in principle, could invade the germ cells, and co-evolve with human beings (much the way mitochondria have co-evolved with eukaryotes), leading to a "super-human" with increased survival characteristics. Or alternatively, the organism might integrate with the human genome within the germ cells and add new genes to the human complement, in the same way that retroviral genes have already entered the genome of a number of mammalian species, including humans.

Is this an unlikely scenario? You may be surprised to learn that genetically modified cytotoxic viruses and bacteria are already used experimentally to kill cancer cells. Similar vectors are used to transfer genes to correct genetic defects in human clinical experiments. Generally speaking, extreme care is taken to make sure that these genes cannot be passed horizontally by infection. But so far there has been relatively little thought given to the possibility that therapeutic genetic tampering might be transmitted vertically through the genome. Though this is certainly an improbable event, the existence of retrovirus genes in the human genome is an existence proof that such vertical transmission is possible.

If you are dying of cancer or a genetic disease, the last thing on your mind, surely, is whether that virus might end up in your germ cells. Individually, it is kind of a moot point. But as a society, is it something we should worry about? This is an ethical debate that may impinge upon some areas of nanobiomedicine.

16.7.3
Environmental Disaster Due to Inhalable or Ingestible Nanoparticles

We have already referred to the fact that little toxicology has been done with respect to nanoparticles. On the face of it, this is a particular problem with carbon nanotubes, as industry is already gearing up to produce multiton quantities of these. Uses are expected in electronics, as field emission devices for cathode-ray tubes for instance. The fear is that nanotubes or other nanoparticles will contaminate the air or waterways, resulting in large-scale environmental damage. Comparisons are made with the miner's black lung disease, from coal tar, lung cancer caused by cigarette smoking, or mesothelioma caused by asbestos fibers. These concerns cannot easily be reasoned away. Carbon nanotubes, for instance, penetrate cells with ease. They are harder than diamonds, stronger than steel, and not biodegradable.

Nanoparticles already exist in large quantities in the environment in the form of carbon black, also known as acetylene black, channel black, furnace black, lampblack, and thermal black. This is used in tires, inks, lacquers, carbon brushes, electrical conductors, and insulating materials. Typically, carbon black comes in 10- to 40-nm particles of elementary carbon with various chemicals adsorbed to the surface. Workers in tire plants may have this nanoparticle adsorbed into their skin, such that they sweat it out on their sheets and personal clothing even weeks after they leave their employment. So nanoparticles are not really a new thing. Carbon nanoparticles have been around as a product of incomplete combustion since Prometheus brought fire to mankind.

We will most likely deal with nanotech environmental hazards in the way that we always have, by ignoring them until they have become disastrous in an obvious way. Even that level of control will develop on a case-by-case basis. Nanobiomedicine, by its nature, is not likely to be a major cause of environmental damage, since nanoparticles and nanoscale devices will be used sparingly compared with other uses – the manufacture of tires as an obvious example.

16.7.4
End of Shortage-Based Economics

One of the claims of the more utopian nanotech enthusiasts is that we will be able to control matter at will, and that desired objects will be "instantiated" by nanotech assemblers drawing material out of the environment. Nanotech disaster theorists claim that this apparent feature is actually a bug; it will mean the end of economics as we know it, and therefore the end of the capitalist system. Or whatever.

This is not a problem that I am prepared to worry about. The world today produces an over-abundance of food and yet people go hungry, even in very rich nations. So the economics of shortage is probably safe for quite awhile.

16.7.5
"People Will Live for Ever, Leading to Overpopulation"

This is another feature vs. bug dichotomy. Nanomedicine, it is thought, will progress to the point where we can fix all diseases of aging so that people will outlive their usefulness, but refuse to move on to their reward. Meanwhile, the younger generation will struggle and the world will become overpopulated with aged but still healthy folks.

Overpopulation has been with us since Thomas Malthus [17] apparently invented the concept in 1798. Overpopulation turns out to be a relative concept. The Indian population has successfully resisting forced sterilization and less coercive measures by the state to control population. This already crowded nation is on track to add 50% to its population by 2050 according to a report by the US Census Bureau. On the other hand, Europeans have reduced their fertility to the point that some countries are well below replacement values. Italy, once renowned for its large happily families, is now among the lowest in the world with respect to fertility.

Serious people now worry about a crash in human populations. Such a crash appears to be well in progress in sub-Saharan Africa where life expectancies are back under 40, an incredible reversal due to the AIDS epidemic. New epidemics, due to Nipah virus or the SARS virus or a pathogen yet to be discovered, are threatened constantly in the headlines.

Population concerns aside, the more relevant question is whether nanobiomedicine can fulfill the expectation of adding substantially longevity. After all, modern medicine to this point has added significantly to the average life expectancy of those in industrialized countries but precious little to the prospective longevity of any individual. The bell curve has shifted, but the outer bounds remain the same. People can live a productive life for their biblical three score years and ten; maybe they can be active into their 80s, but still only a lucky few reach the century mark. Nanobiomedicine may in fact contribute to substantially greater longevity, in combination with other technologies, such as therapeutic cloning, that seem to have even a greater current potential, in this regard.

Therapeutic cloning has already engendered a fire-storm of political controversy. Ostensibly, this has been about the necessity of destroying "embryos" to create patient-specific embryonic stem cells. In fact, these embryos would result from the placement of a nucleus from the patient into the environment of an egg cytoplasm. There is no fertilization involved between a sperm and egg cell, and therefore no creation of a new individual. But the destruction of an embryo – no matter that it is only a ball of cells at this point – is seen by some as tantamount to murder.

Even if you do not subscribe to the extreme conservative view, it is true that the process has the potential to violate what has been termed "normative medicine." Normative medicine seeks to return the patient from a condition of disease to one of "normal" health. To add attributes or performance beyond what could be considered normal is not medicine, in the eyes of some observers.

Of course, medicine as practiced is full of violations of normative medicine. Some are rather trivial, like plastic surgery to remove signs of aging. Others are really drastic, like sex-change surgery, for example. Most plastic surgery, at some level, actually endangers the patient's health to achieve a cosmetic end. Sex-change surgery is an attempt to transcend the bounds of nature. Similarly, conservatives would say, therapeutic cloning seeks to transcend the barriers of nature with respect to aging. If it is possible, for instance, to replace heart tissues with new cardiomyocytes that behave as if they are newborn, then medicine is no longer normative. We are seeking to create something that has never existed before, an aging human with a newborn heart.

Nanobiomedicine, as it is currently constituted, is ethically benign compared to the promise of tissue engineering and therapeutic cloning. However, it is likely that as our control of matter at the nanoscale level increases, our ability to repair aging tissues will increase as well. A nanotech-enabled artificial retina, as yet a primitive instrument, could be constructed such that people could see into the infrared or ultraviolet ranges, for instance. Such an individual might have certain advantages, as a soldier in battlefield situations, for instance. Thus, capabilities introduced by nanobiomedicine will eventually affect the debate over "normative" medicine as an ethical ideal.

16.7.6
"Only Rich People Will Live Forever": Nanotech Benefits Accrue Only to Those in Charge

Nanobiomedicine will surely benefit the rich first. This is the way of things in this world.

This is part of a larger debate about who should control the benefits of nanotechnology. Some have claimed that since nanotech research has been paid for by taxpayers, there should be some mechanism to spread the benefits in an equitable fashion. A gaping hole in this logic, to date, is that the more visible benefits of nanotech have actually been the result of private investment. The atomic force microscope is the result of efforts at IBM's Zurich laboratory. Carbon nanotubes were an accidental observation of a researcher at Nippon Electric Company. Dendrimers came from years of research at Dow Chemical.

New advances in medicine have always benefited the rich first. Implantable defibrillators, to cite a recent example, are incredibly expensive items, especially when surgical expenses are considered. Neither Medicare nor most insurance carriers will pay to implant them in all that could potentially benefit; nor will they pay for the most advanced models of these devices. However, Medicare will pay to implant these devices in certain patients – people with arrhythmia who have already suffered a heart attack; at least some patients who could not afford these devices from their own resources are granted the benefits of this technology. As technology improves, the bells and whistles in the advanced devices eventually become standard on all models. This "trickle-down" approach to medical distribution is not ideal, but it is the system currently in place.

Ideally, technology should benefit all who could benefit; however, the lack of a societal mechanism to insure such a distribution should not be regarded as challenge to the development of the technology in the first instance. Nanotechnologists do not bear any special responsibility for the organization of society.

16.7.7
"Nanotech Will Turn Us Into Cyborgs"

This objection can be restated to read, "Nanotechnology will accelerate the trends that are turning us all into cyborgs ("cyborg" was originally a contraction of "cybernetic organism," but we will take it in its more extended Star Trek-informed popular meaning as a partially mechanical organism)." Ever since George Washington was fitted with wooden teeth, we have become increasingly beings who were part mechanical, part organic. In our mouths, we have fillings, crowns, caps, bridges, or artificial teeth. In order to see, we have spectacles or contact lenses or lasik surgery that transforms our eyeballs. To hear better, we have hearing aids or cochlear implants. Already, there are people walking around with retinal implants giving them a modicum of vision who would otherwise be profoundly blind. We have artificial shoulders, hips, elbows, finger joints, knees, elbows, and intervertebral discs. Some people require implanted insulin pumps to maintain their blood sugar levels or morphine pumps to keep at bay the pain of cancer. Dialysis machines have taken the place of kidneys for an unfortunate few.

On an extraorganismal level, many, if not most, of us have become tethered to a computer or at least a telephone to accomplish our daily tasks. Some of us have been known to circle a parking lot for 30 min to avoid walking an extra hundred yards, substituting the automobile for legs. Because of paraplegia, breathing difficulties, or extreme weight, many people require wheelchairs for locomotion.

Nanobiomedicine may indeed accelerate this cyborgian trend. Resistance is futile. Face it. Few among us will reject the nanotech-enabled retinal implant when faced with the alternative of blindness. Few of us will care that we have become dependent upon nanomachines if they turn out to be efficient in keeping cancer at bay. It is not difficult to imagine a future in which we are still born naked into the world but are fitted with more and more electronic and mechanical parts as we age, to the point where our original organic origins are completely abandoned, at some point.

16.7.8
"Nanotechnology Can Be Used to Create Incredible Weapons of Mass Destruction"

Beyond Michael Crichton's *Prey*, there is a sci-fi menagerie of potential destructive agents that could be created using nanotechnology. Tiny mechanical creatures could recognize certain people based on their histocompatibility antigens and proceed to their brainstem and take control of their neural functions – to create an instant premise for a novel. We should not be facetious about these fears, however.

It is not impossible that weapons of mass destruction, too small to see, could eventually be created by a malign power with expertise in nanotechnology.

It is not impossible that a similar malign power armed with biotechnology could bring back smallpox, or create a new bioterror agent, based on existing viruses or bacteria, to kill millions of people. The Department of Homeland Security, in fact, fears such an event, and puts a lot of money and effort in trying to prevent it.

It is not impossible that a sufficiently powerful group armed with nuclear technology could destroy much of the known world. When I was a child, we were taught to believe that this scenario was a likely event.

There is no end to the number of disaster scenarios that could be created. Eternal vigilance, it is said, is the cost of freedom, and it is up to us and our governmental institutions to prevent the misuse of technology. It would be a mistake, however, to forgo a particular technology altogether because of such fear. It is not impossible that technology invented in the interest of nanobiomedicine, for instance, will hasten the day when a malign power will use that same technology destructively. On the other hand, if we do not develop this technology as a society, then nonsocietal powers may be the only ones to understand such a technology and to use it. At that point, we would be helpless.

16.8
Summary

Nanobiomedicine, to a small extent, already exists in the form of nanoparticles used to deliver medicine and nanoscale devices used for diagnostic purposes. Biotechnology, which has already made a large contribution to medicine in the form of protein-based drugs and diagnostic technologies, has been described as "wet nanotechnology."

Existing regulations and policies that govern the introduction of drugs and medical devices apply to nanobiomedical products as well.

The principal impediments to the further development of nanobiomedicine at this point are financial and operational. Financial constraints to small entrepreneurial firms have developed because of the shortage of venture capital. This is due primarily to a hangover effect of the dot.com debacle, but venture capital players are also scared off by some of the excessive claims being made for nanotechnology. Operationally, there exists a shortage of people who can successfully navigate all the various disciplines required to put together nanobiomedical products. In the US, federal agencies that play a role in the implementation of nanobiomedicine, like the Food and Drug Administration, the Environmental Protection Agency, and the US Patent Office, feel this shortage most acutely.

Some of the social and ethical concerns expressed concerning the development of nanotechnology are premature and do not necessarily apply to nanobiomedicine in particular. However, it is likely, in the long run, that nanobiomedicine will contribute to a lengthening of the functional lifespans of human beings, leading to societal stress. There are also certain nanobio products that could be objectionable on ethical grounds. Finally, like any advanced technology, nanobiomedicine has the potential for abuse by terrorists or other malcontents. Political institutions will have to develop methods to guard against this abuse.

Abbreviations

ASR – analyte-specific reagent
BCC – Business Communications Company
CEO – chief executive officer
DNA – deoxyribonucleic acid
DARPA – Defense Advanced Research Projects Agency
EPA – Environmental Protection Agency
FDA – Food and Drug Administration
IDE – investigational device exemption
IND – investigational new drug
IP – intelectual property
IPO – initial public offering
MIT – Massachusetts Institute of Technology
NSF – National Science Foundation
PMA – premarket approval

References

1 Perkel, J. M., Nanoscience is out of the bottle, *Scientist 17*, 15, **2003**.

2 Personal communication, A. Watson, Quantum Dot Corp.; also *Technology Review, Emerging Technologies*, Wednesday update, May 12, **2004**.

3 Personal communication, R. Ellson, Picoliter, Inc., **2003**.

4 Freitas, R., *Nanomedicine: Basic Capabilities*, Landes Biosciences, Georgetown, TX, USA, **1999**.

5 Gordon, N., Sagman, U., *Nanomedicine Taxonomy*, Canadian Nanobusiness Alliance,www.regenerativemedicine.ca/nanomed/ Nanomedicine Taxonomy (Feb 2003). PDF, **2003**.

6 Edwards, S. A., *Biomedical Applications of Nanoscale Devices*, BCC, Inc., Norwalk, CT, **2003**.

7 Drexler, K. E., *Engines of Creation: The Coming Era of Nanotechnology*, Anchor Books, New York, NY, **1986**.

8 Crichton, M., *Prey*, Harper Collins, New York, **2002**.

9 Smalley, R., *Of Chemistry, Love, and Nanobots*, *Sci. Am.* **2001**, Sep, 285 (3), 78–83.

10 Interview by John Roy from Merrill Lynch, February 2, **2004**.

11 Rocco, M., Bainbridge, W., eds., *Converging Technologies for Improving Human Performance*.National Science Foundation report, www.technology.gov/reports/2002/NBIC/ Part1.pdf, **2002**.

12 Reynolds, G. H., *Forward to the Future: Nanotechnology and Regulatory Policy*, Pacific Research Insitute, www.pacificresearch.org/ pub/sab/techno/forward_to_nanotech.pdf, **2002**.

14 Hsieh, T., *Emerging IP Issues in Nanotechnology*, Presented at the NanoBiotech IP Landscape Conference, July, Boston, MA, USA, **2003**.

15 Joy, W., *Why the Future Doesn't Need Us*, *Wired* 8.04, April 2000.

16 Kurzweil, R., *The Age of Spiritual Machines*, Penguin Books, Harmondsworth, **1999**.

17 *The Big Down – Atomtech: Technologies Converging at the Nanoscale*, Etc Group, www.etc-group.org/documents/TheBigDown.pdf.

18 Malthus, T., *An Essay on the Principle of Population*. Printed for J. Johnson in St. Paul's Churchyard, London, **1798**.

Index

Nanofabrication Towards Biomedical Applications. C. S. S. R. Kumar, J. Hormes, C. Leuschner (Eds.)
Copyright © 2005 WILEY-VCH Verlag GmbH & Co. KGaA, Weinheim
ISBN 3-527-31115-7